S0-BHD-143

Regional Trends in the Geology of the Appalachian-Caledonian-Hercynian-Mauritanide Orogen

NATO ASI Series

Advanced Science Institutes Series

A series presenting the results of activities sponsored by the NATO Science Committee, which aims at the dissemation of advanced scientific and technological knowledge, with a view to strengthening links between scientific communities.

The series is published by an international board of publishers in conjunction with the NATO Scientific Affairs Division

A	Life Sciences	Plenum Publishing Corporation
B	Physics	London and New York
C	Mathematical and Physical Sciences	D. Reidel Publishing Company Dordrecht, Boston and Lancaster
D	Behavioural and Social Sciences	Martinus Nijhoff Publishers
E	Engineering and Materials Sciences	The Hague, Boston and Lancaster
F	Computer and Systems Sciences	Springer-Verlag
G	Ecological Sciences	Berlin, Heidelberg, New York and Tokyo

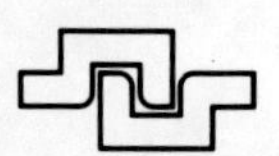

Series C: Mathematical and Physical Sciences Vol. 116

Regional Trends in the Geology of the Appalachian-Caledonian-Hercynian-Mauritanide Orogen

edited by

Paul E. Schenk

Department of Geology, Dalhousie University,
Halifax, Nova Scotia, Canada

Associate editors

R. T. Haworth

Geological Survey of Canada,
Dartmouth, Nova Scotia,
Canada

J. D. Keppie

Nova Scotia Department of Mines and Energy,
Halifax, Nova Scotia, Canada

W. E. Trzcienski

Department of Geology,
University of Montreal,
Montreal, Quebec, Canada

P. F. Williams

Department of Geology,
University of New Brunswick,
Fredericton, New Brunswick,
Canada

G. Kelling

Department of Geology,
University of Keele,
Keele, England

D. Reidel Publishing Company

Dordrecht / Boston / Lancaster

Published in cooperation with NATO Scientific Affairs Division

Proceedings of the NATO Advanced Study Institute on
Regional Trends in the Geology of the Appalachian-Caledonian-
Hercynian-Mauritanide Orogen
Atlantic Canada and Fredericton, New Brunswick, Canada
August 2-17, 1982

Library of Congress Cataloging in Publication Data

NATO Advanced Study Institute on Regional Trends in the Geology of the Appalachian-Caledonian-Hercynian-Mauritanide Orogen (1982 : Fredericton, N.B.)
Regional trends in the geology of the Appalachian-Caledonian-Hercynian-Mauritanide orogen.

(NATO ASI series. Series C, Mathematical and physical sciences ; Vol. 116)
Proceedings of the NATO Advanced Study Institute on Regional Trends in the Geology of the Appalachian-Caledonian-Hercynian-Mauritanide Orogen, held in Fredericton, N.B., preceded by a pre-conference (10-day field excursion) across southeastern Atlantic Canada.
Includes index.
1. Orogeny—North Atlantic Region—Congresses. 2. Geology, Stratigraphic—Paleozoic—Congresses. I. Schenk, Paul E., 1937- . II. Title. III. Series.
QE621.5.N75N37 1982 551.7'2'091821 83-19113
ISBN 90-277-1679-X

Published by D. Reidel Publishing Company
P.O. Box 17, 3300 AA Dordrecht, Holland

Sold and distributed in the U.S.A. and Canada
by Kluwer Academic Publishers,
190 Old Derby Street, Hingham, MA 02043, U.S.A.

In all other countries, sold and distributed
by Kluwer Academic Publishers Group,
P.O. Box 322, 3300 AH Dordrecht, Holland

D. Reidel Publishing Company is a member of the Kluwer Academic Publishers Group

---

Printed in The Netherlands

CONTENTS

PREFACE

The classical Appalachian, Caledonian, Hercynian, and Mauritanide orogens are now only segments of a once-continuous Paleozoic mountain belt which has been fragmented during Mesozoic-Cenozoic formation of the North Atlantic Ocean. These segments are major parts of the countries surrounding the North Atlantic - most of which are members of NATO. The aim of this NATO conference was to evaluate these fragments in terms of their pre-Mesozoic positions, and to attempt a synthesis of their geologic evolution on an international and orogen-wide scale.

Geologists who have studied these scattered remnants have been separated by both geography and discipline. Orogen-wide syntheses have beeen attempted in the past by individuals who are specialists not only in discipline but also in geography; therefore, these attempts have not been satisfactory to everyone. This conference brought together the foremost specialists in different disciplines from each country. They attempted to teach other specialists, not only in their own fields, but in other disciplines, about regional variations and particular problems. The resulting international cross-fertilization, both within and between specialties, enriched individual workers and helped to provide a multi-disciplinary overview of the orogen. The gathering included specialists in plutonics, volcanics, metamorphism, deformation, sedimentology/stratigraphy, paleontology, basement relations, and geophysics. Participants came from Norway, Sweden, Great Britain, Ireland, France, Spain, Morocco, the United States, and Canada. In this way, a volcanologist working in Portugal (for example) saw his/her own geological problem place into its context both within the entire, continuous orogen as well as amongst other disciplines of Earth Science.

Regional trends along the Orogen were shown by special geological maps. Many of the participants were members of Project 27 - "The Caledonide Orogen" - of the International Geological Correlation Programme. In preparation, compilation maps by discipline had been prepared at the same scale and legend for national segments of the Orogen. These were assembled, discussed, and corrected at the meeting. In most

P. E. Schenk (ed.),
Regional Trends in the Geology of the Appalachian-Caledonian-Hercynian-Mauritanide Orogen, xi–xiii.

cases, trends of specific variables between countries were detected for the first time.

The conference was organized into three parts.

1) Approximately 40 people participated in a pre-conference, 10 day field excursion across southeastern Atlantic Canada. We surveyed the two largest suspected terranes in the Canadian Appalachians - the Avalon and the Meguma. Both have been interpreted as being exotic to North America - i.e. as chips of Western Europe and Northwestern Africa. As such, they offer greatest hope in trans-Atlantic correlation. A guidebook of 308 pages with separate maps, entitled: A. F. KING, compiler, "Guidebook for Avalon and Meguma Zones" is available from the Department of Earth Sciences, Memorial University, St. John's, Newfoundland, A1B 3X5.

2) Two days were spent in Fredericton, New Brunswick in preparation for a four-day plenary symposium there. The papers presented at the symposium and printed in this volume were prepared or modified during discussions on the field trip or around the compilation maps. At the plenary session, one half-day was assigned to each specialist group in the order of this volume. Each group attempted to synthesize regional trends in their discipline down the entire orogen. Two special meetings were held in the evening: one on new and possible contoversial papers by usually younger people; and one on metalogenesis in the Orogen. The last day of the Symposium was spent on multidisciplinary summaries of each participating country, based on the preceding analyses of regional trends. Two general summaries highlighted the Symposium - an introduction by Dr. John Rogers, and a grand synthesis by Dr. H. R. Williams.

3) In the week following the Symposium six field trips organized for specialist groups were led over Atlantic Canada, Guidebooks for each of these are available from the Department of Geology, Dalhousie University, Halifax, Nova Scotia, B3H 2J5. They are:

1) ST. JULIEN, P. "External and Internal Domains between Quebec City and Thetford Mines"
2) DAVIES, L. and FYFFE, L. R. "Sulphide Deposits of the Bathurst Area, New Brunswick"
3) RUITENBERG, A. A. and McCUTCHEON, S. "Mount Pleasant Burnt Hill Tungsten- Tin- Molybdenum-Antimony Deposit"
4) STRONG, D. F., WILLIAMS, P. F., DEAN, P. L., HIBBARD, J., and BLACKWOOD, F. "Notre Dame Bay Island Arc, Ophiolite and Melange Complexes"
5) WILLIAMS, H., HIBBARD, J., MALPAS, J., and BLACKWOOD, F. "Trans-Newfoundland Trip"

6) STRINGER, P. "Polyorogenic deformation in Coastal Southern New Brunswick".

In order that the formal results of the conference be available to all, most of the papers prepared at the Symposium are published in the present proceedings, for which I am much obliged to the publishers. All of the participants are grateful to NATO Brussels who sponsored this meeting and made the Advanced Studies Institute possible. Further help was supplied by a Conference Grant from the National Sciences and Engineering Research Council of Canada, and the International Geological Correlation Programme. Assistance was also given freely from Dalhousie University, Memorial University, the University of New Brunswick, and the Departments of Energy and Mines of the provinces of Newfoundland, Nova Scotia, and New Brunswick. Finally I thank the Canadian working group leaders of IGCP Project 27 for their continued support, Dr. P. F. Williams for his hospitality at the University of New Brunswick, and to Mrs. D. Crouse for her able administration.

Halifax, June 1983

Paul E. Schenk

GEOPHYSICS AND GEOLOGICAL CORRELATION WITHIN THE APPALACHIAN-CALEDONIDE-HERCYNIAN-MAURITANIDE OROGENS - AN INTRODUCTION

Richard T. Haworth

Institute of Geological Sciences, Nicker Hill,
Keyworth, Nottingham, England, BG125GG

INTRODUCTION

A geophysical study group within IGCP Project 27 "The Caledonide Orogen" was informally organized in 1978. Its initial objective was to foster the compilation of geophysical data within individual countries on a common scale so that international compilations and correlations between separate elements of the orogen might be possible. Initially, data were most readily available for the Appalachian, Caledonide and Hercynian orogens so that compilations of magnetic and gravity data have been produced (1,2) and their results interpreted (3,4). From these beginnings more recent work has concentrated on compiling new data for the Caledonides and the Mauritanides, so that correlation can be extended over greater distances, and so that the correlations suggested by potential field data can be evaluated in the light of other geological and geophysical data. The geophysical session of the NATO/IGCP meeting in Fredericton, Canada in August 1982 concentrated primarily on newer aspects of the work of the study ground, including an introduction to the difficulties in deciding the reliability of paleomagnetic results. Although I presented an overview of geophysical work and interpretation in the orogen, much of that work has been published elsewhere (5,6). This note therefore serves primarily to introduce the papers that follow and refers the reader to significant papers elsewhere that provide an introduction to the geophysical literature regarding the orogen.

Gravity

Woollard and Joesting's compilation of gravity data for

P. E. Schenk (ed.),
Regional Trends in the Geology of the Appalachian-Caledonian-Hercynian-Mauritanide Orogen, 1–9.

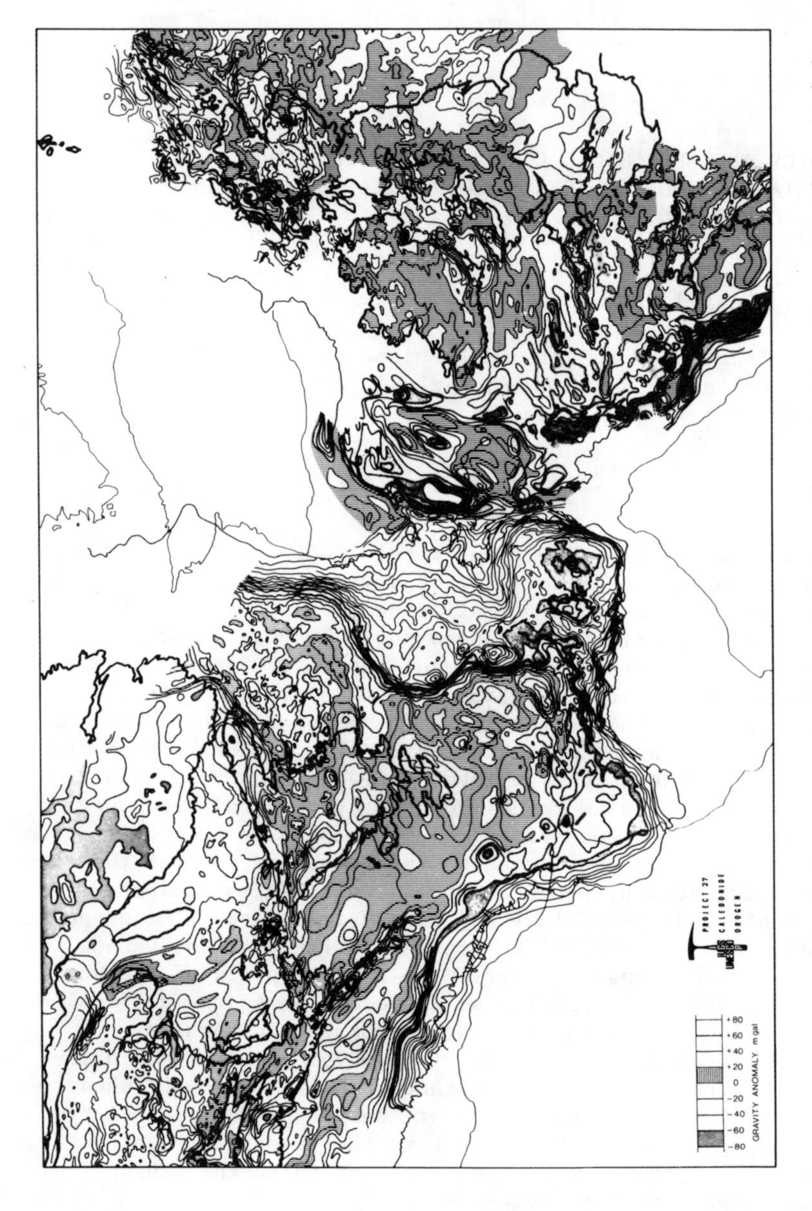

Figure 1: Compilation of gravity anomaly data for the Canadian, British, Irish and French continental margins (3) Courtesy of the Geological Society of America.

the United States (7) demonstrated the continuity of a line of gravity gradient associated with the Appalachians. Speculation about its significance (e.g. 8) continued until recent deep seismic reflection studies (9) confirmed that it marks a significant lateral change in crustal structure across the Appalachians. This recent interest in the deep structure has had two consequences: The publication of a new gravity map of the Appalachian Orogen (2) and the preparation of a new digital data file of U.S. gravity data (10) on which to base quantitative interpretation. Kane reports (this volume) on the crustal structure of the U.S. Appalachians as interpreted from these data.

The success of the digital analysis techniques developed by the U.S. Geological Survey has attracted other scientists who have been offered the use of their facilities. Wolff, for example, has been able to prepare a more detailed gravity map of Scandinavia (this volume). Independently, Hipkin (Edinburgh University) has put together a digital data base of gravity data for the northern half of the British Isles and the adjacent continental margins, and produced a series of similar derivative maps (e.g. 11) with an accompanying interpretative text (12). These activities will serve to expand the scope of present Appalachian-Caledonide gravity compilations (2;figure 1) and focus attention on those areas immediately adjacent to the coast (e.g. Miller, this volume) and at the edge of the continental margin where gravity data are sparse on both sides of the Atlantic.

It has always been difficult to make trans-Atlantic correlations between the U.S. east coast and West Africa because of the dearth of African data. Roussel, Lecorche, Liger and their colleagues have now produced a gravity map of western Mauritania and Senegal on the basis of which they interpret the crustal structure of the area (see Roussel and Lecorche, this volume) and by which comparisons can be made with the structure of the eastern United States (Lefort and Klitgord, abstract, this volume).

Magnetics and Paleomagnetism

Many countries having the orogen within their boundaries have carried out aeromagnetic surveys on a routine basis, but the proprietory nature of some of the data, the lack of lines that "knit" together the existing surveys, and the extensive marine areas separating different elements of the orogen have delayed the production of regional compilations. For example, the U.S. Geological Survey has only recently published its first magnetic anomaly map of the entire conterminous United States (13) despite having published a considerable number of maps for discrete parts of the United States, including the east coast offshore (14). However, the Geological Survey of Canada is now in the third edi-

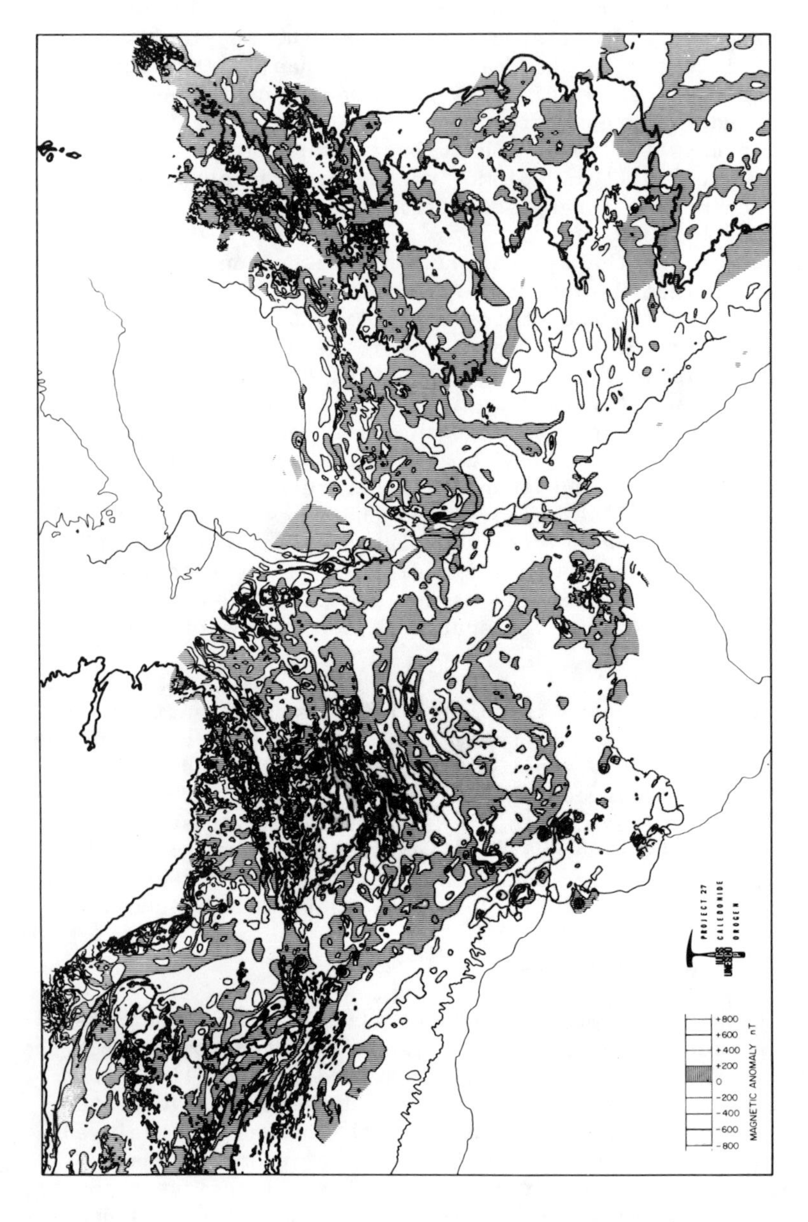

Figure 2: Compilation of magnetic anomaly data for the Canadian, British, Irish and French continental margins (3) Courtesy of the Geological Society of America.

tion of its magnetic anomaly map (15) with good Appalachian coverage (see Hood, this volume). The aeromagnetic map of the British Isles (16) has recently been supplemented by a compilation of magnetic data over Ireland and its adjacent continental margin (17). The magnetic map of France (18) and the Bay of Biscay (19) were both prepared from aeromagnetic data. Aeromagnetic surveys have also been completed for parts of Norway, Sweden, and the Norwegian continental shelf.

Compilations of all the available magnetic data for the Appalachians (1,20) and the British Caledonides (3) have been published (Fig. 2). Together with the published gravity data these have been the basis for attempts to correlate the geological zonation of the northern termination of the Appalachian Orogen with that of the British Caledonides (5,6, Fig. 3), and the structure of the eastern United States with that of west Africa (Roussel and Lecorche; Lefort and Klitgord; both this volume). The early Mesozoic configuration of the land masses surrounding the North Atlantic is well controlled according to the interpretation of oceanic magnetic lineations (e.g. 21,22). The relative positions of some of the structural elements of the orogen are likely to have changed between Hercynian and early Mesozoic time. Some such minor adjustments have been incorporated in the reconstructions of LePichon _et al._ (23) and Lefort (24). There are, however, suggestions from paleomagnetic data (e.g. 25; Seguin, this volume) of major displacements along the axis of the Appalachian-Caledonide system. Whereas this proposition is not necessarily incorrect, Roy _et al._ (this volume) caution geologists to recognize the uncertainties that are taken into account by paleomagnetists, but that may not be apparent to geologists using their data. These uncertainties are primarily associated with the potential difference between the age of a rock unit and the age of the magnetization that rock acquired. Paleomagnetists associated with the study group are currently trying to define the geological reliability of many of the published paleomagnetic poles.

Seismic Reflection and Refraction

Initially, subsurface structural models for the Appalachians and Caledonides were controlled by scattered seismic refraction lines interpreted to give average crustal sections in different areas. Extrapolation of these crustal sections was generally based on gravity data. Recent experiments such as the Blue Road Traverse across Scandinavia (26) and the LISP-B line along the length of the British Isles (27) have provided more detail for these crustal sections of the Caledonides. These sections have been compared, somewhat unsuccessfully (5), with sections based on compilations of seismic refraction data off Canada (28).

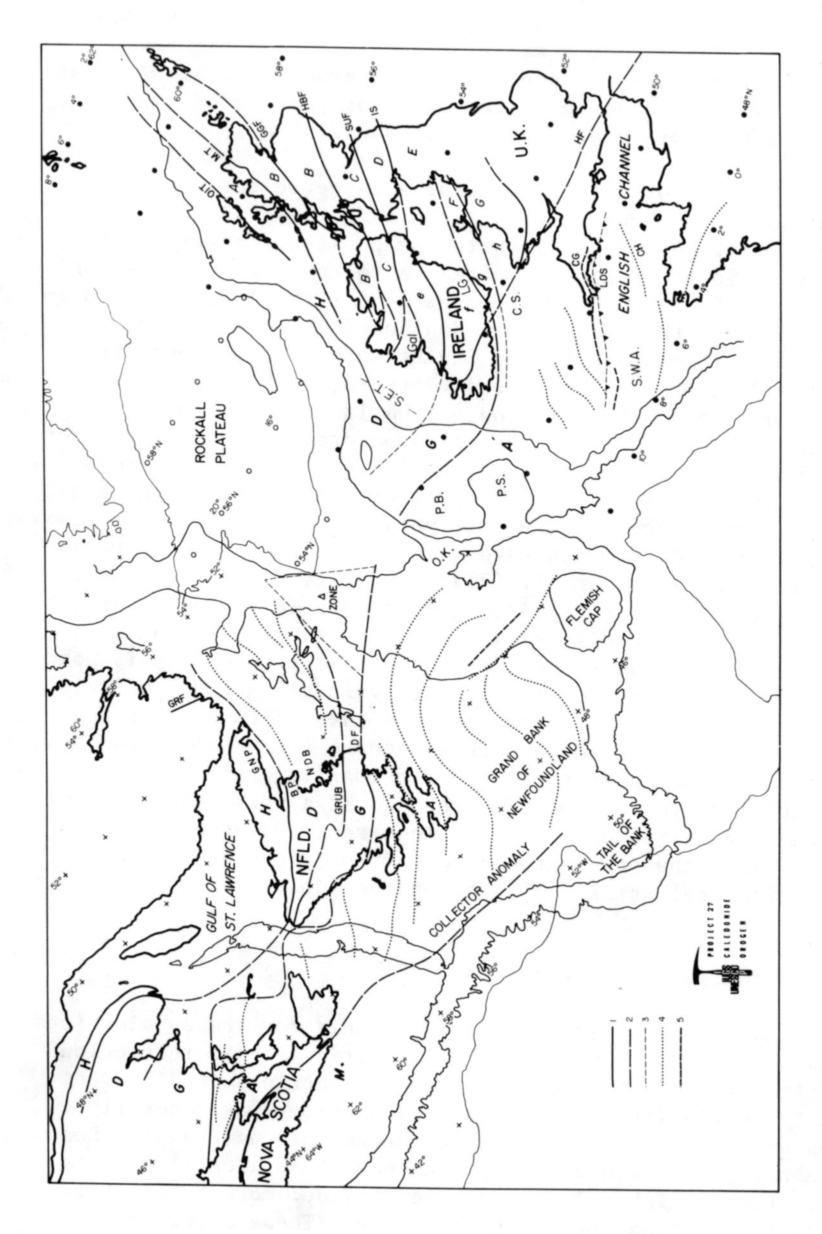

Figure 3: The geological zonation of Newfoundland and the British Isles and its offshore extension as defined geophysically (3) Courtesy of the Geological Society of America.

Meanwhile, the COCORP (9) and USGS (29) groups in the United States have been having spectacular results in their seismic reflection surveys indicating the allochthonous nature of much of the southern Appalachians. This has spurred activity in the northern Appalachians (30,31) where similar allochthony has been interpreted. An active program of deep marine seismic reflection profiling is also underway across the Caledonides of the British Isles (32). These results have focussed much interest in the direct observation of the deep structure of the Appalachian/Caledonide system, but considerably more work will have to be carried out to interpret the physical properties of the layers observed, followed by deep drilling confirmation, before the debate over interpretation of the profiles abates.

Seismicity

The Appalachian and Caledonides are relatively inactive seismically, but coordinated seismic networks in the northern (33) and southern Appalachians (e.g. 34) are beginning to show seismicity patterns that were previously hidden (35). The occurrence of relatively strong motion in New Brunswick in early 1982 (36) created an increased awareness of the seismic potential of the region, although the epicentre was not located on one of the geologically defined zone boundaries.

Magnetotellurics

Information on the electrical conductivity of the earth can be deduced by its susceptibility to imposed or induced electrical or magnetic fields. Syntheses of the irregularly distributed data in the Appalachians (37) point to the existence of a highly conductive zone crossing the northern Appalachians at a high angle to the Taconic trends. These interpretations are more difficult to accept than the results in the Caledonides of the British Isles where Hutton (38) has interpreted her results in terms of the structural layers observed on LISP-B (27).

CONCLUSION

Undoubtedly, deep seismic reflection profiling will remain the *prima donna* geophysical tool for several years until the arguments over the nature of deep seismic reflectors within the crust reach such a pitch that deep drilling must be given priority. By that time it is hoped that the argument that rages over the validity of paleomagnetically interpreted major longitudinal movements along the Appalachians will have been resolved. This will pave the way towards deciding how well we might expect there to be trans-Atlantic correlation between elements of the Appalachian, Caledonide, Hercynian and Mauritanide orogens. That path will, however, be more easily trod as the result of the compilation of all available geophysical data, the task to which this study group is dedicated.

REFERENCES

(1) Zietz, I., Haworth, R.T., Williams, H., and Daniels, D.L.: 1980, Memorial University of Newfoundland Map No. 2.
(2) Haworth, R.T., Daniels, D.L., Williams, H., and Zietz, I.: 1980, Memorial University of Newfoundland Map No. 3.
(3) Haworth, R.T., and Jacobi, R.D.: 1983, in Geol. Soc. America Mem. 158.
(4) Lefort, J.-P., and Haworth, R.T.: 1978, Can. J. Earth Sci. 15, pp. 397-404.
(5) Haworth, R.T.: 1981, Can. Soc. Petrol. Geol. Memoir 7, pp. 429-446.
(6) Jacobi, R.D., and Kristoffersen, Y.: 1981, Can. Soc. Petrol. Geol. Memoir 7, pp. 197-229.
(7) Woollard, G.P., and Joesting, H.R.: 1964, Am. Geophys. Union and U.S. Geol. Survey Map G64121.
(8) Diment, W.H., Urban, T.C., and Revetta, F.A.: 1972, in Nature of the Solid Earth (ed. E.C. Robertson) McGraw Hill pp. 544-572.
(9) Cook, F.A., Albaugh, D.S., Brown, D.L., Kaufman, S., Oliver, J.E., and Hatcher, R.D.Jr.: 1979, Geology 7,pp. 563-567.
(10) Simpson, R.W., Hildenbrand, T.H., Godson, R.H., and Kane, M.F.: 1982, U.S. Geol. Survey Open File Rept. 82-477.
(11) Hussain, A., and Hipkin, R.G.: 1982, Regional gravity map of the British Isles (Northern Sheet).
(12) Hipkin, R.G., and Hussain, A.: 1982, Inst. Geol. Sci. (U.K.) Rept. 82/10.
(13) Zietz, I. et al: 1982, U.S. Geol. Survey Map GP-954-A.
(14) Behrendt, J.C., and Klitgord, K.D.: 1979, U.S. Geol. Survey Map GP-931.
(15) McGrath, P.H., Hood, P.J., and Darnley, A.G.: 1977, Geol. Survey of Canada Map 1255A.
(16) Institute of Geological Sciences : 1965 and 1972, Aeromagnetic map of Great Britain, 2 sheets, 1:625,000.
(17) Max, M.D., Inamdar, D.D., and McIntyre, T.: 1982, Geol. Survey of Ireland Report 82/2 (Geophysics), 7 pp.
(18) Autran, A., and Weber, C.: 1971, Pub. Inst. Francais du Petr., Coll. Coll. et Semin. 22, IV.10
(19) Le Mouel, J.L., and Le Bourgne, E.: 1971, Pub. Inst. Francais du Petr., Coll. Coll. et Semin. 22, VI.3.
(20) Haworth, R.T., and MacIntyre, J.B.: 1975, Geol. Surv. Canada Paper 75-9, 22 pp.
(21) Sclater, J.C., Hellinger, S., and Tapscott, C.: 1977, J. Geology 85, pp. 509-552.

(22) Srivastava, S.P.: 1978, Geophys. J. Roy. Astron. Soc. 52, pp. 313-357.
(23) LePichon, X., Sibuet, J.-C., and Francheteau, J.: 1977, Tectonophysics 38, pp. 169-209.
(24) Lefort, J.-P.: 1980, Mar. Geol. 37, pp. 355-369.
(25) Kent, D.V., and Opdyke, N.E.: 1978, J. Geophys. Res. 83, pp. 4441-4450.
(26) Lund, C.E.: 1979, Geol. Foeren. Stockh., Foerh. 101, pp. 191-204.
(27) Bamford, D., Nunn, K., Prodehl, C., and Jacob, B.: 1977, J. Geol. Soc. Lond. 133, pp. 481-488.
(28) Haworth, R.T, and Keen, C.E.: 1979, Tectonophysics, 59, pp. 83-126.
(29) Harris, L.D., and Milici, R.C.: 1977, U.S. Geol. Survey Prof. Paper 1018, 40 pp.
(30) Ando, C.J., Cook, F.A., Oliver, J.E., Brown, L.D., Kaufman, S., Klemperer, S., Czuchra, B., and Walsh, T.: 1981, EOS (Trans. Am. Geophys. Union) 62, pp. 1046-1047.
(31) Ministere des Richesses Naturelles du Quebec : 1979, Open File (DP-) 721 and 722.
(32) Smythe, D.K., Dobinson, A., McQuillin, R., Brewer, J.A., Matthews, D.H., Blundell, D.J., and Kelk, B.: 1982, Nature Lond. 299, pp. 338-340.
(33) Vudler, V. and Celata, M.A. (eds): Northeastern U.S. Seismic Network Bulletins, Weston Observatory, Boston College, Massachusetts, U.S.A.
(34) Carver, D., Turner, L.M., and Tarr, A.C.: 1977, U.S. Geol. Survey Open File Rept. 77-429, 66 pp.
(35) Sbar, M.L., and Sykes, L.R.: 1973, Geol. Soc. Am. Bull. 84, pp. 1861-1882.
(36) Hasegawa, H.S., Adams, J., Wetmiller, R.J., and Basham, P.W.: 1982, EOS (Trans. Am. Geophys. Un.) 63, p. 1249.
(37) Bailey, R.C., Edwards, R.N., Garland, G.D., and Greenhouse, J.P.: 1978, Geophys. J. R. Astr. Soc. 55, pp. 499-502.
(38) Hutton, V.R.S., Ingham, M.R., and Mbipom, E.W.: 1980, Nature Lond. 287, pp. 30-33.

THE PALEOMAGNETIC RECORD OF THE APPALACHIANS

J. L. Roy, E. Tanczyk and P. Lapointe

EARTH PHYSICS BRANCH;
Energy, Mines and Resources
Ottawa, Ontario, Canada, K1A 0Y7
Contribution No. 1043

The present study synthesizes paleomagnetic results from the Paleozoic of North America, revealing important features of the tectonic history of the Appalachians. The far-reaching effects of the Hercynian Orogeny are apparent, as its magnetic signature can be detected on both sides of the Appalachian structural front. Appalachian rocks usually carry multi-remanences composed of the initial remanence acquired during rock formation, and later remanences produced during subsequent tectonic events. In some instances the initial remanence may have been erased and replaced by one of orogenic age. Many poles are believed to be signatures of orogenic episodes of Hercynian or Acadian age. Thus pole ages and rock unit ages can by no means be equated. Hypotheses based on the assumption of the contemporaneity of pole and rock unit ages are unsubstantiated unless it is proven conclusively, that poles of the same age are being compared. Only by recognizing the signatures particular to specific tectonic events is it possible to reconstruct, on the basis of the paleomagnetic record, the regional extent of orogenic episodes, and to further our knowledge of Appalachian tectonics throughout the Paleozoic.

INTRODUCTION

In the last decade, detailed paleomagnetic studies have shown that the natural remanent magnetization (NRM) of rocks, is commonly composed of several remanences acquired at different times during the rock's history. Consequently, the interpretation of paleomagnetic data has been complicated by the need to resolve several magnetic components, and to identify the geologic age at which a given remanence was acquired. This two-fold problem is

P. E. Schenk (ed.),
Regional Trends in the Geology of the Appalachian-Caledonian-Hercynian-Mauritanide Orogen, 11–26.

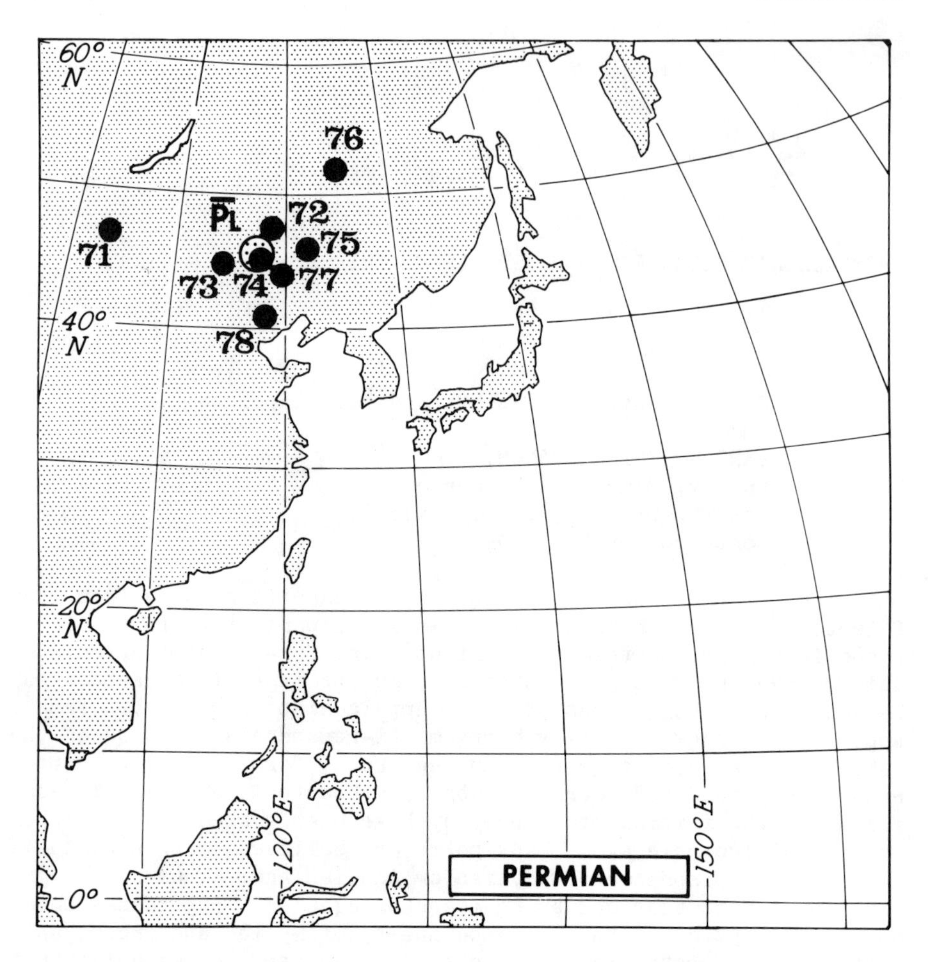

Fig.1. Paleopoles derived from Permian rock units: nos. 71 to 78 in Table 1 of Irving and Irving, [24] - Pl is the calculated mean pole.

summarized by Buchan et al., [1]: "paleomagnetic techniques for separating multiply-overprinted NRM's are just being developed. Determining reliable paleomagnetic directions for the components --- is not enough. Determining the origins of the components --- is an obligatory first step in assigning relative and ultimately absolute ages to the components. Age determination is almost taken for granted in single-component studies where a primary NRM can be assumed, but it is far from trivial in rocks that have been repeatedly remagnetized".

Multi-remanences have been found in a large number of North

American Paleozoic rocks (e.g. Van der Voo and French, [2], and especially in Appalachian rocks which have been affected by various orogenic episodes (Taconic, Acadian, Hercynian). Paleomagnetic results can be and have been used to formulate hypotheses about relative motions during the Paleozoic among different parts of North America in general and about the evolution of the Appalachians in particular. The validity of some such hypotheses will be assessed in the light of age determinations on poles from the presently available paleomagnetic data base.

THE SIGNIFICANCE OF PALEOMAGNETIC POLES

Paleomagnetic results are often expressed in terms of paleomagnetic poles which are not measured directly, but are a convenient means for comparing results from different localities. The significance of a paleomagnetic pole is evaluated by assessing both the paleomagnetic reliability and the age control of the remanence from which the pole was derived.

A remanence is paleomagnetically reliable if its direction is a representation of any true paleofield. This is accomplished by isolating it from all other remanences contained in the rock. Often, the total remanence carried by a rock unit is the vector sum of a number of remanence vectors, each representing a true paleofield direction, but at different points in time. Such a paleomagnetic record, often present in rock units of fold belts, can be most informative about the tectonic development of orogenic terranes. The interpretation of these records requires that all remanences be recognized. This usually necessitates extensive experimental work using a variety of techniques [3,4]. If one of the remanences remains undetected, then none of the <u>observed</u> remanence vectors, being the resultant of unresolved vectors, will represent a paleofield accurately. The magnitude of the divergence between the observed direction and the paleofield direction is dependent upon the vectorial differences among the undetected and detected remanences.

Therefore, the divergence, whose real cause remains unrecognized, may be insignificant, in which case the observed vectors are still reasonable estimates of paleofield directions, or it may be appreciable, in which case the divergence may be erroneously attributed to other causes (e.g. apparent polar wander, relative motions between sampling localities, tilting, etc.) Even worse, if the undetected remanence happens to be the initial remanence, another remanence may be mistaken for the initial one, leading to an incorrect tectonic interpretation as discussed below. Therefore, if the paleomagnetic record is to be interpreted correctly, one must ensure that all remanences have

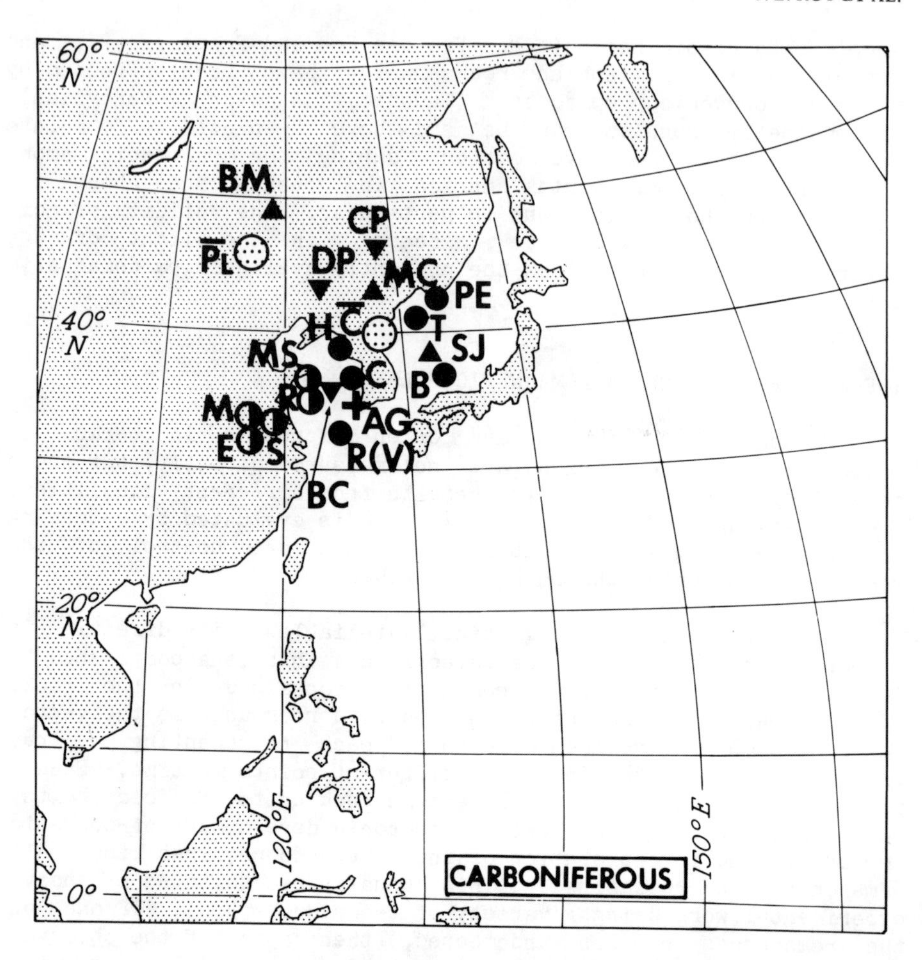

Fig.2. Paleopoles derived from Carboniferous rock units. Re-drawn from Roy and Morris, [9]; original references are: 25, 26, 27, 28, 29, 30, 31, 8, 32, 33, 10, 11, 34 and 35. C is the calculated mean pole. The pole mean (stippled circle) for the Early Permian is from Irving and Irving, [24].

been at least uncovered and, whenever possible, isolated by experimental and/or vectorial analyses, because only an isolated remanence can provide one with a truly reliable pole.

All reliable poles are valid representations of events that have affected a given rock unit, but they are of limited value unless their ages are properly established. The initial pole derived from the remanence acquired during rock formation can be associated with the rock age, but all others represent subsequent events. The assignment of ages to poles is not an easy task [1,5]. In many instances, the initial remanence has been erased by subsequent overprints, or has not been uncovered owing to insufficient experimental work. In the case of a sedimentary rock unit, a pole may be derived from a remanence shown to be pre-folding. That evidence by itself only brackets an age interval (rock age to folding age) for the pole. Similarly, for an extensive igneous body, the age of the remanence can be confidently associated with the body only if both radiometric dating and paleomagnetic results originate from the same localities. Unfortunately, this has rarely been achieved. Authors of an original paper are usually careful to point out the known age limitations. However, when results are summarized in review papers, the age uncertainties associated with ages are often neglected.

Because of the difficulties in pole age assignments, it has been common practice in paleomagnetism to identify poles according to the rock unit from which they were derived. Irving et al., [6], carefully point out in their catalogue listing that even the most reliable determination should "NOT be taken to mean the magnetization necessarily records the field at the time the rock unit was formed. Detailed studies are needed before the age of magnetization can be determined". Thus, for example, the listing of a pole under a Devonian heading means that it was derived from a Devonian rock unit. It should not be construed that the pole is necessarily of Devonian age.

It is the principal purpose of this article to assess paleomagnetic data as a means of studying the evolution of the Appalachian Orogen during the Paleozoic; in particular, to recognize possible relative motions of blocks within the Appalachians and of the entire mobile belt in relation to the North American craton. However, because the evolutionary processes of the Earth's crust change with time, it is often difficult to define the extent of the craton for a given epoch. The Appalachians have been subjected to at least three orogenic episodes (Taconic, Acadian and Hercynian). Hence, it is most probable that the eastern limits of the craton were quite different from Cambrian to Permian times. In this brief review, for simplicity, the Appalachian structural front, [7], will be used as the demarcation line for comparing results from the Appalachians with results from Interior North America. We stress that this line is used only to locate geographically the paleomagnetic sampling localities. It should not be taken to indicate the boundary of any of the orogenic episodes, as some may have extended past that line, affecting rock units to the west. Only well documented and generally accepted poles are considered in the following.

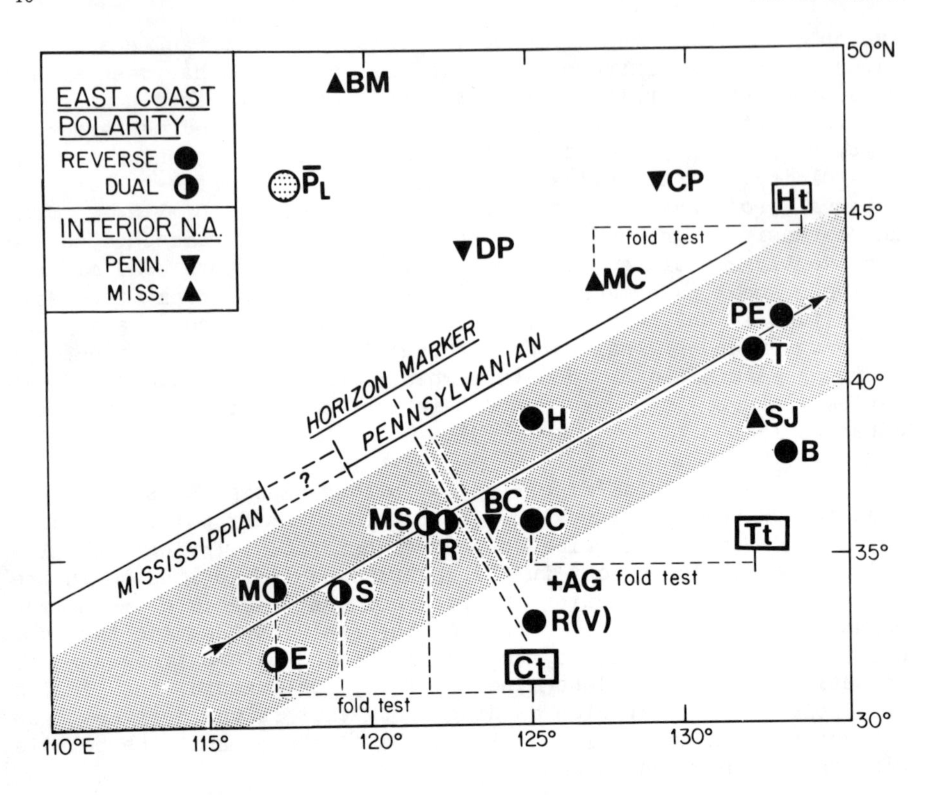

Fig.3. Paleopoles derived from Carboniferous rock units shown on expanded scale (modified from Roy and Morris, [9]). The probable location of the horizon marker is shown. Paleopoles derived from remanences acquired prior to folding are shown with the upper limit of the folding event indicated: Ct (Cumberland time). Tt (Tormentine time), Ht (Hercynian time). Symbol (+) refers to the undetermined age of the AG pole (Austell gneiss). It may range from Devonian to Mississippian (Ellwood and Abrams, [35]).

Permian

Poles from Lower Permian rock units shown in Fig.1 have all been derived from Interior North America, and are shown principally for completeness and comparison with other Paleozoic poles. Diehl and Shieve, [8], have suggested that the East Coast and Interior North America may have been separate entities until the Permian. Their analysis was based on equating pole age to rock age, an unacceptable procedure for reasons discussed above. The broad range of Late Pennsylvanian to Permian, assigned to the rock units themselves, casts further doubt on the age control of the remanences. Thus paleomagnetic data are still inadequate for drawing conclusions on Appalachian tectonics in the Permian.

Carboniferous

Poles derived from Carboniferous rock units are shown in Figs.2 and 3. The circle symbols indicate that the poles were derived from nine red bed formations from the Carboniferous basin of the Maritime Provinces in Canada. They show a well defined apparent polar displacement of at least 16° during the Carboniferous period (Fig.3). The pole sequence running from southwest to northeast conforms to the stratigraphic column indicating that pole ages are closely related to rock ages. Further age control is provided by positive fold tests as indicated (Fig.3). For example, all remanences yielding poles E, M, S and MS were acquired prior to the folding that occurred before the deposition of the Cumberland Group (Ct) from which pole C was derived. Most importantly, the results define a time stratigraphic horizon marker that could be of global significance. All Carboniferous units postdating this time line yield a single reverse polarity (full circles), while all units predating the time line yield dual polarity remanences (half open circles), (Fig.3). This horizon marker has been pinpointed at the Riversdale Formation level (Early Pennsylvanian; Namurian-Westphalian boundary) (Roy and Morris, [9]).

Six poles have been derived from sedimentary units from Interior North America. The assignment - Mississippian or Pennsylvanian - of the <u>rock unit age</u> is identified by upright or inverted triangles respectively. However, the pole ages are not accurately defined. Only the Late Mississippian Mauch Chunk pole (MC) is derived from a remanence acquired prior to folding, that apparently took place during the Permian. Thus, as stated by Knowles and Opdyke [10], "The actual age of magnetization could be anywhere from Upper Mississippian to Lower Permian". The remaining five poles are not constrained by a fold test and could be of any (post-depositional) age, except that their locations indicate that they are probably Permo-Carboniferous. This is supported by the fact that all six poles are derived from a single reverse polarity that characterizes the Late Carboniferous-Permian interval. The BM pole derived from the Barnett Formation of Texas, lies some 15°to the north of the Mississippian poles from the Northeastern Appalachians. Assuming that this pole represents the location of the North American craton in Mississippian time, Kent and Opdyke, [11], have argued that the 15°difference provides one with further evidence for their earlier suggestion, [7], based on Devonian results (discussed below) that the Northeastern Appalachians did not become attached to North America until Late Carboniferous. There are several reasons to suggest that the BM pole is not of Mississippian age aside from the fact that its reliability as a paleofield indicator is questionable, [9].

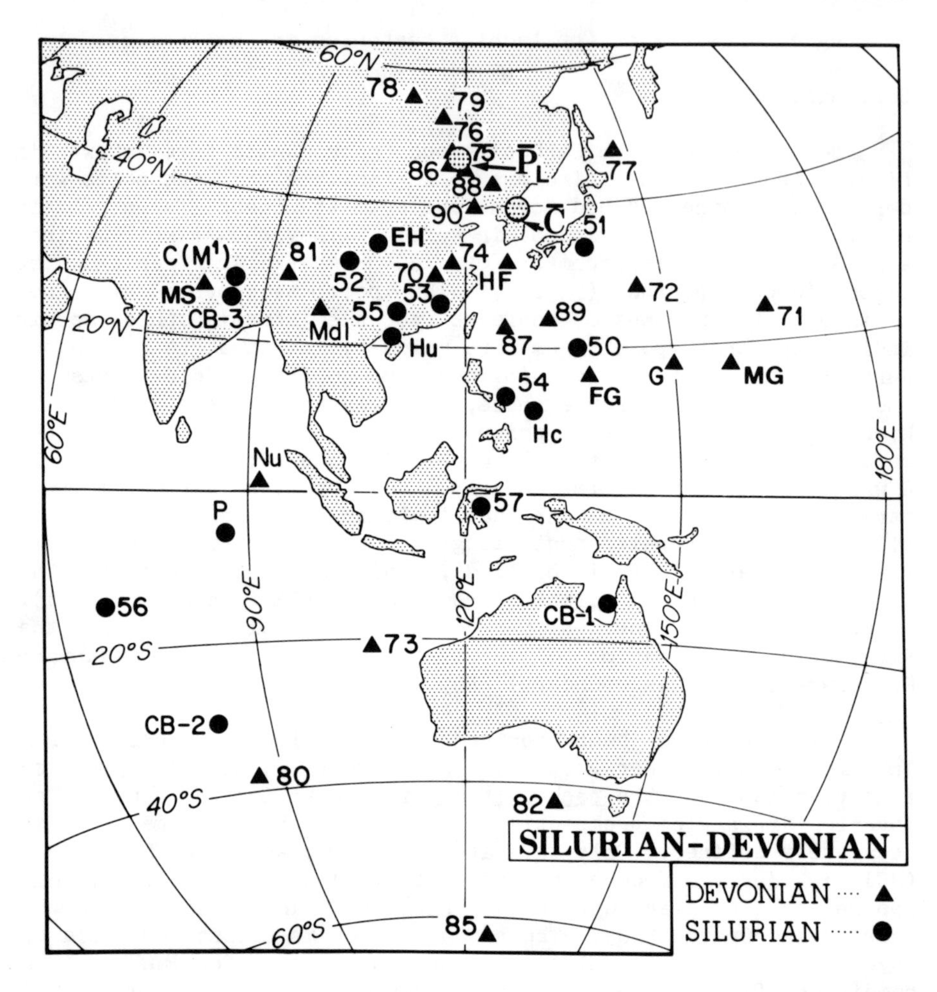

Fig.4. Paleopoles derived from rock units of Devonian to Silurian age. Re-drawn from Roy and Anderson, [4]. See also Roy et al., [36], for list of poles. Additions: CB-1, CB-2, CB-3, MS, HF, FG, G and MG from Seguin, Rao and Pineault, [13], Hu, Hc and EH from Roy, [5] (recalculated from Kent and Opdyke, [37]); Mdl and Nu from Dankers, [38].

Firstly, it lies very close to the Early Permian mean (Pl, Fig.2), indicating that it could be of Permian age. Secondly, the sampling locality may not have been part of Interior North America at that time, [12]. Thirdly, a dual polarity, expected if the remanence was of Mississippian age (Fig.3), was not detected. Using the horizon marker, Roy and Morris [9], have suggested that all the Interior North American poles shown in Fig.3 are post time-line in age, and that none of them is representative of the position of North America prior to the time-line.

Devonian-Silurian

Poles derived from Devonian and Silurian rock units are shown together (Fig.4), because the rock unit age in these time intervals is not always accurately known; e.g. the age of the St.George pluton yielding poles 80 and 81 is near the Devonian-Silurian boundary. The scatter of poles illustrates the complexities that can arise when dealing with multi-remanences. It should be realized that most of the rock units studied have been subjected to more than one orogenic episode: primarily the Acadian and the Hercynian. Consequently, many of them have yielded multi-poles; e.g. CB-1, CB-2 and CB-3 were derived from the same rock unit (Fig.4, Seguin et al., [13]. It is impractical to review all of the results in detail here; readers are referred to the original articles listed in the figure captions.

The much discussed hypothesis, [14,7,15,4,13], that the northeastern Appalachians were detached from Interior North America during the latter part of the Paleozoic era is based, in essence, on the observation that poles (75, 76, 78, 79, 86 and 88 referred to as group K; Fig.4) derived from Interior North America lie some 15°to the north of pole 74 derived from the Late Devonian Perry Formation [16]. However, pole 74 was derived from two only partially resolved remanences, and the authors quoted that it "may be in error by 15°or more" [16], p.1175. Such an uncertainty obviously precludes detection of latitudinal discrepancies of the same magnitude. Furthermore, the pole ages of group K are poorly defined. Only one of them (88, (Catskill Fm., [15]), was derived from a remanence acquired prior to what seems to be Hercynian folding. Therefore owing to the near perfect agreement between group K poles and the Early Permian mean pole (Pl, Fig.4), the possibility that those poles are of Permian age is not easily discounted. This possibility is further enhanced by the fact that pole 90 (Onondaga Limestone; [17]), derived from a remanence acquired after a folding event of probable Hercynian age, falls into that group. It should be realized that although the sampling localities are west of the demarcation line adopted in this article, most of them nonetheless, are located in parts of the Appalachians which could have been affected by the Hercynian Orogeny. Thus there exists considerable evidence to suggest that poles in group K were derived from remanences imprinted by the Hercynian Orogeny.

The remaining poles of Fig.4 have largely been derived from localities east of the demarcation line, and the significance of their scatter is not fully understood. Most of them are the result of extensive studies, and the isolated remanences can be

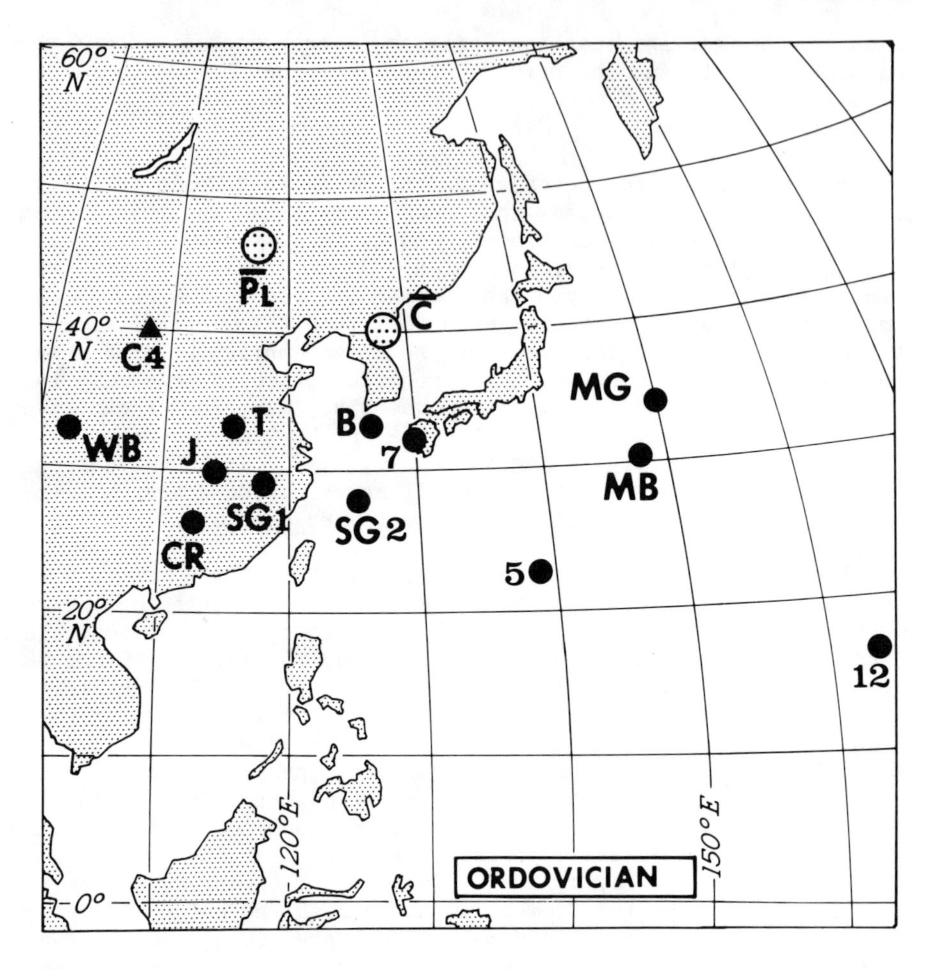

Fig.5. Paleopoles derived from Ordovician rock units. Re-drawn from Seguin et al., [22]. Additions: 5, 7, 12 and WB from Buchan and Hodych, [20].

considered reliable representations of the paleofield at the time of remanence acquisition. However, a fully accurate paleomagnetic result is achieved only if the studied rock unit is restored to its original attitude. This is not easily accomplished, as explained in detail by MacDonald, [18]. Gillett [19], has also shown that differences between contemporaneous poles may be attributable to block rotation. Considering that the rock units have been subjected to two orogenic episodes, and that in many instances, the rock age itself is not accurately defined, it is not surprising to find a scatter among poles derived from different localities.

In spite of these uncertainties we believe that these scattered poles lying a considerable distance away from the Carboniferous pole (Fig.4), are derived from remanences older than Carboniferous and indicate that the Devonian and Silurian paleopoles are different from the Carboniferous or Permian paleopoles. Whether the difference is attributable to polar wandering or to plate or block motions cannot yet be established. If plate motion is involved, it is not clear whether it is applicable to the Appalachians only, or to North America as a whole, since no data, which can positively be identified with the Silurian-Devonian, are available from Interior North America.

Worth noting is a broad band of poles about 30° N latitude. Such a pattern could be a characteristic paleomagnetic signature of local rotations, [4].

Ordovician

Poles derived from Ordovician rock units are shown in Fig.5. Of special interest is the fact that poles from Newfoundland, (WB; 5,7, SG1 and SG2), are intermingled with the poles from Interior North America. Buchan and Hodych [20] have suggested that, if these poles are of Ordovician age, then the paleomagnetic evidence indicates that the Avalon zone of Newfoundland and Interior North America were located on a similar paleolatitude in Ordovician times. The similarity between the distribution of poles about 30° N latitude in Fig.5, and a similar pattern of poles in Fig.4 is remarkable. The possibility that the poles derived from Ordovician rock units may be of younger age cannot therefore be ruled out.

Cambrian

Fig.6 shows, for completeness, poles derived from Cambrian rock units, though none of these were obtained from the Appalachians. However, results obtained from the St.Lawrence Lowlands, [21,22], indicate that this area was an integral part of Interior North America in Cambrian times. The results also indicate that during the Cambrian, the apparent pole moved by some 80°-90° in a north-westerly direction.

CONCLUSIONS

The paleopoles shown in the figures represent data points of a paleomagnetic record produced during the Paleozoic era. The record is far from complete, and unless the significance (reliability and age) of each data point is carefully analysed, a variety of sometimes conflicting hypotheses can be advanced resulting in increasing confusion. Much of this can be avoided if

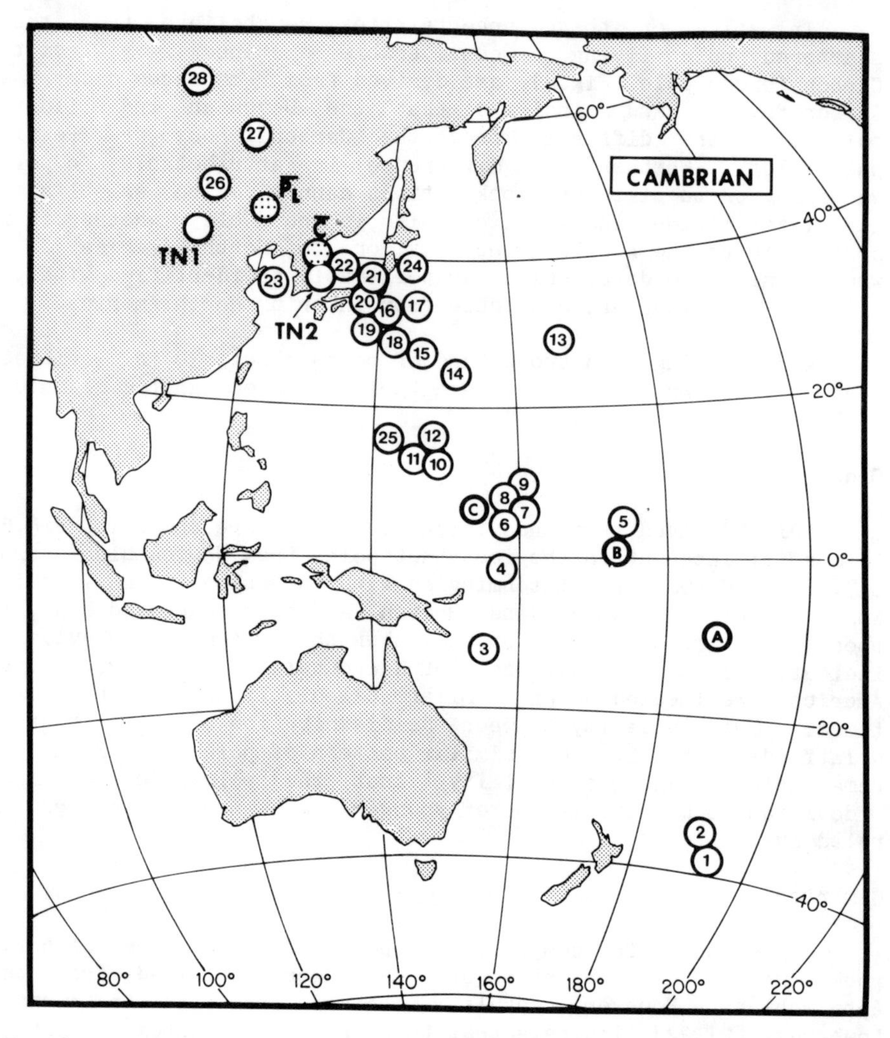

Fig.6. Paleopoles derived from Cambrian rock units. Re-drawn from Dankers and Lapointe, [21]. Additions: TN1 and TN2 from Gillett, [9].

certain misconceptions are dispelled and certain rules followed. In assessing the significance of a paleopole, particular attention must be given to the age of remanence. It cannot be assumed a priori that this age is equivalent to the age of the rock unit, as more than one reliable paleopole can be obtained from a single rock body. This means that a distinction must be made between the contemporaneity of poles and the contemporaneity of rock units from which poles are derived. It is absolutely essential to establish the contemporaneity of poles before attempting to demonstrate relative motion between terranes. Ignoring such basic rules can artificially increase the accreditation of "tentative" hypotheses which are initially based on data of doubtful quality. The credibility or gradual acceptance of a hypothesis should not be, but unfortunately often is, upgraded by adding more questionable evidence (the band-wagon syndrome). A case in point is the hypothesis regarding the possible Paleozoic displacement of the northern Appalachians with respect to Interior North America. Lately, abundant results in support of the hypothesis are apparently being discovered from Devonian, Carboniferous and even Permian rocks. To non-paleomagnetists, it might appear that the quantity of favorable evidence is overwhelming. However, as the original papers reveal, the "evidence" is based in most cases on the assumption that the pole age and rock unit age are contemporaneous. Drawing parallels between poles from Permian formations of Interior North America and Appalachian poles from Carboniferous rock units, [8], is hardly a convincing procedure. As the difference in poles can, for instance, be attributable to apparent polar wander, comparison of non-contemporaneous poles as an argument for relative motion is clearly unacceptable.

In order to circumvent the pitfalls caused by uncertainties in pole age determinations, and to avoid improbable scenarios, the Paleozoic data base as a whole can be inspected for internal consistency. A series of interesting patterns are revealed:

Results from Interior North America

The bulk of the poles obtained from Late Cambrian to Permian rock units plot near longitude 120°E and latitude 30°-40°N (Figs.1 to 6). Two interpretations are compatible with this observation: either North America remained stationary for some 300 Ma, or the poles are derived from remanences acquired during the Late Paleozoic. We believe that the latter is more plausible as, apparently, all other cratons moved during this interval [23]. It is noteworthy that several of the sampling areas are located in Appalachian terranes (west of the structural front).

Results From Appalachian Terranes

Poles derived from Ordovician to Devonian rock units (Figs.4 and 5) can be subdivided into three groups: 1) poles falling close to the Carboniferous and Permian mean poles, 2) poles strung along latitude 30°N, and 3) poles in lower latitudes (20°N-60°S).

The following suggestions can be advanced: poles in group 1) are derived from remanences imprinted by the Hercynian Orogeny.

The linear pattern of poles in group 2) may be explained by a local block rotation at the time of the Acadian Orogeny. Poles in group 3) represent remanences pre-dating the Acadian Orogeny.

If our assessment is correct that poles in groups 1), 2) and 3) are Hercynian, Acadian and Pre-Acadian signatures respectively, then paleomagnetic results provide a very powerful tool for delineating different periods, or even epochs, of the Paleozoic. As more data points become available, the regional extent and severity of each orogenic episode could be contoured. Only then will it become possible to distinguish between magnetic overprints caused by intraplate deformation, and well-defined pole divergences caused by interplate motions between the Appalachians and the North American craton. The essential condition for drawing such conclusions on a regional level is comparison of contemporaneous paleopoles.

REFERENCES

[1] Buchan, K.L., Berger, G.W., McWilliams, M.O., York, D., and Dunlop, D.J.: 1977, Adv. Earth Planet. Sci.1, pp. 167-178.
[2] Van der Voo, R., and French, R.B.: 1977, J. Geophys. Res. 82, pp. 5796-5802.
[3] Roy, J.L. and Lapointe, P.L.: 1978, Phys. Earth Planet. Int. 16, pp. 20-36.
[4] Roy, J.L. and Anderson, P.: 1981, J. Geophys. Res. 86, pp. 6351-6368.
[5] Roy, J.L.: 1982, Can. J. Earth Sci. 19, pp. 225-232.
[6] Irving, E., Tanczyk, E. and Hastie, J.: 1976, Earth Physics Branch, Energy, Mines and Resources Canada, pp 99.
[7] Kent, D.V. and Opdyke, N.E.: 1978, J.Geophys. Res. 83, pp. 4441-4450.
[8] Diehl, J.F. and Shieve, P.N.: 1981, Earth Planet. Sci. Letters 54, pp. 281-292.
[9] Roy, J.L. and Morris, W.A. (In preparation)
[10] Knowles, R.R. and Opdyke, N.D.: 1968, J.Geophys. Res. 73, pp. 6515-6526.
[11] Kent, D.V. and Opdyke, N.D.: 1979, Earth Planet. Sci. Letters 44, pp 365-372.
[12] Irving, E.: 1979, Can.J.Earth Sci. 16, pp 669-694.
[13] Seguin, J.K., Rao, K.V. and Pineault, R.: 1982, J. Geophys. Res. 87, pp. 7853-7864.

[14] Schutts, L.D., Brecker, A., Hurley, P.M., Montgomery, C.W. and Krueger, H.W.: 1976 Can.J.Earth Sci. 13, pp. 898-907.
[15] Van der Voo, R., French, A.N., and French, R.B.: 1979, Geology 7, pp. 345-348.
[16] Robertson, W.A., Roy, J.L. and Park, J.K.: 1968, Can.J.Earth Sci., 5, pp. 1175-1181.
[17] Kent, D.V.: 1979, J.Geophys. Res. 84, pp. 3576-3588.
[18] MacDonald, W.D.: 1980, J.Geophys. Res. 85, pp. 3659-3669.
[19] Gillett, S.L.: 1982, Earth Planet. Sci. Letters 58, pp. 383-394.
[20] Buchan, K.L. and Hodych, J.P.: 1982, Can.J.Earth Sci.19, pp. 1055-1069.
[21] Dankers, P. and Lapointe, P.: 1981, Can.J.Earth Sci.18, pp. 1174-1186.
[22] Seguin, M.K., Rao, K.V., and Arnal, P.: 1981, Earth Planet. Sci. Letters 55, pp. 433-449.
[23] Morel, P. and Irving, E.: 1978, Jour. Geol. 86, 5, pp. 535-561.
[24] Irving, E. and Irving, G.A.: 1982, Geophysical Surveys 5, pp. 141-188.
[25] Roy, J.L.: 1966, Can. J. Earth Sci. 3, pp. 139-161.
[26] Roy, J.L., Robertson, W.A., and Park, J.K.: 1968, J. Geophys. Res. 73, pp. 697-702.
[27] Roy, J.L.: 1969, Can. J. Earth Sci. 6, pp. 663-669.
[28] Roy, J.L.: 1977, Can. J. Earth Sci., 14, pp. 1116-1127.
[29] Roy, J.L. and Park, J.K.: 1969, J. Geophys. Res. 74, pp. 594-604.
[30] Roy, J.L. and Robertson, W.A.: 1968, Can. J. Earth Sci. 5, pp. 276-285.
[31] Roy, J.L. and Park, J.K.: 1974, Can. J. Earth Sci. 11, pp. 437-471.
[32] Helsley, C.E.: 1965, J. Geophys. Res. 70, pp. 413-424.
[33] Payne, M.A., Shulik, S.J. Donahue, J., Rollins, H.B. and Schmidt, V.A.: 1981, Phys. Earth Planet. Int. 25, pp. 113-118.
[34] Scott, G.R.: 1979, J. Geophys. Res. 84, pp. 6277-6285.
[35] Ellwood, B.B. and Abrams, C.: 1982, J.Geophys. Res. 87, pp. 3033-3043.
[36] Roy, J.L., Anderson, P. and Lapointe, P.L.: 1979, Can.J.Earth Sci., 16, pp. 1210-1227.

[37] Kent, D.V. and Opdyke, N.D.: 1980, Can.J.Earth Sci. 17, pp. 1653-1665.
[38] Dankers, P.: 1982, Can.J.Earth Sci. 19, pp. 1802-1809.

PALEOMAGNETISM OF LOWER CARBONIFEROUS (TOURNAISIAN TO NAMURIAN) VOLCANIC ROCKS OF NEW BRUNSWICK, CANADA, AND ITS TECTONIC IMPLICATIONS

Maurice K.-Seguin, Paleomagnetic Laboratory, Université Laval, Quebec, P.Q., G1K 7P4 Canada &
L. R. Fyffe, Department of Natural Resources of New Brunswick, Geological Surveys Branch, P. O. Box 6000, Fredericton, N. B., E3B 5H1, Canada.

ABSTRACT ONLY

As part of a major study of the northern Appalachians, the paleomagnetism of well dated lower Carboniferous volcanic flows is described. The tectonostratigraphic zones of New Brunswick and the distribution of paleomagnetic sampling sites are shown in Figure 1. Detailed alternating field and thermal experiments performed on the New Brunswick volcanic rocks indicate that they are characterized by two different mean directions of magnetization in the 20-50 mT coercivity and 400-600°C temperature spectra: ESE (D=160°, I=+15°, K=9, N=11 sites) and W (D=264°, I=+34°, K=9, N=3 sites) for the formation <u>in-situ</u> and ESE (D=158°, I=+16°, K=8.5) and W (D=267°, I=+38°, K=10) after tilt correction. After exclusion of site C-7, an altered acidic tuf, the predominent cleaned ESE component is D=159°, I=+12°, K=9. The antiparallel direction is D=166°, I=+11°. K=25, and the corresponding palepole position 37°N, 131°E; it is not significantly different from early Carboniferous poles from the stable North American craton.

Since there was little apparent polar motion in the early Carboniferous interval, the results obtained from this study indicate that the North American craton and northern Appalachians shared a common apparent polar wander path in this time interval. We conclude that Kent and Opdyke's hypothesis (1980) for a sinistral motion north of the Maine coastal volcanic belt bringing the New England-Canadian Maritimes region to its present position with respect to cratonic North America during the Carboniferous is

P. E. Schenk (ed.),
Regional Trends in the Geology of the Appalachian-Caledonian-Hercynian-Mauritanide Orogen, 27–28.

not warranted by this investigation.

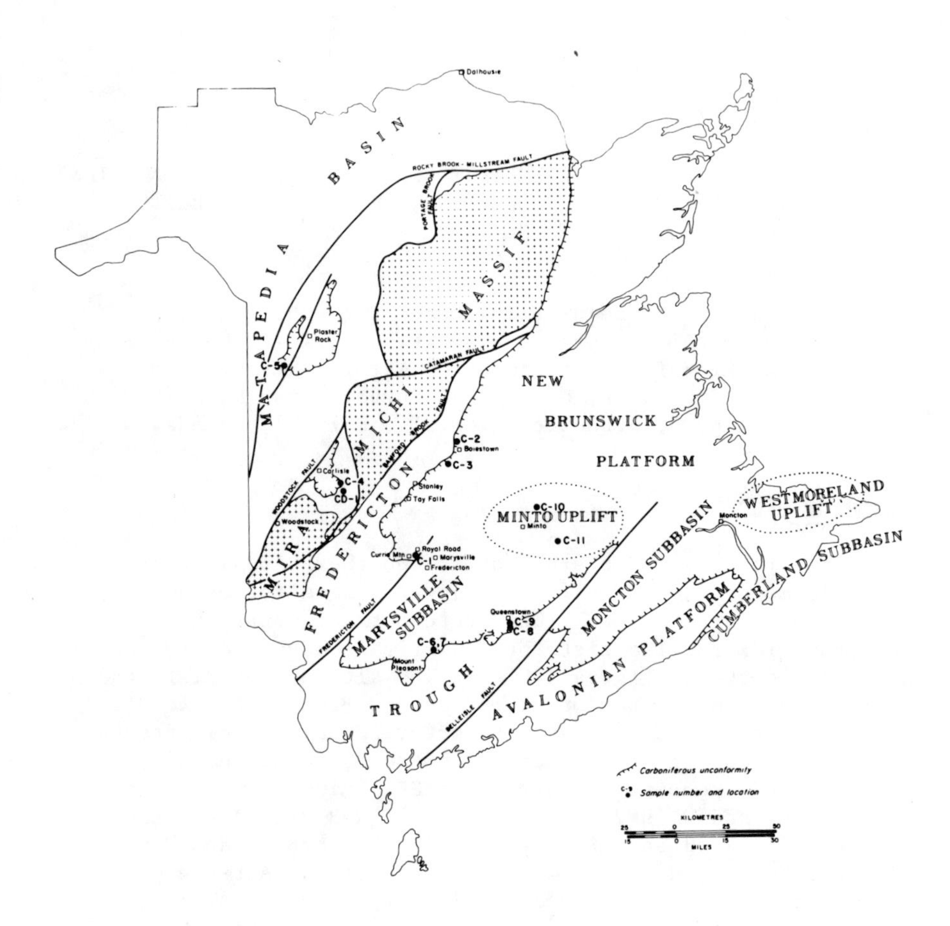

Fig. 1: Distribution of paleomagnetic sampling sites with respect to the tectonostratigraphic zones of New Brunswick

Reference:

Kent, D. V., and Opdyke, N. D. (1980). Paleomagnetism of Siluro-Devonian rocks from eastern Maine. Can. J. Earth Sci., 17, pp. 1653-1665.

GRAVITY AND MAGNETIC STUDIES OF CRUSTAL STRUCTURE IN NEWFOUNDLAND

MILLER , H.G.

Department of Earth Sciences,
Memorial University of Newfoundland,
St. John's, Newfoundland, Canada A1B 3X5

INTRODUCTION

The concept of geological zonation in the Appalachian Caledonian orogen has been developed and refined over the past two decades (Williams, 1964; Williams, et al, 1972, 1974; Williams, 1979) (Fig. 1). Concurrent with the development of this concept geophysical surveys were being conducted in the Newfoundland portion of the Appalachians which resulted in the refinement of the zone boundaries (Miller and Deutsch, 1976; Miller, 1977; Miller and Weir, 1982). In addition, the geophysical data have enabled the geological zones to be extended to offshore areas (Haworth, 1980; Haworth & Jacobi, 1982; Haworth & Miller, 1982; Miller, 1982).

This paper reviews the results from the systematic gravity surveys conducted specifically to deduce the crustal structure of Newfoundland. The objective is to summarize the data bases from which the geophysical interpretations are derived, to indicate the geophysical contributions to our understanding of the three dimensional nature of the crustal structure in the key Dunnage and Avalon Zones, and to discuss the direction of future geophysical studies.

The first published gravity survey undertaken to ascertain the broad crustal structure of Newfoundland was conducted by the Earth Physics Branch in 1964 (Weaver, 1967) while underwater surveys were conducted in the Gulf of St. Lawrence (Goodacre, 1964; Goodacre et al, 1971). Subsequent studies were conducted by the Earth Physics Branch, the Atlantic Geoscience Centre, and Memorial University as both separate and joint investigations.

P. E. Schenk (ed.),
Regional Trends in the Geology of the Appalachian-Caledonian-Hercynian-Mauritanide Orogen, 29–38.

Aeromagnetic surveys conducted for the Geological Survey of Canada commenced in the late 1950's in portions of central Newfoundland.

The details of the geology of the island have been published in numerous papers dating from reports by Cormack (1824) to those of the modern era typified by Williams (1964, 1979) and in others too numerous to enumerate separately. Several such papers appear in the current volume.

EXISTING DATA BASE

1) Gravity

The existing data base extends from the Gulf of St. Lawrence in the west to the edge of the continental margin east of the island, and from the Grand Banks in the south to the Labrador Sea (Fig. 1).

The only survey encompassing the whole island is the Earth Physics Branch survey at a mean station spacing of 13 km. (Weaver, 1967, 1968). More detailed surveys commenced with Weir's (1971) work at 0.8 km. spacing along 800 km. of the Trans Canada highway which traverses the island in an arcuate route. This was followed by 2.5 km. spacing work in the eastern Notre Dame Bay area (Miller, 1970; Miller and Deutsch, 1973; Miller and Weir, 1982) to delineate in more detail several features initially mapped by Weaver (1967, 1968). Later surveys were undertaken in the western Notre Dame Bay area (Miller and Deutsch, 1976) which tied in with the initial surface meter work north of the island conducted by Atlantic Geoscience Centre, Bedford Institute (Haworth *et al*., 1976; Haworth, 1977; Folinsbee, *et al*., 1978). Prior to this surface meter work had been conducted in the Gulf of St. Lawrence and on the Grand Banks (Haworth and McIntyre, 1975).

Even with the intense surveying conducted from 1964 to 1975 several significant gaps in the coverage existed, (Fig. 1), most notably in the bays of the Avalon Zone where surface meter data and navigation were unreliable. As a result of joint Memorial/ Earth Physics Branch underwater surveys which commenced in 1982 these gaps should be filled by 1986 according to present plans. Detailed mapping at 2.5 km. spacing is being conducted to provide the necessary land coverage in parallel to these underwater surveys (Pittman, 1981; Miller and Pittman, 1982; Miller, 1982).

2) Magnetic

Magnetic data for the whole island are available in the

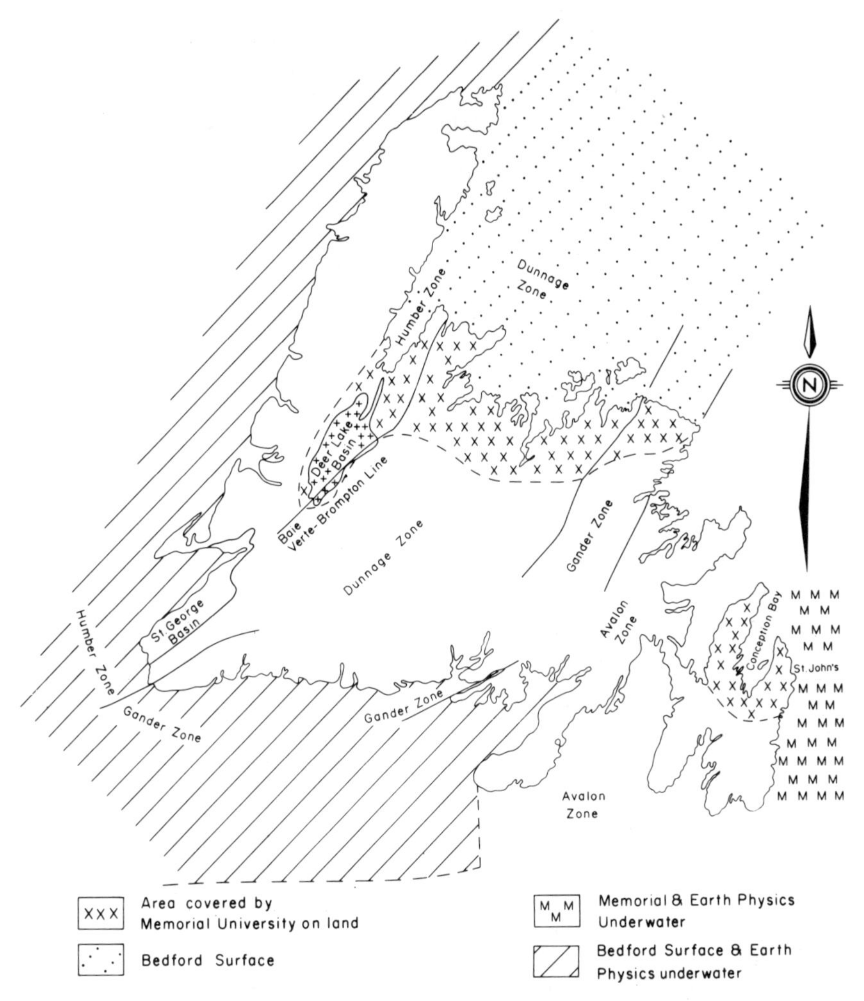

Figure 1 Generalized geological zone map after Williams 1979 on which the gravity survey boundaries are shown. Explanation of designation: Bedford Surface depicts area in which gravity data was collected by Bedford Institute ships using shipborne gravity meters, Bedford Surface and Earth Physics underwater denotes areas of partially overlapping coverage by shipborne surveys of Bedford Institute and underwater coverage by sea bottom meters by the Earth Physics Branch, Memorial University on land represents areas surveyed by Memorial personnel using 2.5 km spacing on the island. Memorial and Earth Physics underwater denotes area covered in 1982 by a joint project using underwater meters undertaken by Memorial University and Earth Physics Branch. The remainder of the island not designated by any shading has been covered by the Earth Physics Branch survey at 13 km spacing (Weaver, 1967, 1968)

form of one inch to one mile (1:63,360) total field aeromagnetic maps produced from data collected along flight lines separated by one-half mile. These are published by the Geological Survey of Canada and are currently being digitized by them at half-mile intervals. The maps for the Notre Dame Bay and Avalon Peninsula areas have been digitized and have had the IGRF 1965.0 removed by Memorial University.

The magnetic data for the offshore are available in the Natural Resource Series Maps published by the Canadian Hydrographic Service at a scale of 1:250,000. These maps are published as total field anomaly maps from which the IGRF field for the epoch appropriate to the collection date has been removed. The magnetic coverage does not have the same gaps in coverage as the gravity data base.

SUMMARY OF INTERPRETATIONS

1) Completed Surveys

The nature of the crustal structure in Newfoundland has been the subject of numerous gravity and magnetic studies beginning with Weaver's (1967) in which the presence of positive anomalies was ascribed to diorite and gabbro bodies in the upper 15 km. of the crust. The most pronounced feature on the gravity map (Fig. 2) of the island and its adjacent offshore areas is the zone of positive Bouguer anomalies which can be traced continuously northeastward from the southwest portion of the island. North of the island the western boundary of this zone has a due northerly trend and passes close to the northern portion of the island near Hare Bay (Fig. 1). The eastern boundary of this zone follows an arcuate path along the southern portion of the island and then veers northeastward. It continues this general trend well offshore until it becomes more northerly near the continental margin. An inspection of the geological zoning (Fig. 1) reveals that this zone of positive Bouguer anomalies is coincident with Williams (1979) Dunnage Zone. Miller and Deutsch (1976) and Miller (1977) correlated this zone of positive anomalies with the ophiolitic material from Iapetus which had been identified in the Betts Cove area (Williams, Kennedy and Neale, 1972).

Early detailed studies in the northeastern portion of the zone substantiated Weaver's results (Miller, 1970; Miller and Deutsch, 1973) and had revealed that this denser crustal material was outcropping along the coast of Notre Dame Bay and lay beneath the Ordovician and Silurian sediments of the rest of the area. This material was tentatively identified by Miller and Deutsch (1973) as similar to the intermediate layer identified on seismic

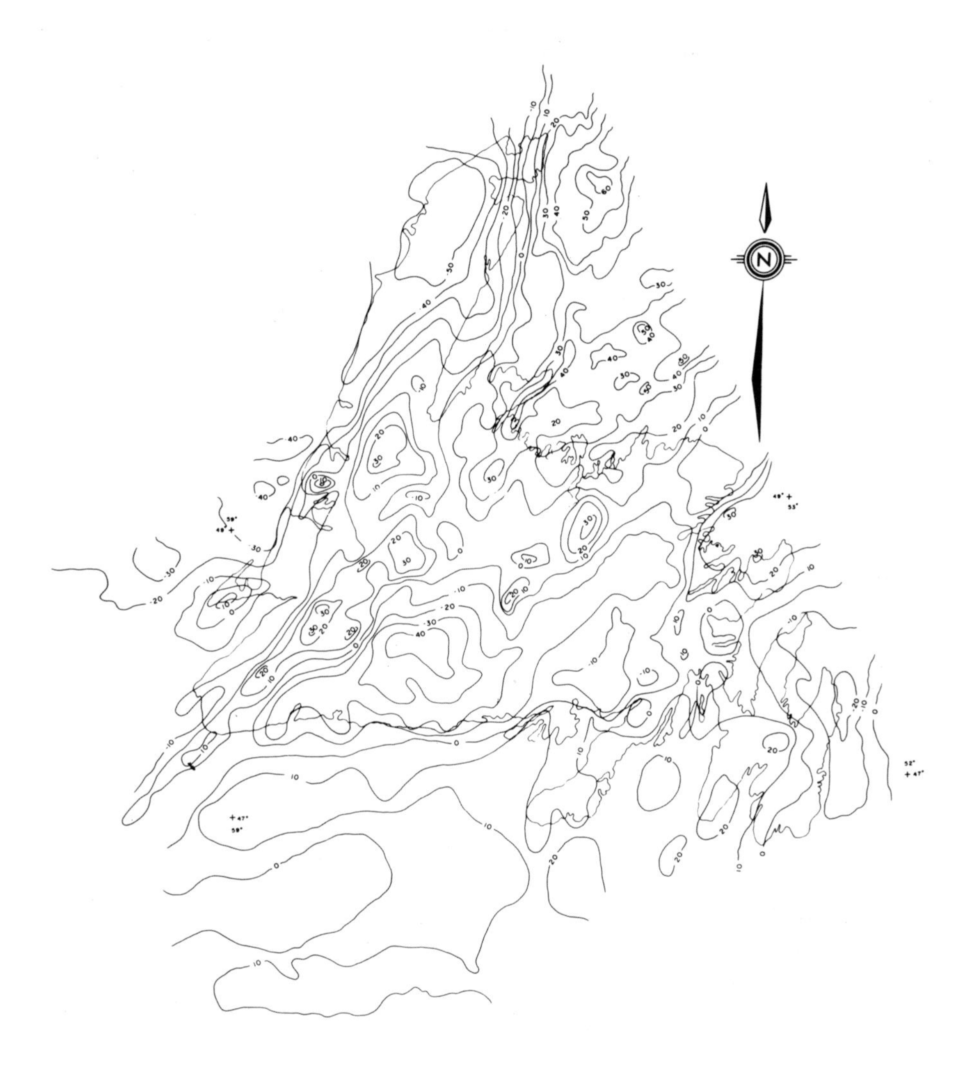

Figure 2 Generalized Bouguer anomaly map for Newfoundland and adjacent offshore region from Haworth, Daniels, Williams and Zietz, 1980. Gravity anomaly map of the Appalachian Orogen.

profiles north of their study area (Sheridan and Drake (1968). It is now recognized that the seismic lines on which the intermediate layer was detected lie within the Dunnage Zone offshore (Haworth, et al., 1978).

Subsequent studies focussed on understanding the nature of the margins of Iapetus. Miller and Deutsch (1976) showed using geophysical evidence that the true western boundary of Iapetus lay not at the coast of Notre Dame Bay near Betts Cove but farther to the west at Baie Verte as is now recognized by Williams (1979). Offshore studies to the north (Haworth and Miller, 1982) have shown that this western boundary signature is recognizable at sea and can be followed as far north as 52°N (Folinsbee, et al., 1978; Haworth, 1980).

The eastern margin of Iapetus has been studied on land by the Canadian Plutonics Working Group of this IGCP project (Pajari et al., 1979). The geophysical signature of this eastern region has demonstrated that the boundary is farther east than the trace of the Gander River Ultramafic Belt (Miller and Weir, 1982). Its signature can also be traced seaward (Jacobi & Kristofferson, 1976; Haworth and Miller, 1982) to the northeast of the island.

All of these studies on the Dunnage Zone and its boundaries have led to the conclusion that the Dunnage Zone is a synclinorium of Iapetus oceanic crustal material (Haworth and Miller, 1982) which served as a source zone for the ophiolites obducted westward (Williams, 1979). In the north at the coast of Notre Dame Bay the Dunnage is in excess of 180 km. wide whereas in the south it is only 20-30 km. wide (Miller, 1977). Throughout the zone the anomalously high density material ascribed to the oceanic crust of Iapetus is less than 15 km. thick. The nature of the crust beneath this material has not yet been deduced. This will be the subject of seismic work proposed for Newfoundland.

3) Surveys in Progress

Avalon Zone. Significant areas of positive gravity anomalies occur in the Avalon Zone of Newfoundland (Fig. 2). The crustal structure of this zone has not been elucidated as well as that in the Dunnage Zone. Haworth and Lefort (1979) have noted the correlation of positive magnetic and gravity anomalies with Precambrian volcanics outcropping in the zone on land (Fig. 1). They have delineated belts of similar geophysical signature on the Grand Banks and have concluded that the geophysical anomalies are caused by paleotopographic highs bounding the Paleozoic and younger basins of the area. The correlation between the onshore exposures of mafic volcanics and their geophysical signature has been used to deduce the detailed subsurface structure of Conception Bay (Miller, 1982). The underwater gravity data base has been extended in 1982 to include Conception Bay and the area immediately to the east of it. The interpretation of the eastern data is in progress. Initial modelling indicates that the mafic volcanics underlie significant portions of the Avalon Zone and outcrop on seabottom or are covered by a thin veneer of recent sediments to the north of Conception Bay. The total thickness of sediments and volcanics is less than 15 km. The nature of the crust beneath these volcanics cannot be deduced from gravity and magnetic data.

Humber Zone. Land gravity surveys at 2.5 km. spacing are being conducted in the Deer Lake Carboniferous Basin of the Humber Zone. The proposed extension of the Baie Verte-Brompton Line (Williams, 1979; Williams and St. Julien, 1982) passes through the study area in which Carboniferous sediments obscure the older relationships. Preliminary results indicate that it will be possible to delineate the trace of the Humber/Dunnage boundary i.e. the Baie Verte/Brompton line in this area.

4) Future Studies

The success of the gravity and magnetic interpretation in deducing the presence of buried mafic volcanics, ultramafics, and ophiolitic material has led to plans by the three groups for further studies, jointly and independently, onshore and offshore.

Over the next five years the St. George's Carboniferous basin in the Humber Zone of Western Newfoundland and the major bays and adjoining peninsulas of the Avalon Zone will be surveyed using 2.5 km. spacing on land and 6 km. at sea. With the conclusion of these surveys, major portions of the island will be covered at 1:250,000 scale and the nearshore area will have similar coverage enabling reasonable correlations between onshore units and their extensions offshore to be made.

SUMMARY

The gravity and magnetic methods have been successfully applied over the last 20 years to the study of the crustal structure of the Newfoundland Appalachians. The conclusions from these studies have made significant contributions towards developing a regional geological framework for the island and extending that framework offshore.

The major conclusion from the interpretation of the gravity and magnetic fields of Newfoundland and the adjoining offshore region is that the material causing the anomalies is confined to the upper fifteen kilometres of the crust. In regions onshore where the geology is well know, positive Bouguer and magnetic anomalies correlate well with outcrops of mafic volcanics. This correlation has been used to map areas offshore underlain by such rocks and has been used to deduce the presence of such rocks beneath younger sediments onshore.

REFERENCES

Cormack, W.E. 1824. Account of a journey across the island of Newfoundland. Edinburgh Philosophical Journal, 10, pp. 156-164.

Folinsbee, R.A., Haworth, R.T., and MacIntyre, J.B. 1978. Marine gravity and magnetic maps, northeast of Newfoundland. Geol. Survey Canada, Open File 525.

Goodacre, A.K. 1964. Preliminary Results of Underwater Gravity Surveys in the Gulf of St. Lawrence, Gravity Map Series No. 46, Earth Physics Branch, Ottawa.

Goodacre, A.K., Stephens, L.E. and Cooper, R.V. 1971. Results of Underwater Gravity Surveys over the Nova Scotia Continental Shelf. Gravity Map Series No. 123, Earth Physics Branch, Ottawa.

Haworth, R.T. and MacIntyre, J.B. 1975. Gravity and Magnetic Fields of Atlantic Offshore Canada, Geol. Survey of Canada. Paper 75-9.

Haworth, R.T., Poole, W.H., Grant, A.C., and Sanford, B.V. 1976. Marine geoscience survey northeast of Newfoundland. Geol. Survey of Canada, Paper 76-1A, pp. 7-15.

Haworth, R.T. 1977. The continental crust northeast of Newfoundland and its ancestral relationship to the Charlie Fracture Zone. Nature, 266, pp. 246-249.

Haworth, R.T., Lefort, J.P. and Miller, H.G. 1978. Geophysical evidence for an east dipping Appalachian subduction zone beneath Newfoundland. Geology, 6, pp. 522-526.

Haworth, R.T. and Lefort, J.P. 1979. Geophysical evidence for the extent of the Avalon Zone in Atlantic Canada. Can. Jour. Earth Science, 16, pp. 552-567.

Haworth, R.T. 1980. Appalachian structural trends northeast of Newfoundland and their trans-Atlantic correlation. Tectonophysics, 64, pp. 111-130.

Haworth, R.T. and Jacobi, R.D. Geophysical correlation between the geological zonation of Newfoundland and the British Isles. In press, GSA Special Paper (Preprint only available to author, 1982).

Haworth, R.T. and Miller, H.G. 1982. The structure of Paleozoic oceanic rocks beneath Notre Dame Bay, Newfoundland. Geol. Assoc. Canada Special Paper No. 24, P. St. Julien and J. Beland, eds., pp. 150-173.

Jacobi, R.D. and Kristoffersen, Y. 1976. Geophysical and geological trends on the continental shelf off northeastern Newfoundland. Can. Jour. Earth Sciences, pp. 1039-1051.

Miller, H.G. 1970. A gravity survey of eastern Notre Dame Bay, Newfoundland. Unpublished M.Sc. Thesis, Memorial University of Newfoundland 84 p.

Miller, H.G. and Deutsch, E.R. 1973. A gravity survey of eastern Notre Dame Bay, Newfoundland. In Earth Science Symposium on Offshore Eastern Canada, P. Hood (ed.). Geol. Surv. Can., Paper 71-23, pp. 389-406.

Miller, H.G. and Deutsch, E.R. 1976. New gravitation evidence for the subsurface extent of oceanic crust in north-central Newfoundland. Can. Jour. Earth Sciences, 13, pp. 459-469.

Miller, H.G. 1977. Gravity zoning in Newfoundland. Tectonophysics, 38, pp. 317-326.

Miller, H.G. and Weir, H.C. 1982. The northwestern portion of the Gander Zone - a geophysical interpretation. Can. Jour. Earth Sciences, 19, pp. 1371-1381.

Miller, H.G. and Pittman, D.A. 1982. Geophysical constraints on the thickness of the Holyrood pluton, Newfoundland. Maritime Sediments and Atlantic Geology, 18, pp. 75-82.

Miller, H.G. 1982. A geophysical interpretation of the geology of Conception Bay, Newfoundland. Submitted Can. Jour. Earth Sciences, Sept. 1982.

Pajari, G.E., Jr., Pickerill, R.K. and Currie, K.L. 1979. The nature, origin, and significance of the Carmanville ophiolitic melange, northeastern Newfoundland. Can. Jour. Earth Sciences, 16, pp. 1439-1451.

Pittman, D.A. 1981. The gravity and magnetic fields of the northeastern Avalon Peninsula, Newfoundland. Unpublished B.Sc. Dissertation. Memorial University of Newfoundland, 56 p.

Sheridan, R.E. and Drake, C.L. 1968. Seaward extension of the Canadian Appalachians. Can. Jour. Earth Sciences, 5, pp. 337-373.

Weaver, D.F. 1967. A geological interpretation of the Bouguer anomaly field of Newfoundland. Earth Physics Branch, EMR, Ottawa, Vol. XXXV, No. 5, pp. 223-251.

Weaver, D.F. 1968. Preliminary results of the gravity survey of the island of Newfoundland with maps. No. 53, 54, 55, 56, 57. Earth Physics Branch, EMR, Ottawa.

Weir, H.C. 1971. A gravity profile across Newfoundland. M.Sc. Thesis Memorial University of Newfoundland, St. John's, 164 p.

Williams, H. 1964. The Appalachians in northeastern Newfoundland - a two-sided symmetry. Amer. J. Science, 262, pp. 1137-1158.

Williams, H., Kennedy, M. and Neale, E.R.W. 1972. The Appalachian structural province. In Tectonics Styles in Canada. R.A. Price and R.J.W. Douglas (eds.). Geol. Assoc. Can. Spec. Paper 11, pp. 181-262.

Williams, H., Kennedy, M., and Neale, E.R.W. 1974. The northeastern termination of the Appalachian Orogen. In The Ocean basins and margins, Vol. 2, A.E. Nairn and F.G. Stehli, eds. Plenum Publishing, New York, pp. 79-123.

Williams, H. 1979. Appalachian Orogen in Canada. Can. J. Earth Sciences, 16, pp. 792-807.

Williams, H. and St.Julien, P. 1982. The Baie Verte-Brompton Line: Early Paleozoic continent-ocean interface in the Canadian Appalachians. Geol. Assoc. Can. Special Paper No. 24, P. St. Julien and J. Beland (eds.), pp. 177-208.

COMPILATION OF CANADIAN MAGNETIC ANOMALY MAPS: A STATUS REPORT

Peter J. Hood

Geological Survey of Canada, Ottawa
601 Booth Street, Ottawa, Canada K1A 0E8

ABSTRACT

The first 1:5 M national magnetic anomaly map (MAM) was published by the Geological Survey of Canada in 1967 and subsequent versions were published in 1972 and 1977. The coloured MAM of Canada demonstrated that features of large areal extent such as geological provinces were readily delineated and proved the usefulness of producing such regional compilations. In 1981, the first four digitally-produced coloured 1:1 M MAM's were published for an area west of Hudson Bay, and these show much more detail than the 1:5 M MAM's. It is abundantly clear that such MAM's should be made a standard end product of regional aeromagnetic surveys because of their several uses. They serve not only as an index map to the aeromagnetic survey coverage available on a national basis (as well as the gaps in the coverage), but also stimulate comparison of the regional magnetic features with those appearing on similar scale geological and geophysical maps and with Landsat imagery.

1:5 M Magnetic Anomaly Map Series

Aeromagnetic surveying by Canadian Government agencies commenced in 1947 and has resulted in the systematic coverage of the mineral-rich Canadian Precambrian Shield, part of the Cordillera, the Arctic Islands and Canadian Appalachia (see Figure 1) using an essentially common set of specifications. Important areas in British Columbia, the Yukon, the Arctic Islands and offshore, however, remain to be flown. The main end product of the aeromagnetic survey of Canada has been line contour maps of the total magnetic field at the one inch to one mile (1:63,360) scale,

P. E. Schenk (ed.),
Regional Trends in the Geology of the Appalachian-Caledonian-Hercynian-Mauritanide Orogen, 39–44.

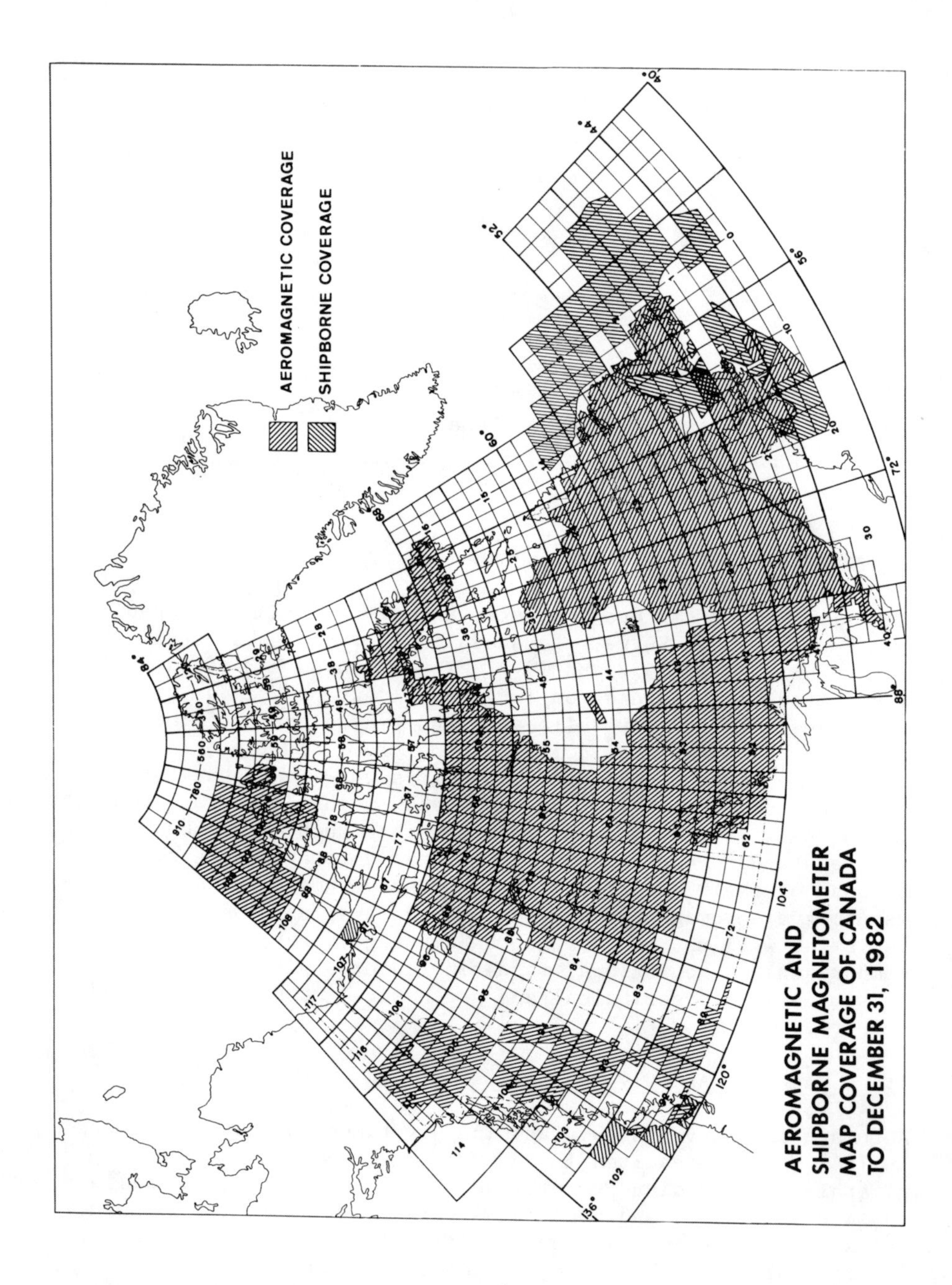
AEROMAGNETIC COVERAGE
SHIPBORNE COVERAGE
AEROMAGNETIC AND
SHIPBORNE MAGNETOMETER
MAP COVERAGE OF CANADA
TO DECEMBER 31, 1982

FIG. 1

although when Canada adopted the metric system in the middle 70's this scale was changed to 1:50,000 for compatibility reasons. Composite aeromagnetic maps at the one inch to four mile (1:253,440) and after metrication the 1:250,000 scale have also been issued. By 1967 a sufficient area of the country had been covered that it was feasible to produce a Magnetic Anomaly Map (MAM) of Canada.

The first edition of the 1:5,000,000 Magnetic Anomaly Map of Canada (GSC Map 1255A) was published in time for the 1967 Canadian Centennial Conference on Mining and Groundwater Geophysics held in Niagara Falls. Because the first 1:5 M national MAM was published by the Geological Survey before the IGRF (1965.01) was agreed upon in Washington in October 1968, the total field (F) map produced by the EMR Earth Physics Branch was utilized to remove the earth's core field. A graphical subtraction technique was used in its compilation so that a considerable amount of fine detail was eliminated. A further edition of the 100-gamma colour contour map was published in time for the 1972 International Geological Congress in Montreal and the third edition was published in time for the Exploration '77 Symposium held in Ottawa.

Many interesting features seen on the Magnetic Anomaly Map of Canada correlate with features on the Tectonic Map of Canada. In general the boundaries of the various age provinces can be distinguished because of the change in magnetic character between provinces. For example, the Grenville Province has a very complex short-wavelength bird's eye pattern which tends to reflect the metamorphic grade of the rocks rather than their lithology. A broad featureless magnetic low parallels the boundary between the Superior and Grenville Provinces for about 250 miles and is usually referred to as the Grenville Front Low.

It should be realized that a 200-gamma magnetic anomaly map is a form of filtered map in which the majority of anomalies less than 200 gammas in amplitude are eliminated. Because of the small scale of the 1:5 million map and colour contour interval, the longer wavelength anomalies that have an amplitude in excess of 200 gammas will tend to be emphasized. Features that produce anomalies less than about 0.5 millimetre in width at the map scale cannot be physically drawn on the map; 0.5 mm at the 1:5,000,000 scale represents 2,500 metres but only 500 metres at the 1:1,000,000 scale. This means that many important narrow geological features, such as dykes or dyke swarms, will not be portrayed on the resultant end product. Thus the short wavelength cut-off is directly proportional to the scale of the map. Hence, the scale and colour contour interval represent graphical filtering parameters which may be varied to produce various filtering effects on the magnetic anomaly map.

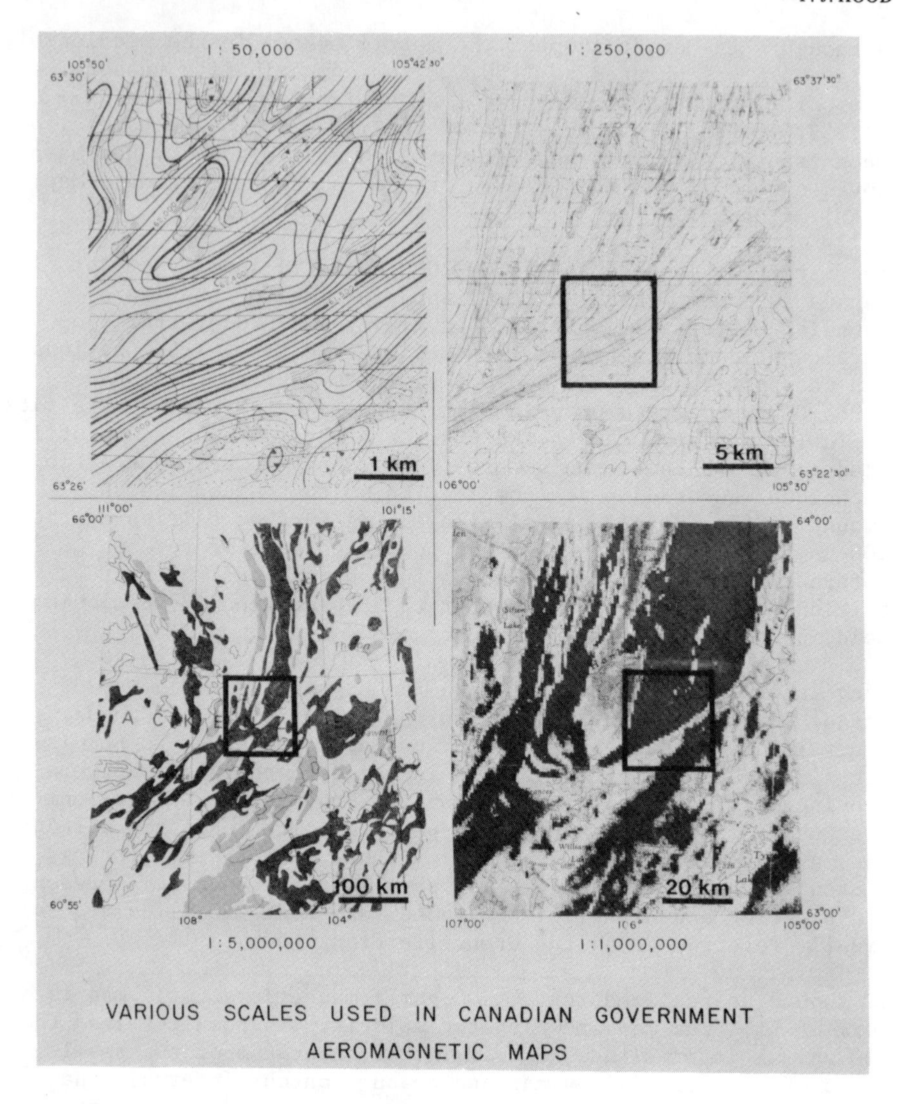

FIG. 2

1:1 M Magnetic Anomaly Map Series

Thus although the 1:5 million Magnetic Anomaly Map of Canada series serves a useful purpose, the scale is not the optimum one for regional studies. The 1:1 M magnetic anomaly map series is much better suited for that purpose because the greater degree of detail possible at the larger scale permits the portrayal of features that are not otherwise seen at the smaller scale. Figure 2 illustrates the various scales utilized in the publication of Canadian Government aeromagnetic maps. The areas outlined on Figure 2 indicate in turn the area covered by the adjacent larger scale map. The first four coloured 1:1 M magnetic anomaly maps (GSC Maps 1566A, 1567A, 1568A and 1569A) were issued by the Geological Survey of Canada in November 1981 for an area that straddled the T-junction of Manitoba, Saskatchewan and the Northwest Territories.

Because most of the aeromagnetic survey data in Canada was obtained prior to the general adoption of digital recording techniques, a great deal of the original data is in analog form. As a result, because it would be very expensive to go back to the original chart records, the data on the resultant contour maps have been digitized to produce the digital data set from which the 1:1 M maps have been prepared. This procedure also avoids a re-levelling of the data. Several digitizing techniques have been experimented with and the one that has been finally adopted is to digitize the location of the contour intercepts along the flight lines. Because the values along the flight lines have the greatest accuracy in comparison to the interpolated contours between the flight lines, the technique retains the accuracy of the original compilation prior to contouring. Moreover there is close compatibility between the 0.8128 mm square size of the 16 dot matrix pixel and the flight line spacing which is .805 mm at the 1:1 M scale utilized on the final digitally-produced colour map.

An Applicon plotter is utilized to print the final map. The plotter utilizes three colour ink jets, one for each primary colour (namely red, yellow and blue) and these jets are controlled by a digital tape. A combination of ink blobs corresponding to the correct shade is fired at a rotating drum around which is wound a sheet of paper. The ink jet system moves progressively down the rotating paper which is consequently painted by the ink raster to produce the final end product. It is possible by utilizing one ink jet at a time to produce three colour separations for printing purposes. Unfortunately stable-base plastic material cannot be utilized by the Applicon plotter because the ink is not absorbed and consequently smears. The three colour separations therefore have to be photographed the same day they are produced to avoid the effects of differential expansion along the length and across the width of the map. The resultant end products are utilized together with a base map showing planimetry etc. in printing the final coloured 1:1 M magnetic anomaly map.

The colour scheme used in the 1:1 M magnetic anomaly maps is essentially the spectrum of white light. Thirty-nine shades are employed ranging from red at the positive end to blue at the negative end. The zero value coincides with pure yellow. Twenty-six of the shades in the central part of the spectrum represent 25 gamma contour intervals with the contour interval of the remainder coarsening in discrete steps to cover the necessary range required.

It is intended that the 1:1 M magnetic anomaly maps will be utilized essentially as basic building blocks in compiling future editions of the 1:5 M Magnetic Anomaly Map of Canada. In turn the Canadian map will be utilized for the Magnetic Anomaly Map of North America being compiled as part of the Decade of North American Geology to celebrate the centenary of the Geological Society of America in 1988. However an end product of the 1:1 M MAM compilation will be a digital aeromagnetic data base for Canada which can be utilized for many purposes. One example might be the production of a variety of filtered maps that typically emphasize the high or low frequency content of the data. Also of much current interest to qualitative interpreters of aeromagnetic data are the simulated shaded-relief maps that emphasize linear features in the magnetic anomaly maps that are not otherwise easily seen. These can also be compiled from the digital data and are quite stunning in appearance. However there is no doubt the GSC will continue to produce a magnetic anomaly map at the 1:5 million scale to accompany the concomitant geological, tectonic, gravity etc. maps of the country at the same scale.

Aeromagnetic coverage of the sedimentary areas of Saskatchewan and Alberta has been obtained by the Petroleum industry. Negotiations to obtain this data is being undertaken at the present time both to complete the Magnetic Anomaly Map of Canada and also the Magnetic Anomaly Map of North America which is being produced as part of the Geological Society of America centenary.

Thus, to summarize, the coloured Magnetic Anomaly Map of Canada demonstrated that features of large areal extent such as geological provinces were readily delineated and proved the usefulness of producing such regional compilations. Consequently in carrying out aeromagnetic surveys overseas as contributions to Canadian aid programs of the Canadian International Development Agency, the preparation of regional magnetic anomaly maps has usually been made part of the contract at the interpretation stage. It is abundantly clear that such magnetic anomaly maps should be made a standard end product of regional aeromagnetic surveys because of their several uses. They serve not only as an index map to the aeromagnetic survey coverage available on a national basis (as well as the gaps in the coverage), but also stimulate comparison of the regional magnetic features with those appearing on similar scale geological and geophysical maps and with Landsat imagery.

GRAVITY EVIDENCE OF CRUSTAL STRUCTURE IN THE UNITED STATES APPALACHIANS

M. F. Kane

U.S. Geological Survey, Box 25046,
Denver Federal Center, Denver, CO 80225

ABSTRACT. The pattern of anomalies on a wavelength-filtered residual gravity map (wavelengths shorter than 250 km) over the Appalachian basin is similar to that ascribed to a rift system in the central United States. The gravity pattern coincides laterally with thick carbonate deposits of Cambrian-Ordovician age that are associated with the continental shelf of late Precambrian-early Paleozoic North America. The anomalies must arise in the crust below the sedimentary sequence implying that development of the shelf was accompanied by subjacent rifting. A similar interpretation may be applied to gravity anomaly patterns in the Piedmont and western New England if allowance is made for rift-system exhumation and compression probably caused by continental collision. Second-order changes in gravity and magnetic trends in a zone extending from central Virginia to south-central Tennessee is interpreted as evidence of a transverse structural zone separating the central and southern Appalachians. A complementary regional gravity map (wavelengths longer than 250 km) is dominated by a northeast-trending gradient extending from northeast Alabama to the southeast coast of Maine. It is interpreted as the trace of the suture zone between the continental plates that converged in Paleozoic time. The distribution of earthquakes in the Appalachians can be associated with structures derived from the gravity maps.

1. INTRODUCTION

With the insight provided by the concept of plate tectonics it is now generally accepted that the Appalachian orogen is the location and in many ways the result of a continent-continent

P. E. Schenk (ed.),
Regional Trends in the Geology of the Appalachian-Caledonian-Hercynian-Mauritanide Orogen, 45–54.

collision and the subsidiary events connected with such a collision (1,2). One of the earlier significant generalizations linking the northern and southern Appalachians was the identification by Rodgers (3) of a carbonate bank that extends the length of the orogen and was part of the continental margin of North America in early Paleozoic time. More recently the work of Harris and Bayer (4) and Cook and others (5) established that the terranes of the southern Appalachians have moved laterally northwestward in broad thin sheets with displacements that may have been as much as 100 km or more. These displacements together with other potentially large lateral displacements along the orogen-long thrust belt effectively restrict to shallow depths the confident projection of surface structures over much of the area. Gravity is one of the geophysical methods that offers a means to "look deep", below many of the structures manifested at the surface. The following sections describe a pair of residual and regional gravity maps developed by wavelength filtering, and an attempt to interpret certain features of these maps as crustal structures seen within the framework of plate tectonics.

2. MAP PARAMETERS

It is well known in the calculation of gravity and magnetic fields that short-wavelength anomalies damp out with increase in depth of source (or increase in height of observation) more rapidly than long-wavelength ones. For this reason sharp anomaly features are usually attributed to shallow sources. It is possible to calculate a minimum damping factor for an anomaly by calculating the damping for its longest wavelength component as the anomaly source is made deeper and deeper. This factor may be used to estimate the depth at which an anomaly is so reduced in amplitude and gradient (gradient falls off with depth at one power higher than amplitude) that it can no longer be effectively observed on a map. For example, the 250 km wavelength component of an anomaly caused by a source at 40 km depth is reduced in amplitude by two thirds compared with that of a similar source at 5 km depth; all shorter wavelengths are reduced at proportionately higher rates. Because of the severity of reduction in amplitude and gradient within a wavepass band of 0-250 km for sources deeper than 40 km, it seems reasonable to state that a map composed of wavelengths 250 km and shorter effectively shows anomalies that arise primarily from sources shallower than 40 km, that is, from within the Earth's crust. A residual map prepared in this way however, will also effectively exclude crustal anomalies whose dominant wavelengths are longer than 250 km. A complementary regional map (wavelengths 250 km and longer) will contain long wavelength anomalies arising from sources in the crust (a broad basin of sedimentary strata is one example), anomalies from relief on the crust-mantle interface, and anomalies from sources within the mantle. Of these, anomalies from the crust-mantle interface are estimated as being the most significant because of its large density contrast and relief.

3. RESIDUAL GRAVITY MAP

Figure 1 is a Bouguer gravity map composed of wavelengths shorter than 250 km. The most prominent continuous feature is a broad linear band of highs extending southwestward from western Pennsylvania to northeastern Alabama. A gravity high within the band in northwestern Pennsylvania is elongate northeastward and flanked by two broad lows; a similar pattern of a northeast-trending high with flanking lows can be seen centered at 40°N, 79°W, in southwestern Pennsylvania. These two sets of anomalies are illustrated more clearly and commented on by Diment and others (6); they are also brought out more clearly in a colored version of a filtered gravity map of the United States (7). A pattern similar to those seen in Pennsylvania is centered in central West Virginia. A similar pattern can also be seen trending across the northern part of the Alabama-Georgia State line. Although the broad band of high gravity is continuous from southern West Virginia to northern Alabama, the picture is somewhat obscured by several more localized anomalies that may be related to rock units of the allochthon underlying this region.

The band of high gravity together with the flanking lows broadly resembles a pattern of gravity anomalies intepreted as expression of a rift system in the central United States (8). The pattern consists of a series of sharply defined northeast-trending sets of gravity highs flanked by lows in which each set is offset one from the other along northwest trends; the offsets are interpreted as transform faults. In particular the offset of the gravity highs and flanking lows in western Pennsylvania resembles those seen in the central United States. Northward from the southern part of West Virginia, the band of high gravity with flanking lows coincides with the thickest part of Cambrian-Ordovician carbonate strata which are identified with the continental shelf of early Paleozoic North America (3,9). Since it seems clear that these anomalies cannot arise from within the sedimentary section they are interpreted as evidence of a rift system in the crust that underlies and may have developed as part of the early Paleozoic continental shelf. This structural model is similar to that recently proposed for the interior basins of the continents (10, 11).

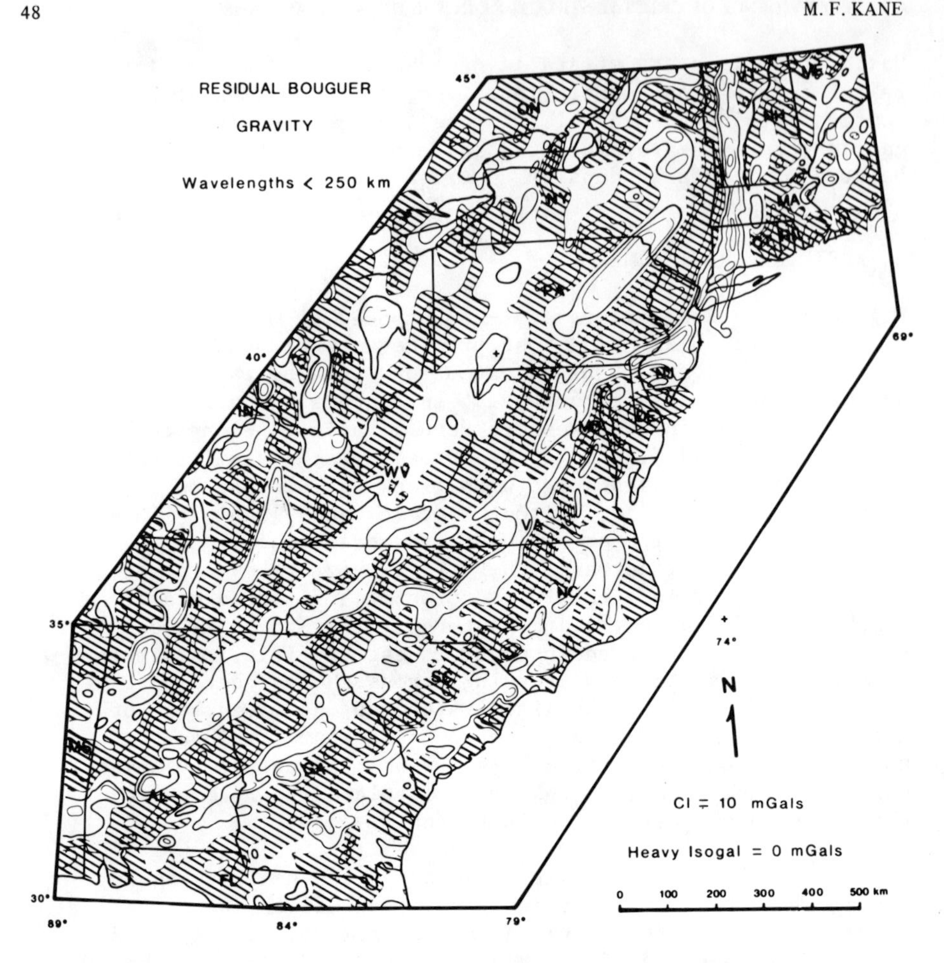

Fig. 1. Bouguer gravity map (residual) of the U.S. Appalachians composed of wavelengths less than 250 km. Lined areas show values less than -10 mGal. Hachures indicate areas of relatively lower gravity. ON, Ontario; VT, Vermont; ME, Maine; NH, New Hampshire; NY, New York; MA, Massachusetts; OH, Ohio; PA, Pennsylvania; CT, Connecticut; RI, Rhode Island; IN, Indiana; KY, Kentucky; WV, West Virginia; MD, Maryland; DE, Delaware; NJ, New Jersey; VA, Virginia; TN, Tennessee; NC, North Carolina; SC, South Carolina; MS, Mississippi; AL, Alabama; GA, Georgia; FL, Florida.

The pronounced linear high trending northeastward through northeast Pennsylvania is associated spatially with thick clastic rocks of Devonian age (9). The anomaly may be the expression of a rift that was related to the emplacement of the clastic rocks. Perhaps the rift was activated in a tensional zone caused by continental collision in a setting much like that described for the Himalayas by Molnar and Tapponier (12).

There is a discordance between gravity trends (north of east) in a zone extending southwest from central Virginia to south-central Tennessee and those (northeast) in the regions north and south of the zone. It is shown most clearly in the trend of the prominent gravity high in southwestern Virginia. The discordance is more pronounced on colored versions of this and other derivative gravity maps (7,13). Although some of the discordant gravity expression may be associated with density contrasts in the allochthon whose northwest margin coincides with the zone, other aspects like the change in the general trend of regional gravity features suggest that it reflects a change in crustal structure. There are changes in the trends of magnetic anomalies along this zone (14) which also indicate that crystalline or igneous basement rocks are involved. The zone is interpreted as a structural break between the central and southern Appalachians.

A second band of gravity highs can be seen extending from southeast Alabama to north-central Maryland. Long (15) interprets the part of the band crossing the Georgia-South Carolina State line as the axis of a rift. The band of highs continues as a much narrower feature east from northern Maryland and then north through western New England (CT, MA, VT). These narrower highs might be interpreted as the central gravity highs associated with rifts if allowance is made for compression of their sources perhaps as a result of the collision between the continental plates.

4. REGIONAL MAP

Figure 2 is a regional Bouguer gravity map composed of wavelengths longer than 250 km. Its most striking feature is the gravity gradient that extends from Maine to Alabama and separates a continental interior region characterized by strongly negative values from a region along the continental margin that exhibits more positive values (16). As stated in a previous section a likely cause for much of the gravity expression shown by the map is relief on the crust-mantle interface. Support for this interpretation is found in a crustal cross-section of the southern Appalachians by Long (15) and in a map of mantle depths of the New England region by Taylor and Toksoz (17) where contours strongly resemble those exhibited in the northeast part of Figure 2. One probable crustal anomaly source is the thick sequence of sedimentary strata of the Appalachian basin centered in eastern

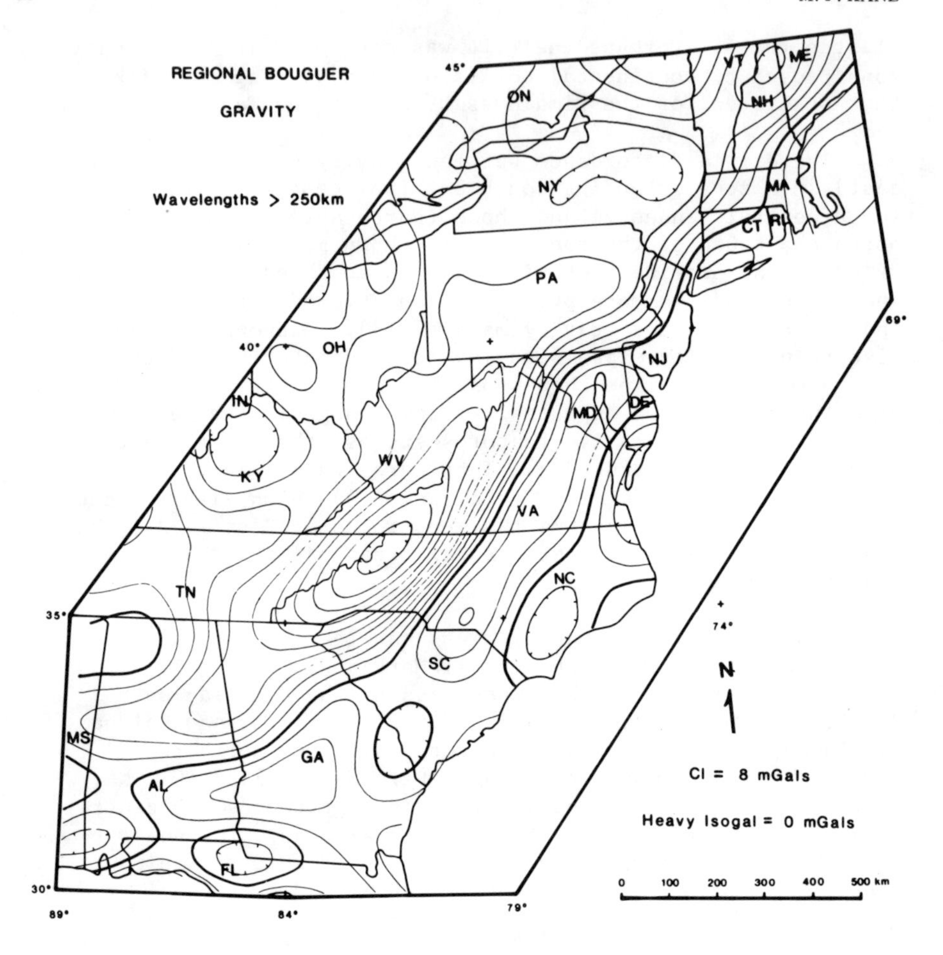

Fig. 2. Bouguer gravity map (regional) of U.S. Appalachians composed of wavelengths more than 250 km. Hachures indicate areas of relatively lower gravity. State abbreviations explained in Figure 1.

Pennsylvania (9). The profound low centered in northwest North Carolina however, coincides most closely with the allochthon of the southern Appalachians suggesting that emplacement of the allochthon may have caused a substantial depression of the crust.

A plausible explanation of the gravity gradient is that it marks the zone of collision between continental plates in mid and late Paleozoic time. The interpretation of thinner crust southeast of the gradient is consistent with the deeper level of crustal rocks exposed there compared to those exposed on the northwest. The linking of the carbonate bank to a rift system underlying the Appalachian basin implies that the continental margin of early Paleozoic North America is still essentially in place. The interpretation of the Piedmont as an exhumed and deformed shelf rift system that once underlay the continental plate margin that converged from the southeast places the zone of convergence in the vicinity of the gradient. Presumably this plate margin was elevated during collision and shed its shelf strata in thin-skin thrusts from successively deeper levels until it exposed and finally shed parts of the underlying crystalline crust, perhaps according to the model of Elliot (18). The preferential uplift of the southeast plate may have been caused by subduction of the oceanic plate beneath it. The extension of the gradient and the interpreted suture across southern New England and along the coastal region of Maine (Figs. 2 and 3) leaves a gap between it and the interpreted continental margin rift zone in western New England. This implies that New England is underlain by accreted terrain as suggested previously by Osberg, (19).

5. Structure Map

Figure 3 shows the structures derived from the gravity maps in the preceding sections together with an outline of the seismicity distribution modified after Hadley and Devine (20). The crustal structure beneath the Appalachian basin is depicted as a series of rifts offset by transform faults. In analogy to the central U.S. rift system (8) the gravity highs are interpreted as rift elements which probably include a substantial amount of volcanic rocks and a subjacent lower-crustal part composed of intruded mantle rock; the flanking lows are attributed to clastic sediments emplaced in a graben phase. If the interpretation is correct, the rift system outlines the entire longitudinal extent of the continental shelf of North America in early Paleozoic time (basin sediments are hidden by allochthonous rocks southeast of the transverse zone indicated by the dot-dash line). A similar but exhumed rift system that underlay the margin of the continental plate that converged from the southeast is shown by dashed symbols to indicate the less clear and in many places, incomplete gravity expression. The approximate position of the collision zone is interpreted from the mid-value of the gradient shown in Figure 2. Transform faults, interpreted from offsets in the gravity gradient and interruptions in the linear gravity high that lies south and southeast of the gradient (Fig. 2) are shown as dashed medium weight lines but it should be emphasized that the trends of these suggested transforms are not well-constrained.

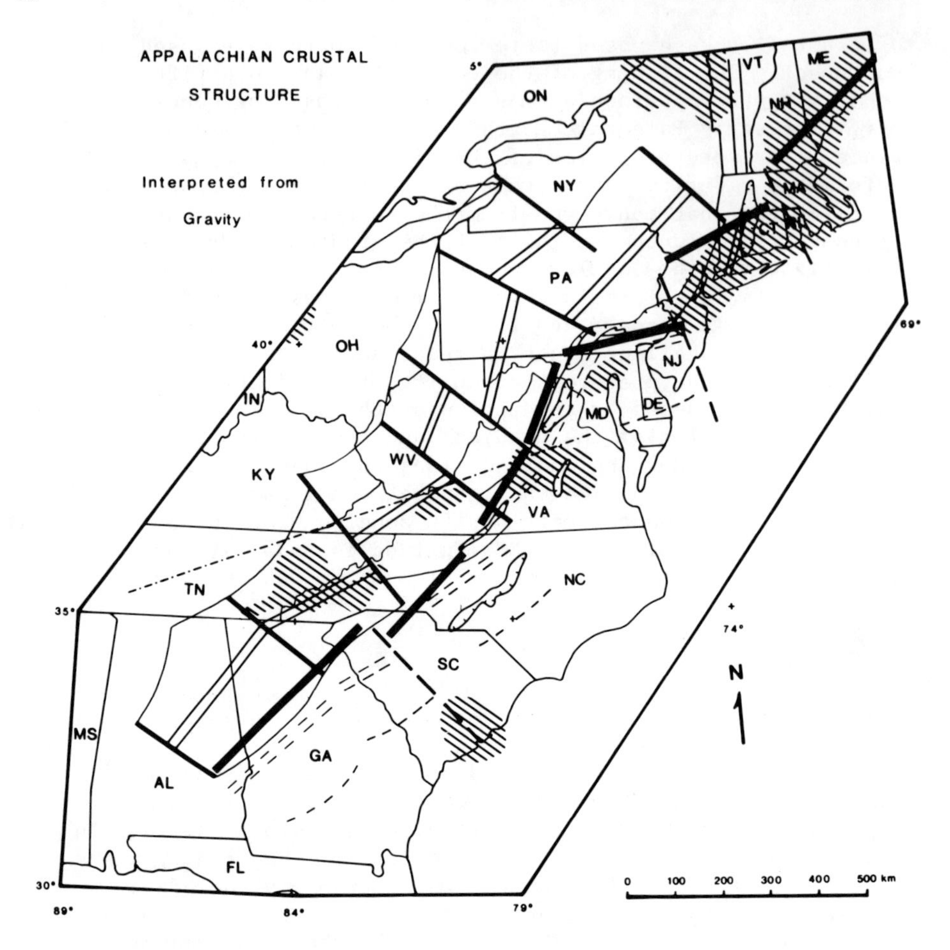

Fig. 3. Crustal structure map of the U.S. Appalachians derived from regional and residual (wavelength-filtered) gravity maps. Heavy lines indicate suture zones; medium lines indicate transform faults--dashed where less certain; light lines indicate outer boundaries of basins flanking central features of rifts--dashed where less certain; double lines indicate central features of rifts--dashed where less certain; dot-dash line indicates transverse structural zone; hachured irregular closures indicate Triassic basins; lined areas indicate zones of seismicity. State abbreviations explained in Figure 1.

The rift model suggested herein offers a possible explanation for the Triassic basins that trend the length of the crystalline belt of the Appalachians. R. W. Bromery (oral commun., 1966) observed earlier that the basins tend to occur on either side of the gravity highs that make up the band of high gravity in the eastern part of the map (Fig. 1). Kane and others (21) have suggested that the flanks of the central features of rifts are principal regions of deformation when rift systems undergo compressional or tensional phases subsequent to their original formation. According to this model the basins would be secondary relaxation structures that formed along boundaries of central rift features during the tensional regime that gave rise to the opening of the Atlantic.

As shown in Figure 3 seismicity in the eastern region from Maine to Virginia is concentrated in the area of the suggested suture zone and particularly near the inferred transform faults. Except for the cluster at Charleston, SC, the seismicity from Virgina southward trends across the Appalachians along and south of the indicated transverse zone. The seismicity at Charleston, SC, is near the proposed location of a transform fault but as noted above, the location of the interpreted fault is not well-constrained.

ACKNOWLEDGMENT

This report would not have been possible without the efforts of my colleagues R. W. Simpson, Jr., T. G. Hildenbrand and R. H. Godson in developing the maps.

REFERENCES

1. Bird, J. M. and Dewey, J. F.: 1970, Geol. Soc. America Bull. 81, pp. 1031-1060.
2. Hatcher, R. D., Jr.: 1978, in Tozer, E. T., and Schenk, P. E., eds. "Caledonian-Appalachian orogen of the North Atlantic region", Geol. Surv. Canada Pap. 78-13, pp. 149-157.
3. Rodgers, John: 1968, in Zen, E-an and others, eds., "Studies of Appalachian Geology: Northern and Maritime", New York, John Wiley, pp. 141-149.
4. Harris, L. D., and Bayer, K. C.: 1979, Geology 7, pp. 568-572.
5. Cook, F. A., Albaugh, D. S., Brown, L. D., Kaufman, Sidney, and Oliver, J. E.: 1979, Geology 7, pp. 563-567.
6. Diment, W. H., Muller, Otto, and Lavin, P. M.: 1972, in Wones, D. R., ed., "The Caledonides in the U.S.A.", Blacksburg, VA, Virginia Polytechnic and State Univ., pp. 221-227.
7. Hildenbrand, T. G., Simpson, R. W., Jr., Godson, R. H., and Kane, M. F.: 1982, U.S. Geological Survey Geophys. Inv. Map GP-953A, scale 1:7,500,000.

8. King, E. R., and Zietz, Isidore: 1971, Geol. Soc. America Bull. 82, pp. 2187-2208.
9. Colton, G. W.: 1971, in Fisher, G. W., and others, eds., "Studies of Appalachian Geology: Central and Southern", New York, John Wiley, pp. 5-47.
10. McKenzie, Dan: 1978, Earth Planet. Sci. Lett. 44, pp. 25-32.
11. Sclater, J. C., and Christie, P.A.F.: 1980, Jour. Geophys. Res. 85, pp. 3711-3739.
12. Molnar, Peter, and Tapponier, Paul: 1977, Geology 5, pp. 212-216.
13. Simpson, R. W., Jr., Hildenbrand, T. G., Godson, R. H., and Kane, M. F.: 1982, U.S. Geological Survey Open-File Report 82-477.
14. Zietz, Isidore, Haworth, R. T., Williams, Harold, and Daniels, D. L.: 1980, St. Johns. Nfld., Mem. Univ. of Nfld., Map No. 2, scale 1:1,000,000.
15. Long, T. L.: 1979, Geology 7, pp. 180-184.
16. Diment, W. H., Urban, T. C., and Revetta, F. A.: 1972, in Robertson, E. C. and others, eds., "The nature of the solid earth", New York, McGraw-Hill, pp. 544-572.
17. Taylor, S. R., and Toksoz, M. N.: 1979, Jour. Geophys. Res. 84, pp. 7627-7644.
18. Elliot, D. W.: 1976, Jour. Geophys. Res. 81, pp. 946-963.
19. Osberg, P. H.: 1978, in Tozer, E. T. and Schenk, P. E., eds. "Caledonian Appalachian orogen of the North Atlantic region", Geol. Surv. Canada Pap. 78-13, pp. 137-147.
20. Hadley, J. B., and Devine, J. F., 1974, U.S. Geological Survey Misc. Field Studies Map MF-620, scale 1:5,000,000.
21. Kane, M. F., Hildenbrand, T. G., Simpson, R. W., Jr., Godson, R. H., and Bracken, R. E., 1982, Soc. Explor. Geophys. Tech. Prog. Abstr. and Biograph., 52nd Ann. Mtg., Dallas, pp. 232-234.

CRUSTAL STRUCTURE OF THE SCANDINAVIAN PENINSULA AS DEDUCED BY WAVELENGTH FILTERING OF GRAVITY DATA

Fredrik Chr. Wolff

Geological Survey of Norway
Leiv Eriksons vei 39
N-7000 Trondheim

ABSTRACT

The general gravity map prepared for the IGCP Caledonide meeting in Uppsala 1981 has been digitized and processed by wavelength filtering using the computer facilities at the U.S. Geological Survey in Denver.

A 250 kilometer low-pass regional map is believed to reflect primarily the MOHO topography and mantle sources. The extremely low gravity field below the Caledonian mountain chain is thought to be due to a thickening of the light Precambrian rocks in this region, and an extensive gravity high along the Norwegian coast is probably caused by a thinning of the crust associated with the opening of present Atlantic Ocean.

A 250 kilometer residual high-pass map which is believed mainly to reflect the more superficial Caledonian nappe lithologies gives new information on the form and thickness of these allochthons. The map also reflects some structural features occurring in the upper parts of the eastern Scandinavian Precambrian region.

INTRODUCTION

Interpretation of standard Bouguer gravity maps is hampered by the complexity caused by the overlap of gravity anomalies of different magnitude originating at different depths. In recent years computers have been used to help to resolve this complexity (1).

P. E. Schenk (ed.),
Regional Trends in the Geology of the Appalachian-Caledonian-Hercynian-Mauritanide Orogen, 55–61.

A new attempt to produce gravity maps by using computers to remove the effects of broad variations in the gravity field was made by a group of geophysicists at the U.S. Geological Survey (R. Godson, T. Hildebrand, M.F. Kane and R. Simpson). The author was introduced to this technique during a year of leave at the USGS in Denver (1981-1982).

PROCESSING OF GRAVITY MAPS

A general Bouguer anomaly gravity map of Scandinavia was compiled by the Geological Survey of Sweden for the 1981 IGCP Project 27 meeting in Uppsala. The data were collected from national geodetic surveys in Denmark, Finland, Norway and Sweden. More recent data from Norway were compiled by the author for the same meeting. The Scandinavian gravity map was digitized by identifying the location of each 5 mgal contour line at increments of 15' of latitude and filtered to produce a data set containing only wavelengths longer than 250 kilometers (regional field). This data set was then subtracted from the original data to produce a set containing wavelengths shorter than 250 km (residual field). Two factors should be considered in any interpretation of these filtered maps: the depth range of the anomaly sources and the distortion of anomalies caused by filtering. Test filtering (2) of calculated anomalies for idealized sources indicates that most of the anomalies on 250 km residual maps arise from depths less than 40 km, i.e. mainly from crustal levels. The regional map composed of wavelengths greater than 250 km is thought to reflect mantle sources, MOHO topography and broad crustal sources.

INTERPRETATION OF THE REGIONAL GRAVITY MAP

On the regional gravity map (Fig. 1), which shows only anomalies of wavelength greater than 250 km, gravity highs (0 to + 50 mgal) occur along the west coast of Norway and around the Oslo Graben, indicating high density rocks in these areas. In a study from northern Norway, Brooks (3) considered the coastal positive anomaly to embrace two prominent anomalies; firstly, the source area of the late Cambrian Seiland plutonic province situated near Soroy, and secondly the anomaly relating to Proterozoic basic granulites in the Lofoten archipelago. Talwani and Eldholm (4) subsequently showed that the Lofoten anomaly is not connected with the Soroy high but can be traced southwestwards across the Norwegian Sea to northeast Scotland. Furthermore this anomaly is attributed primarily to intrabasement density contrasts in rocks of Precambrian age and could well be related to mafic granulites as in Lofoten. The data used in the present study do not cover the area of the Norwegian Sea west of the Lofoten archipelago and

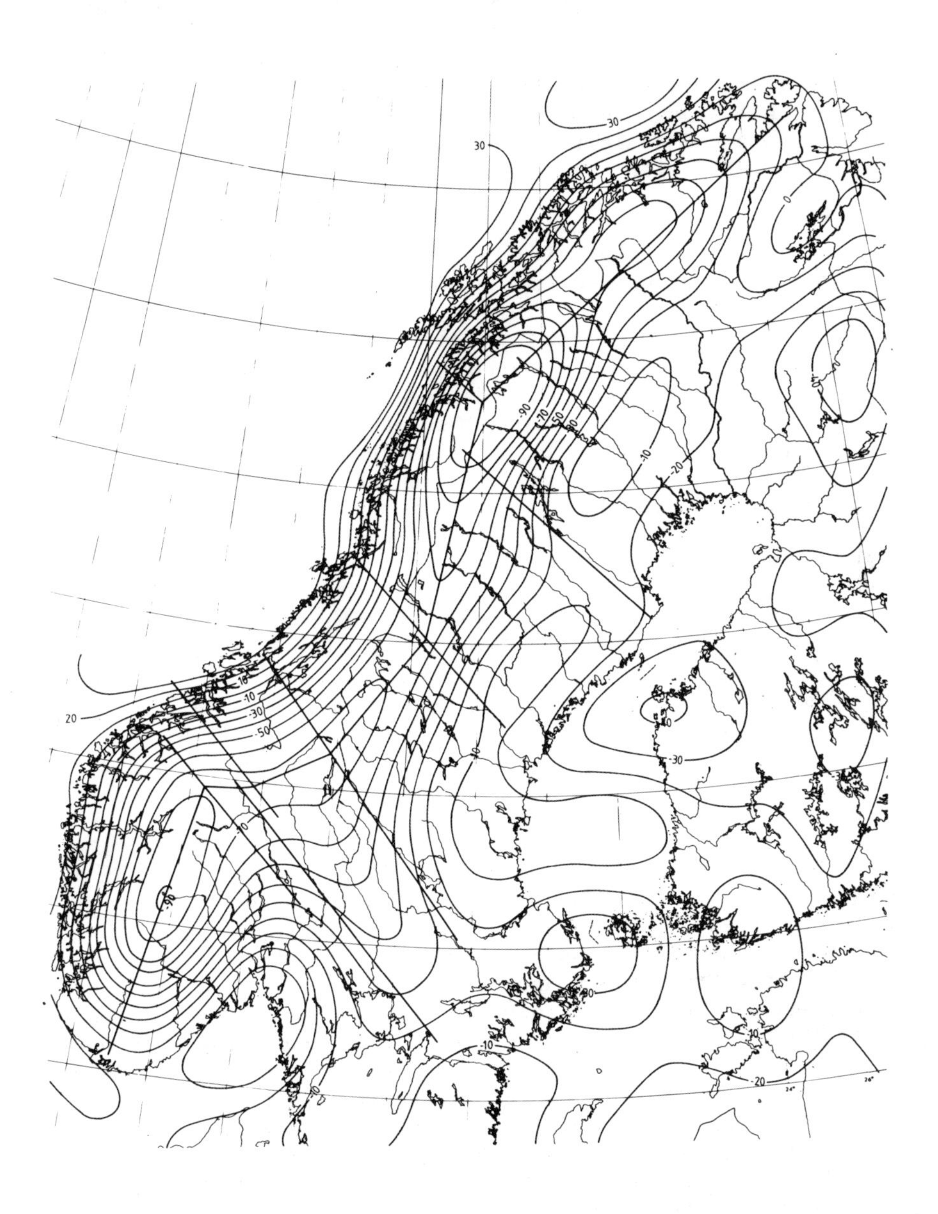

Figure 1. Regional, low-pass gravity map of Scandinavia produced by computer processing of the standard Bouguer gravity map by filtering, leaving only broad-scale gravity features with wavelengths greater than 250 kilometers. Main negative anomaly trend offset by paleofractures trending northwest - southwest.

therefore add no new information to this interesting trend across the Norwegian Sea. The regional low-pass gravity map would probably not have reflected this kind of narrow high gradient anomaly anyway. It shows, however, the marked general increase in gravity values all along the west coast of Norway which may be interpreted as a broad-scale thinning of the continental Precambrian crust associated with the opening of the present Atlantic Ocean.

Earthquakes with magnitude 3 to 4 are located at the eastern edge of this main gravity high. Similar seismic features have also been recognized in the Appalachians (5).

Relatively low gravity anomalies occur over wide areas across the Baltic Shield craton from Central Sweden to Finland and the western Soviet Union. Extremely low gravity anomalies (-50 to -110 mgal) appear along the central part of the Scandinavian Peninsula. A thickening of light Precambrian basement is interpreted to be the reason for this, while the depth to MOHO under the Baltic Shield proper is assumed to be much shallower. Sellevoll (6), who contoured the MOHO topography in Fennoscandia, reached similar conclusions.

Based on the apparent offsets of the main negative anomaly trend in North and Central Norway, major paleofractures may be inferred trending northwest - southeast. These fractures coincide closely with some of the paleofractures described and interpreted by Stromberg (7).

INTERPRETATION OF THE RESIDUAL GRAVITY MAP

The residual gravity map (Fig. 2) depicts anomalies with wavelengths less than 250 km which are attributed primarily to crustal sources. Low gravity anomalies (-10 to -40 mgal) reflecting Precambrian lithologies of roughly granitoid composition with low specific gravity appear in three large areas in South Norway and one area in the northeasternmost part of North Norway. These areas coincide with (1) the central-southern Precambrian area (the Egersund anorthosite province, the Kongsberg-Bamble complex, the Telemark supracrustal suite and the southern basement gneisses and post-tectonic granites); (2) the southeastern Precambrian area (gneisses and post-tectonic granites); (3) the Namsos-Bergen coastal gneiss area; and (4) the Varanger area of Finmark (gneisses, supracrustals and granites with minor amounts of basic rocks and an iron ore formation).

Another striking feature of this map is the presence of several gravity lows above domes and windows of Precambrian crystalline rocks close to the southeastern front of the Caledonian thrust belt (AT, SY, TO, GR, BO, NA, RO, KV and KO in Fig. 2). A

Figure 2. Residual, gravity map of Scandinavia produced by computer processing of the standard Bouguer gravity map by removing broad-scale gravity features with wavelengths greater than 250 kilometers. CS - Central-South Precambrian area, V - Varanger Precambrian area, AT - Atnsjo window, SY - Sylene window, TO - Tommeras window, GR - Grong culmination, BO - Borgefjell window, NA - Nasafjell window, RO - Rombak window, KV - Kvaenangen window and KO - Komagfjord window. GT - Gothian trench, SF - Svecofennian block, BB - Botnian basin and SK - Svecokarelian fault.

published residual gravity map from Nordland county, North Norway (8) also reveals some of these windows.

Caledonian rocks in Scandinavia generally have a higher specific gravity than Precambrian lithologies (9) and the Caledonian thrust sheets are therefore reflected by gravity highs on the residual map. Thin thrust nappes occur in the eastern parts of the Caledonides in Sweden and southeastern Norway (10) causing medium-high anomalies (0 to +10 mgal). Highs in the range +10 to +30 mgal correspond to thick piles, up to 10 km (11), of intensely folded nappes which occur in the central parts of the orogen in central Norway (Jotunheimen, Trondheim region and areas in Nordland and Troms).

The residual map also provides some information on the Precambrian Baltic shield in Norway, Sweden and Finland which coincides quite well with the structural model advocated by Anna Hietanen (11) elements of which are indicated in Figure 2 (GT, SF, BB and SK).

REFERENCES

(1) Kerr, R.A.: 1982, New gravity anomalies mapped from old data. Science Vol. 215, pp. 1220-1222.

(2) Simpson, R.W., Hildenbrand, T., Godson, T.G., and Kane, M.F.: 1982, A description of colored gravity and terrain maps for the United States and adjacent Canada east of 104°. U.S. Geol. Surv. Open file report 82.

(3) Brooks, M.: 1970, A gravity survey of coastal areas of West Finnmark, northern Norway. Q. Jl. geol. Soc. Lond. Vol. 125, pp. 171-192.

(4) Talwani, M., and Eldholm, O.: 1972, Continental margin off Norway: a geophysical study. Geol. Soc. of America. Bull. Vol. 83, pp. 3575-3606.

(5) Kane, M.F.: 1982, Gravity evidence of crustal structure in the United States Appalachians. This volume, p. 45.

(6) Sellevoll, M.A.: 1973, Mohorovicic Discontinuity beneath Fennoscandia and adjacent parts of the Norwegian Sea and the North Sea, Tectonophysics 20, pp. 359-366.

(7) Stromberg, A.G.B.: 1976, A pattern of tectonic zones in the western part of the East European Platform. Geologiska Föreningens i Stockholms Förhandlingar, Vol. 98, pp. 227-243.

(8) Gabrielsen, R.H., Ramberg, I.B., Mork, M.B.E., and Tveiten, B.: 1981, Regional geological, tectonic and geophysical features of Nordland, Norway. Earth Evolution Sciences Vol. 1, No. 1, pp. 14-26.

(9) Roberts, D., and Wolff, F.C.: 1981, Tectonostratigraphic development of the Trondheim region Caledonides, Central Norway. Journ. of Structural Geology, Vol. 3, No. 4, pp. 487-494.

(10) Gee, D.G.: 1975, A tectonic model for the central part of the Scandinavian Caledonides. Am. Jour. Sci. 275, pp. 468-515.

(11) Hietanen, A.: 1975, Generation of potassium-poor magmas in northern Sierra Nevada and the Svecofennian of Finland. J. Res. U.S. Geol. Surv. Vol. 3, No. 6, pp. 631-645.

REGIONAL GRAVITY TRENDS ASSOCIATED WITH THE MAURITANIDES OROGEN (WEST AFRICA)

J. Roussel* and J.-P. Lecorche**

* Laboratoire de Géophysique
** Laboratoire de Géologie dynamique
L.A. CNRS n° 132. Faculté des Sciences et Techniques de St-Jérôme, F - 13397 Marseille Cedex 13

INTRODUCTION

The Mauritanides orogenic belt (1); (2); (3); (4); (5); is exposed between latitudes 12°N and 23°N as a long north-south belt in contact eastward with the West African Precambrian craton and its sedimentary cover (Fig. 1). The Belt is interrupted only between 20°N and 21°N by the Reguibat uplift which protrudes westward in that area. Westward, the internal margin of the orogen is hidden beneath the thick Mesozoic-Cenozoic deposits of the Mauritania-Senegal coastal basin. Southward, the prolongation of the chain seems to be the Rokelides belt of Sierra Leone (6).

The main outlines of the geology and the present interpretation of the Mauritanides orogen is developed in Lécorché's article (elsewhere in this volume). In this paper we seek to discuss on a regional scale the interpretation of the orogen on the basis of available gravity data.

GRAVITY DATA

Figure 2 presents a Bouguer gravity anomaly map corresponding to the area marked in Figure 1 and prepared for the IGCP Caledonide project meeting held in Uppsala in 1981. This gravity map has been compiled with a contour interval of 5 mGal largely from data supplied by the Office de la Recherche Scientifique et Technique Outre-Mer (7). For the Senegal basin area west of 15°W, the map has been compiled on data from oil exploration surveys (8).

P. E. Schenk (ed.),
Regional Trends in the Geology of the Appalachian-Caledonian-Hercynian-Mauritanide Orogen, 63–71.

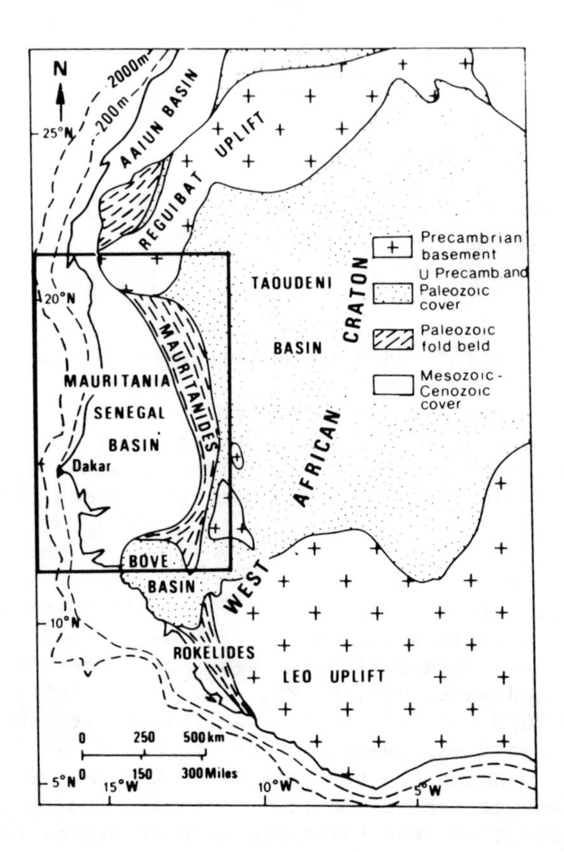

Figure 1. Schematic structural map of West Africa also showing the area of the gravity surveys.

The principal geological subdivisions have been superimposed on the gravity map. Three major tectonic elements can be identified: the western border of the West African craton and its cover (Taoudeni basin), the Mauritanides orogen proper in the area of its outcrop and the Mauritania-Senegal coastal basin.

The map shows gravity patterns that reflect these three main domains: Broad regional low gravity anomalies are associated with the western margin of the craton; in the central area, there is a prominent NNW-SSE trending belt of high gravity anomalies which parallels the main exposed segment of the Mauritanides orogen; West of this large positive anomaly, a relatively positive Bouguer anomaly field occurs over the wide area of the Mesozoic-Cenozoïc coastal basin.

GRAVITY TRENDS ASSOCIATED WITH THE FORELAND OF THE MAURITANIDES BELT

The foreland of the Mauritanides is constituted by the West African craton which crops out in the Reguibat uplift and in the Eastern Senegal and Kayes inliers (Fig. 2). This foreland is generally covered by horizontal epicontinental sediments of the Taoudeni basin.

Areas of Basement Outcrop

The southwestern part of the Reguibat uplift (Tasiast, Tigirit and Amsaga regions) is marked by negative anomalies (RU, in Fig. 2) with trends reflecting the general structural grain of this part of the uplift. The boundary between the southwestern part of the Reguibat uplift and the northernmost extension of the main Mauritanides belt (the Akjoujt portion of the belt which turns towards the West) is characterized by a pronounced gradient in the gravity field which presumably indicates an E-W crustal discontinuity. This feature appears to be aligned with the continentward extension of the Cape Blanc fracture zone (9).

The Eastern Senegal and Kayes basement inliers are mainly marked by SW-NE gravity lows which probably reflect intrusive granites emplaced during the Eburnean orogeny between 2000 and 1600 Ma (10).

Western Margin of the Taoudeni Basin

From Adrar to the south of Assaba, the craton's margin, covered by upper Proterozoic to Palaeozoic sediments of the Taoudeni basin, is gravimetrically marked by a series of gravity lows elongated in a SW-NE direction (SK, Mb, Tj, in Fig. 2) and flanked by relatively positive anomalies of similar trend (GA, Ki, Dk, in Fig. 2). These gravity features are parallel to the structures which appear in the basement inliers and seem to be controlled by major tectonic lineaments in the Precambrian basement. These alternating gravity highs and lows, well defined on the residual anomaly map (11), suggest a horst and graben arrangement which seems to have controlled sedimentation in the upper Precambrian and Palaeozoic along the entire length of the central Mauritanides margin.

The gravity pattern clearly shows that this apparent graben system extends about 30 to 50 km beneath the eastern part of the Mauritanides front (to Selibaby, Mbout, Gadel, .. Fig. 2). This confirms the recent interpretations (2); (3); (4) in which the main part of the outcropping belt forms an allochthonous slice of rocks transported over the craton and its autochthonous sedimentary cover.

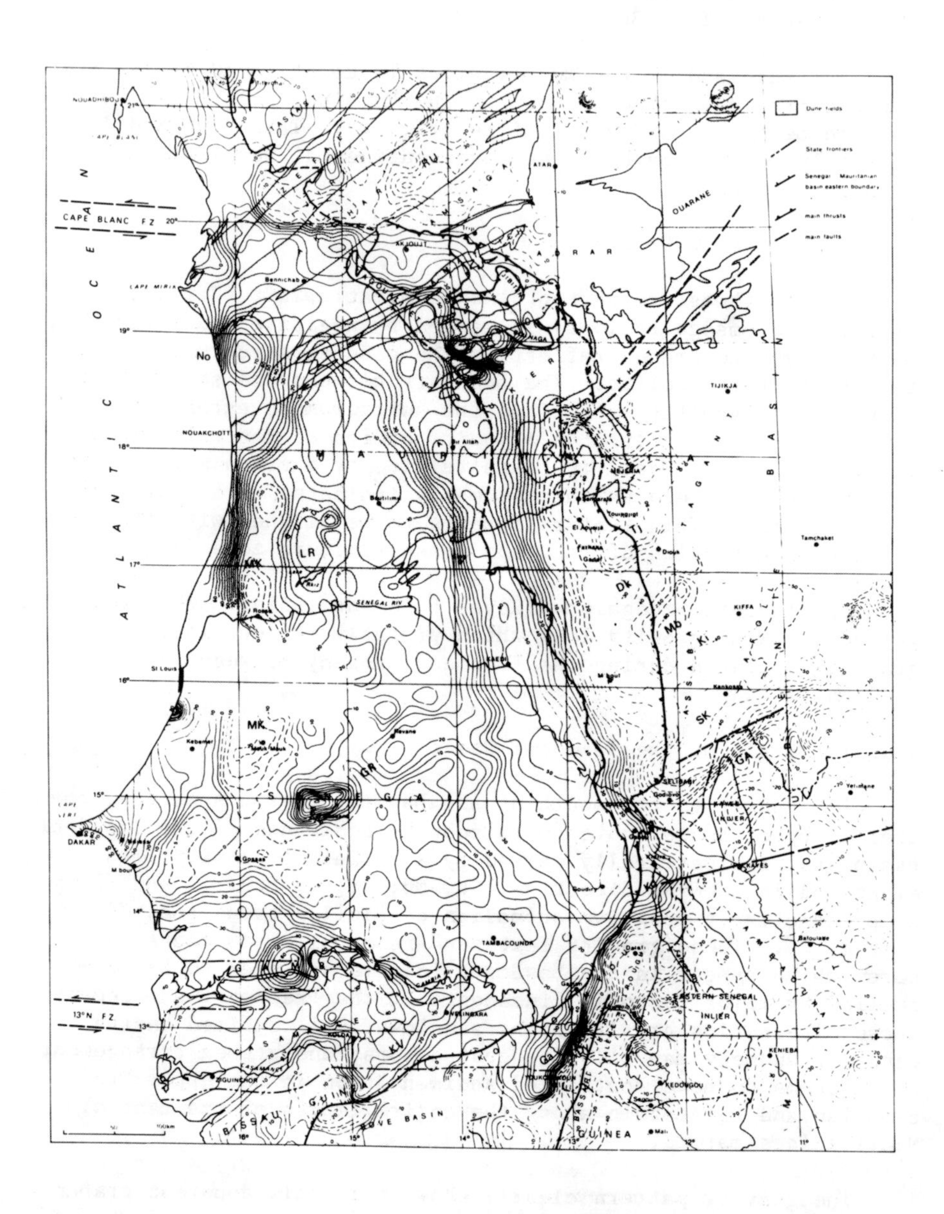

Figure 2. Bouguer gravity map of Western Mauritania and Senegal. The contour interval is 5 mGal. Areas of positive values are shown as continuous lines, negative values as dashed lines. The contours of the Mauritanides orogenic belt (solid lines) are from the map of Lécorché (4).

GRAVITY ANOMALIES OCCURING IN THE MAURITANIA SENEGAL BASIN

In addition to the prominent positive anomaly along the eastern border of the basin, presumably related to the Mauritanides orogen and therefore discussed separately in the next section, relatively positive anomalies occur in the wide area over the Mesozoic-Cenozoic coastal basin. In the vicinity of the coast where post-Palaeozoic cover apparently reaches a thickness of more than 7000 m (12) a large gradient in the gravity field occurs. An implication of the anomaly pattern is that the crust underlying the basin thins westward from the craton to the ocean-continent boundary (13). To the east of a zone of N-S flexure and fault zone along longitude 15°30"W (12) where the depth of the basement is less than 2000 m (14), the dominant NW-SE gravity trend (DV, GR in Fig. 2) is about the same as that observed in the foreland of the belt. This suggests a basement structurally similar to the western border of the West African craton.

INTERPRETATION OF THE MAURITANIDES GRAVITY ANOMALY

In the central region of the gravity map (Fig. 2) a large NNW-SSE trending positive anomaly occurs along the eastern end of the Mauritania-Senegal basin, separating the West Afrcian craton's margin from the thinned coastal basement. This anomaly has an amplitude greater than 50 mGal and is elongated for more than 700 km from Akjoujt in the north to Goudiry in Eastern Senegal. Although its axis is displaced to the west of the Mauritanides outcrop, it remains closely parallel to the trend of the belt. In the region south of the Reguibat uplift the anomaly axis swings westward as do the outcrops of the chain.

This prominent anomaly has been related before to the Mauritanides belt by many authors (15); (7); (16); (17); (2); (4); (18). Recent interpretations (11), have used transformed maps, spectral analysis, inverse theory and linear programming as aid to direct interpretation.

The Mauritanides anomaly may be interpreted by an elongated west dipping dense body along a major crustal discontinuity rooted at between 15 and 30 km within the crust. A schematic geological section across the central Mauritanides from coastal basin to the craton and its cover is presented in Figure 3. This profile shows that the gravity high is clearly but indirectly linked with the outcrop of the allochthonous units of the belt. The form of the dense anomalous mass, asymmetric towards the east suggests major eastward thrusting of a western crustal basement over the craton.

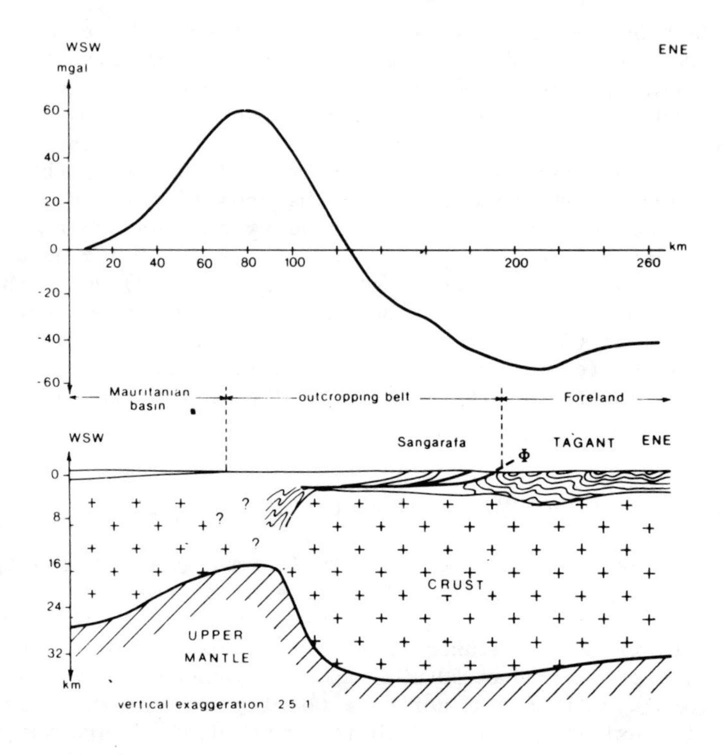

Figure 3. Schematic geological interpretative section across the central Mauritanides at the latitude of Sangarafa, after (5), and showing the indirect but clear relationship between the outcropping belt and the Bouguer anomaly source.

The Mauritanides orogen is a complex mobile belt which is still poorly known in comparison with the Appalachians and a conclusive schema of geodynamic evolution cannot be proposed. According to Lécorché's recent interpretation (4), the Mauritanides gravity anomaly is interpreted as the trace of the suture zone that could be the result of compression directed towards the east during the Ordovician (a "Taconic event"). In this respect, it could reflect continued collision of two continental blocks after the closure of an inter-block trench rather than active margin collision as has been previously suggested (17). A further phase of the chain development, at the end of the Devonian, occurred farther to the west in the coastal region or even farther west (4); (19), but was thrust over the craton, involving the Taconic chain in a late stage of horizontal shearing.

In Eastern Senegal and Guinea, an important elongated NNE-SSW gravity high (GA, in Fig. 2) occurs nearly parallel to the folded formations of the southeastern arm of the Mauritanides belt (Lécorché, this issue). In this region, a teleseismic experiment at stations astride the craton margin (20) has shown a steep seismic boundary in the mantle aligned with the gravity anomaly. Recent magneto-telluric soundings carried out in the same area (21) have shown the existence of a conducting zone at a depth of 30 km in the mobile zone crust. The gravity high, on line with the Mauritanides anomaly but having a different form and strike, would have a different origin. It appears to be related to other similar well-marked positive anomalies occuring in Western Guinea (22) and in Liberia (23) and may be linked to an older (Pan-African) suture corresponding to the northern continuation of the Rokelides of Liberia and Sierra Leone.

REFERENCES

(1) Sougy, J.: 1962, West African fold belt. Geol. Soc. Am. Bull., Vol. 73, pp. 871-876.

(2) Lécorché, J.-P., and Sougy, J.: 1978, Les Mauritanides, Afrique occidentale. Essai de synthèse. In: PICG projet 27, contrib. fr. n° 4, Caledonian-Appalachian Orogen of the North Atlantic Region; Geol. Surv. Canada, paper 78-13, pp. 231-239.

(3) Dia, O., Lécorché, J.-P., et Le Page, A.: 1979, Trois événements orogéniques dans les Mauritanides d"Afrique occidentale. Rev. Géol. dynam. Géogr. Phys., Vol. 21, fasc. 5, pp. 403-409.

(4) Lécorché, J.-P.: 1980, Les Mauritanides face au craton ouest- africain. Structure d'un secteur-clé: la région d"Ijibiten (Est d'Akjoujt, R.I. de Mauritanie). Thèse Univ. Aix-Marseille III, 446 p.

(5) Lécorche, J.-P., Roussel, J., Sougy, J., and Guetat, Z.: (in press), An interpretation of the geology of the Mauritanides orogenic belt (West Africa) in the light of geophysical data. In: Contribution to the Tectonics and Geophysics of mountain chains. R.D. Hatcher, Jr., I. Zietz and H. Williams (editors), Geol. Soc. Am. Memoir, Boulder.

(6) Allen, P.M.: 1969, The Geology of part of an orogenic belt in Western Sierra Leone, West Africa. Geol. Rundsch., Vol. 58, n° 2, pp. 588-620.

(7) Crenn, Y., et Rechenmann, J.: 1965, Mesures gravimétriques et magnétiques au Sénégal et en Mauritanie occidentale. Cah. O.R.S.T.O.M., sér. Géophys., n° 6, p. 59.

(8) Bureau De Recherches Petrolieres (B.R.P.): 1955, Etude géophysique par la méthode gravimétrique du bassin sédimentaire du Sénégal. Unpublished Rep., Comp. Gén. Géophys., Paris.

(9) Le Pichon, X., Sibuet, J.C., and Francheteau, J.: 1977, The fit of the continents around the North-Atlantic ocean. Tectonophysics, Vol. 38, pp. 169-209.

(10) Vachette, M., Rocci, G., Sougy, J., Caron, J.P.H., Marchand, J., Simon, B., et Tempier, C.: 1975, Ages radiométriques Rb/Sr, de 2.000 à 1.700 MA de séries métamorphiques et granites intrusifs précambriens dans la partie N et NE de la dorsale Reguibat (Mauritanie septentrionale). 7ème Coll. inter. Géol. Afr., Firenze, Italie. In: Trav. Lab. Sci. Terre St-Jérôme, Marseille, sér. B, n° 11, pp. 142-143.

(11) Guetat, Z.: 1981, Etude gravimétrique de la bordure occidentale du craton ouest-africain. Thèse 3e cycle, Univ. Sciences et Techniques du Languedoc, Montpellier, p. 185.

(12) De Spengler, A., Castelain, J., Cauvin, J., et Leroy, M.: 1966, Le bassin secondaire-tertiaire du Sénégal. In: "Bassins sédimentaires du littoral africain", lère partie: littoral atlantique, D. Reyre ed., Assoc. Serv. Géol. Afr. Paris, pp. 80-94.

(13) Roussel, J., and Liger, J.L.: (in press). A review of deep structure and ocean-continent transition in Senegal basin (West Africa). Tectonophysics, Vol. 91, p. 29.

(14) ASGA-UNESCO: 1968, International tectonic map of Africa, 1/5.000.000.

(15) Bolt, C., Crenn, Y., et Rechenmann, J.: 1962, Eléments apportés par la Gravimétrie à la connaissance de la tectonique profonde du Sénégal. C.R. Acad. Sci., Paris, t. 254, pp. 1131-1133.

(16) Chiron, J.C.: 1973, Etude géologique de la chaîne des Mauritanides entre le parallèle de Moudjeria et le fleuve Sénégal (Mauritanie). Un exemple de ceinture plissée précambrienne repris à l'Hercynien. Mém. Bur. Rech. Géol. Min., n° 84, p. 284.

(17) Dillon, W.P., and Sougy, J.M.: 1974, Geology of West Africa and Canary and Cape Verde Islands. In: "The ocean basins and margins", Vol. 2, A.E. Nairn and F.G. Stehli ed., Plenum Publish. Corp., New-York, pp. 315-390.

(18) Liger, J.L.: 1980, Structure profonde du bassin côtier sénégalo-Mauritanien. Interprétation de données gravimétriques et magnétiques. Trav. Lab. Sci. Terre St-Jérôme, Marseille, Sér. B., n° 16, p. 158.

(19) Lefort, J.-P., and Klitgord, K.D.: 1982, Geophysical evidence for the existence of Mauritanide Structures beneath the U.S. Atlantic Coastal Plain. Abst., IGCP 27 Caledonian-Appalachian orogen meeting, Fredericton, Canada.

(20) Briden, J.-C., Whitcombe, D.N., Stuart, G.W., Fairhead, J.D., Dorbath, C., and Dorbath, L.: 1981. Depth of geological contrast across the West African craton margin. Nature, Vol. 292, pp. 123-128.

(21) Ritz, M.: 1982, Etude régionale magnéto-tellurique des structures de la conductivité électrique sur la bordure occidentale du craton ouest-africain en République du Sénégal. Can. J. Earth Sci., vol. 19, pp. 1408-1416.

(22) Akhmetjanov, B.S., Loutsenko, N.F., and Dyalo, M.: 1976, Carte des lignes iso-anomales de la gravité en réduction de Bouguer au 1/500.000 de la Guinée occidentale. Technoexport, Moscou.

(23) Behrendt, J.-C., Schlee, J., Robb, J.-M., and Silverstein, M.K.: 1974, Structure of the continental margin of Liberia, West Africa. Geol. Soc. Am. Bull., Vol. 85, pp. 1143-1158.

GEOPHISYCAL EVIDENCE FOR THE EXISTENCE OF MAURITANIDE STRUCTURES BENEATH THE U. S. ATLANTIC COASTAL PLAIN

J. P. Lefort, Universite de Rennes I, Campus de Beaulieu, 35042 RENNES Cedex, France, and
K. D. Klitgord, U. S. Geological Survey, Woods Hole, MA 02543, U.S.A.

ABSTRACT ONLY

Late Paleozoic collision of the North American and African plates produced westward thrusting in the Appalachians, eastward thrusting in the Mauritanides, and a suture zone somewhere between the two zones of thrusting. The reconstruction of the Early Jurassic closure of the Atlantic by Klitgord and Schouten (1982)[1] serves as a framework for examining geophysical evidence of this suture zone. The Precambrian Reguibat Massif protudes west of the West African Craton near Cap Blanc; it is bounded to the south by the Senegal basin and to the north by the Tarfaya basin. The late Paleozoic orogenic belt swings westward near the coast around this massif, which places the possible suture zone west of Cap Blanc. On the United States Atlantic Coastal Plain, a chain of elongate magnetic and gravity highs swings westward around the Salisbury Embayment, mimicking the arcuate shape of the Reguibat Massif, which the Early Jurassic reconstruction places just to the east. These magnetic- and gravity-field anomalies are similar to those associated with the late Paleozoic suture zone farther north and may mark the suture zone beneath the Coastal Plain of New Jersey, Maryland, Virginia, and North Carolina. This interpretation would suggest that a prong of the African plate indented the North American plate in this region during the late Paleozoic collision, and that a part of the Mauritanide orogenic belt was left behind beneath the Coastal Plain by the Mesozoic breakup of Pangea.

[1]Klitgord, K.D. and Schouten, H., 1982 EOS v.63, p.307

P. E. Schenk (ed.),
Regional Trends in the Geology of the Appalachian-Caledonian-Hercynian-Mauritanide Orogen, 73.

A STRATIGRAPHIC SKETCH OF THE CALEDONIDE-APPALACHIAN-HERCYNIAN OROGEN

(1) W. H. Poole, (2) W. S. McKerrow, (3) G. Kelling, (4) P. E. Schenk

(1) Geological Survey of Canada, 601 Booth Street, Ottawa, Ontario K1A 0E8, (2) University of Oxford, Geology Department, Parks Road, Oxford OX1 3PR, England (3) Department of Geology, University of Keele, Keele, England (4) Department of Geology Dalhousie University, Halifax, Nova Scotia, B3H 3J5

INTRODUCTION (W. H. Poole)

During the two-day workshop in Fredericton, August 1983, representatives from national working groups on Stratigraphy, Sedimentology and Faunal Province prepared a series of charts depicting characteristic lithologies for each series from Late Precambrian to Devonian for the presently disrupted orogen. Those contributing to the charts were David L. Bruton, Gilbert Kelling, Jean-Paul Lecorche, S. R. McCutcheon, W. S. McKerrow, R. B. Neuman, Alain Pique, Paul Schenk, John Rodgers and Joel Rolet. Information was drawn from memory and consequently may well be incomplete, inaccurate, and poorly balanced. However, this was done in the spirit of the conference, at which presentations were to be formulated for the sessions on site and on the basis of compilation maps and informed discussion by international authorities. During the subsequent colloquium of specialists in related fields, selected individuals described particular regions within the orogen. The latter included: (1) a western region conceived as the western margin of the Iapetus Ocean; (2) a central region crudely representing a volcanic-arc belt; (3) an eastern region representing the Avalon terrane of North America including the cratonal cover of France and

P. E. Schenk (ed.),
Regional Trends in the Geology of the Appalachian-Caledonian-Hercynian-Mauritanide Orogen, 75–111.

Iberia; (4) the Meguma terrane of southern Nova Scotia, and apparently similar terranes in the southern United States; and (5) the cratonal cover of northwestern Africa.

We have organized this paper in the following manner. First is a presentation of the stratigraphic charts (Fig. 1) accompanied by notes on most of the columns. Next are some comments on most of the above regions. The conclusions consider five episodes of time (Figs. 2-6).

NOTES ON THE COLUMNS

Western and Central Regions (W. S. McKerrow)

Column 1 (Fig. 1) Most of the column is derived from our knowledge of east North Greenland, at around 81°N. But at this latitude there are no known Devonian deposits. The Devonian non-marine deposits shown in the column occur further south in East Greenland (1, 2).

Columns 2, 3, 4 and 5 In Scotland, the Cambrian and Ordovician carbonate sequences (Col. 2) rest directly on Lewisian gneisses (deformed between 2,600 and 1,600 Ma ago); no other stratigraphic record of our time interval (Late Precambrian to Late Devonian) is present in this area to the west of the Moine Thrust.

Between the Great Glen Fault and the Highland Boundary Fault (Col. 3), the Dalradian Supergroup includes shelf deposits above and below the Port Askaig Tillite, but the later Precambrian and Early Cambrian sediments are largely turbidites. The relationship of the Lower Ordovician rocks near the Highland Boundary Fault to the Dalradian Supergroup is uncertain; the recent discovery (Curry et al. 1982) of Arenig fossils at Aberfoyle proves that some of these fine-grained sediments post-date the (500 Ma) Grampian Orogeny. They may be separated from the Grampian Highlands by a major stike-slip fault. The Devonian non-marine clastics are assigned to the Old Red Sandstone and rest unconformably on the metamorphosed Dalradian rocks.

No pre-Ordovician basement is exposed below the Midland Valley (Col. 4) or the Southern Uplands (Col. 5), but there is some geophysical evidence of crystalline rocks below the Midland Valley, where xenoliths in Carboniferous volcanic vents (Upton, 1976) also indicate a metamorphic basement. The Ordovician-Silurian accretionary prism of the Southern Uplands and Ireland indicates one of the very few instances of Early Palaeozoic subduction below the eastern margin of North America.

Figure 1

P = platform

B = basin (these columns refer to sedimentation off the eastern margin of Lower Palaeozoic North America, in slope, trench, and oceanic environments; they also include some regions which may have been island arcs or other island areas situated out in the Iapetus ocean at a distance from North America).

Abbreviations used in columns:

ss	=	sandstone
sh	=	shale
gw	=	greywacke
cong	=	conglomerate
ch	=	chert
olist	=	olistostrome
dol	=	dolomite
ls	=	limestone
Skol.ss	=	*Skolithus* sandstone
bimod	=	bimodal volcanics
ac	=	acid volcanics
bas	=	basic volcanics
c-a	=	calc-alkaline volcanics
calc	=	calcareous

A horizontal line indicates an unconformity; most are not labelled, but the principle orogenies are named.

Geological series: Late P-C = Late Precambrian (700 Ma); L = Lower; Lower Ordovician includes the Tremadocian; Middle Ord. includes the Caradocian and the *Nemagraptus gracilis* zone in part of the Llandeilo; the U. Ord. is the Ashgill.

WESTERN AND CENTRAL REGIONS

	1P	2P	3 P/B	4B
		SCOTLAND		
	EAST GREENLAND	NORTHERN SCOTLAND	GRAMPIANS	MIDLAND VALLEY
U. DEV.	red ss. bimod. vol	-	red ss. cong.	red ss. cong.
M. DEV.	red ss	-	red ss.	-
		CALEDONIAN OROGENY		
L. DEV.	-	-	red ss. c-a. vol.	red ss. c-a. vol.
	CAL. OROG.			
U. SIL.	(No Ludlow) gw	-	-	cong., ss
L. SIL.	sh., ls.	-	-	ss., gw
U. ORD.	dol	-	-	ss., sh
M. ORD.	ls	-	-	gw., sh., ls., cg.
L. ORD.	ls., ss	ls	ls., bas. vol., ch, sh	bas. vol., sh., ch., olist
		GRAMPIAN OROGENY		
U. CAMB.	-			-
M. CAMB.	dol	-	gw?	-
L. CAMB.	dol., ls., Skol.ss	ls., dol., Skol.ss	gw., bas. vol	-
LATE P-C.	ls., sh ss., gw tillite	-	gw., dol., dol., ss tillite	-

WESTERN AND CENTRAL REGIONS

	5B	6B	7 P/B	8 P/B
	SCOTLAND			
	SOUTHERN UPLANDS	NORTHERN ENGLAND	NORTH WALES	S.E IRELAND
U. DEV.	red ss. cong.	-	-	red ss.
M. DEV.	-	-	-	red ss.
	C A L E D O N I A N		O R O G E N Y	
L. DEV.	red ss. c-a. vol.	? red cong.	red ss., cong.	red ss.
	CALEDONIAN OROGENY			
U. SIL.	gw	gw	gw	gw
L. SIL.	gw., sh	sh	sh	sh
U. ORD.	gw., sh	sh., ls	ss., sh	ss
M. ORD.	gw., sh.	c-a. vol	ac., vol, ss., sh., bi., vol.	ac. vol. ls.
L. ORD.	ch., bas. vol	gw	sh., ss sh	sh
U. CAMB.	-	-	ss., sh	
M. CAMB.	-	-	sh	
L. CAM.	-	-	gw., sh ss., ac. vol	sh., gw.
LATE. P-C.	-	-	olist. ls., sh. gw., ss	sh., gw

WESTERN AND CENTRAL REGIONS

	9P	10B	11B	12P
	NEWFOUNDLAND			
	WESTERN IRELAND	GANDER	NEW WORLD ID.	LONG RANGE
U. DEV.	-	-	-	-
M. DEV.	red ss.	-	-	-
	CAL. OROG.			
L. DEV.	-	-	-	-
U. SIL.	ss	gw	-	ss., sh
L. SIL.	sh., ss., cong.	ss., red ss., cong.	cong. ss olist, cong., gw	- -
U. ORD.	-	ss	gw., sh. olist	-
M. ORD.	-	ss., sh., gw	sh., ls., bas. vol gw., ss	ss., sh
				HUMB.OR
L. ORD.	cong. gw., sh., bas. vol.	ss	olist	ls
	GRAMP. ORO.			
L. CAMB.		?	-	ls
M. CAMB.	gw., sh	?ss., sh.	-	ls
L. CAMB.	gw	?ss., sh. ?gw	-	ls., Skol. ss
LATE P-C.	gw., dol., ss., tillite	?ss., sh., ?gw	-	bas., vol.

WESTERN AND CENTRAL REGIONS

	13B	14 P/B	15B	16P	17B
	NEW BRUNSWICK				
	CENTRAL	NORTH	MAINE	N. Y. (ALLOCH)	N.Y. (ALLOCH)
U. DEV.	-	red ss., sh	-	ss., sh.	-
M.DEV.	-	-	-	ss., sh.	-
	A C A D I A N O R O G E N Y				
L. DEV.	-	-	gw., sh.	ls	-
U. SIL.	gw., sh	ss., ls	ss., sh.	-	-
L. SIL.	?gw	ss.	cong., ss., sh.	-	-
U. ORD.	-	gw, ls. cong.	sh	gw	-
	T A C O N I C O R O G E N Y				
M. ORD.	gw., sh. c-a., vol.	gw.	-	-	sh.
L. ORD.	?vol.	gw.	sh., ss., ls., bim. vol., cong.	ls.	sh.
U. CAMB.	?	?gw.	-	dol., ls	sh., dol.
M. CAMB.	-	?gw	-	dol.	ls., sh.
L. CAMB.	-	-	-	dol., ss.	sh., ls.
LATE P-C.	11	11	?sh., ss.		gw., sh.

WESTERN AND CENTRAL REGIONS

	18P	19B	20P
	VA	S. APPAL	ALABAMA
U. DEV.	ss., sh.	-	sh.
M. DEV.	ss., sh.	-	-
L. DEV.	sh., ss., ch.	ch., calc. ss	ss.
U. SIL.	ss.,	?ch., calc. ss.	-
L. SIL.	sh., ss., cong.	ss., cong.	sh., ss., ls., hematite
U. ORD.	ss., gw., sh.	-	ls
M. ORD.	sh., ls.	?calc., ss.	ls., ch., cong.
L. ORD.	dol., ls.	cong.	dol.
U. CAMB.	dol., ls.	-	dol.
M. CAMB.	sh., dol.	-	sh., ss
L. CAMB.	dol., ss.	-	ls., dol., ss
LATE P-C.	vol.	-	vol.

EASTERN REGION

	21P SCANDINAVIA	22P S. BRITAIN
CARB.	-	ss., sh., ls. coal
U. DVE.	-	red ss., sh.
M. DEV.	red ss., congs.	red ss., ls.
	CALEDONIAN OROGENY	
L. DEV.	red ss.	red ss., sh.
U. SIL.	ss.	red ss., sh., ls.
L. SIL.	ls., sh.	sh., ss., volcs.
U. ORD.	ls., sh. (Hirnantia)	sh., ss. (Hirnantia)
M. ORD.	ls., sh. (Remopleurid)	sh., volcs. (Remopleurid)
L. ORD.	blk. sh., ls.	blk. sh., volcs.
U. CAMB.	blk. sh. (Olenids)	blk. sh., (Olenids)
M. CAMB.	blk. sh. (Paradoxides)	blk. sh., ss. (Paradoxides)
L. CAMB.	ss., sh. (Holmia)	ss., sh.
		AVALONIAN OROGENY
LATE P-C.	qtz., tillite	red ss., volcs. (Ediacarid)

EASTERN REGION

	23P CANADA (Avalon)	24 NORTH U.S.A.	25P SOUTH U.S.A.
	red ss., sh., coal	ss., sh., coal	-
CARB.	MARITIME	ALLEGHANIAN	OROGENY
	red ss., ls., sh., volc.	ss. sh.	-
U. DEV.	red ss., volcs.	red ss., volc.	-
M. DEV.	red ss., volc.	red ss., volc.	-
	A C A D I A N	O R O G E N Y	
L. DEV.	red ss., volcs.	sh., volcs.	-
U. SIL.	sh., red ss., acid/basic volcs	ac. volcs., sh.	-
L. SIL.		blk. sh.	-
U. ORD.	-	-	-
M. ORD.	-	-	-
L. ORD.	blk. sh. ss., Fe.	blk. sh.,	?
U. CAMB.	blk. sh. (Olenids)	gw. qtz.	?
M. CAMB.	blk. sh., basic volcs. (Parad).	bl. sh. (Paradoxides)	sh. (Paradoxides)
L. CAMB.	ls., ss., red. sh. (Holmia)	bl. sh., ls.	qtz.? cong.
	A V A L O N I A N	O R O G E N Y	
LATE P-C.	red ss., gw., volcs., tillite (Ediacarid)	Volcs. (ac.+bas.) Flysch	volcs. (ac.+bas.)

EASTERN REGION

	26 P/B	27P
	MEGUMA	N.W. FRANCE
	ss., sh., coal	ss., sh., coal
	(MARITIME) HERCYNIAN OROGENY	
CARB.	red ss., ls., cong. volc.	cong., sh., volc. olisto.
U. DEV.	red cg., ss.	sh., ss.
M. DEV.	red cg., ss.	ls., sh.
	ACADIAN OROGENY	
L. DEV.	ss., sh., ls., Fe	ls., ss.
U. SIL.	blk. sh., qtz., volcs. (ac. + bas)	ss., blk. sh.
L. SIL.	ss., blk. sh., volcs.	ss., blk. sh.
U. ORD.	blk. sh., ? tillite	ls., sh., volcs. ? tillite
M. ORD.	blk. sh., gw.	blk. sh., ss.
L. ORD.	gw., blk. sh.	blk. sh., qtz., Fe, volcs.
U. CAMB.	gw., blk. sh.	volcs.
M. CAMB.	gw., blk. sh.	cong. red sh.
L. CAMB.	?	ls., ss. (Archaeo).
LATE P-C.	?	volcs., ss. (Ediacarid)
		CADOMIAN OROGENY

EASTERN REGION

	28P	29P	30P
	CENTRAL IBERIA	CENTRAL MOROCCO	ANTI-ATLAS
	ss., sh., coal	red ss., cong.	ss., sh., coal
	HERCYNIAN OROGENY		
CARB.	gw., sh., volc. olisto	gw., sh., volc.	
U. DEV.	sh., ls.	gw., sh.	sh., ss., ls.
M. DEV.	-	ls., ss., cong.	ls., sh.
L. DEV.	ss., sh., ls., volc.	ls., sh.	ls., ss., volc.
U. SIL.	blk. sh.	-	sh.
L. SIL.	blk. sh., volc.	blk. sh.	sh., ls.
U. ORD.	sh., ls., ? tillite	sh., gw., tillite	ss., sh., tillite
M. ORD.	blk. sh., qtz.	gw., sh., qtz.,	ss., sh.
L. ORD.	qtz., red ss. sh.	sh., ss.	sh, ss.
U. CAMB.	-	ss., qtz.	ss., ls.
M. CAMB.	-	gw., sh. (Paradox).	gw., sh. (Paradox).
L. CAMB.	ls., cong. volc. (Archaeo).	sh., ls.	ls., ss., sh. (Archaeo)
LATE P-C.	volc. gw.	volc. (ac)	red ss., volc.
	CADOMIAN OROGENY	PAN AFRICAN OROGENY	

EASTERN REGION

31P

MAURETANIA

CARB.	ss., sh.
	HERCYNIAN
U. DEV.	sh., ss.
M. DEV.	ls., sh.
L. DEV.	red ss., sh.
U. SIL.	sh., ls., ss.
L. SIL.	ss., sh.
U. ORD.	tillite, ss.,
M. ORD.	-
L. ORD.	ss., sh. (Skolithus)
U. CAMB.	sh., ss (Westonia)
M. CAMB.	red ss., sh.
L. CAMB.	ls., dol. red ss. (Stromato)
LATE P-C.	volc., sh., tillite
	PAN AFRICAN OROGENY

Turbidite sedimentation did not commence until the Middle Ordovician, perhaps some 40 Ma after the Grampian Orogeny; the source of the Southern Uplands trench sediments is thus still quite uncertain (3-7).

Column 6 The Lake District in northern England shows no rocks older than Ordovician. In contrast to the American Ordovician trilobite and brachipod faunas present in Scotland, the Lake District faunas have European affinities. The north of England thus lay on the opposite side of the Iapetus Ocean from Scotland. Apart from the Middle Ordovician island arc volcanics, most of the Lake District sediments were deposited in deep water close to the continental margin (8).

Column 7 and 8 The Lower Palaeozoic sequences of North Wales and south-eastern Ireland lie unconformably on Late Precambrian rocks exposed in Anglesey and County Wexford. Like the north of England, these sequences contain European faunas. Different parts of this region show many local unconformities at different times. The first regional break common to much of the British Isles occurs towards the end of the Lower Devonian; was this when the Iapetus Ocean finally closed in Britain? (9, 10).

Column 9 The upper parts of the Dalradian Supergroup of western Ireland are very similar to their equivalents in Scotland, but some probable Middle Cambrian fossils have been recorded on Clare Island, while nothing younger than Lower Cambrian is known in Scotland. The Lower Ordovician succession in the South Mayo Trough is not seen in direct contact with the Dalradian rocks. If these beds are separated by a major strike slip fault, the major fault movements must be prior to the Late Llandovery, for the Silurian sequence (which masks the contact) lies unconformably on both the Dalradian and the Ordovician beds (11).

Columns 10, 11 and 12 The Gander belt (Col. 10) and the Dunnage belt (Col. 11) were probably on opposite sides of the Iapetus Ocean until final closure during the Acadian Orogeny. But until the Humberian Orogeny, it has also been postulated that the Dunnage belt lay out in the ocean; it contains Lower Ordovician brachiopods which are distinct both from America and Europe (Neuman, 1972). In Newfoundland, only the sequence on the Long Range (Col. 12) lies clearly on the North American continent. The Humberian Orogeny is known to occur during the N. gracilis Zone, when basin facies Lower Palaeozoic beds were thrust westwards over the continental margin. The allochthonous beds are overlain by the Middle Ordovician Port a Port Formation and the Late Silurian Clam Bank Formation (12-15).

Columns 13 and 14 The Tetagouche volcanic rocks of central New Brunswick (Col. 13) may have been part of an Early Ordovician island arc; they contain similar oceanic brachiopods to New World Island (Col. 11). To the south of these volcanics, Silurian greywackes are present in the Fredericton Trough, which lies close to the Iapetus suture (McKerrow and Ziegler, 1971). The Taconic Orogeny of New Brunswick may be related to collision of the Tetagouche arc with North America. The pre-Taconic sequences of northern New Brunswick (Col. 14) lie very close to the contemporary margin of the North American continent, and the post-Taconic sequences lie unconformably on the Tetagouche rocks (16, 17, 18).

Columns 15, 16 and 17 The Taconic Orogeny in New England has been interpreted as the result of collision of the Bronson Hill island arc with North America (Robinson and Hall, 1980). As in New Brunswick, deeper water facies (Col. 17) now lie allochthonously over shelf facies (Col. 16). It is probable that the Bronson Hill arc extends from Massachusetts across Maine, almost to the New Brunswick border; but it appears to be separated from the Tetagouche massif by the Matapedia Basin. Column 15 shows possible Lower Ordovician arc material, which is overlain unconformably by Late Ordovician and Silurian sediments (12, 19-22).

Columns 18, 19 and 20 In the Central and Southern Appalachains, there is an excellent stratigraphic record on the shelf platform (Columns 18 and 20), but the basin succession is less well known. I am indebted to John Rodgers (personal communications, 1982) for details of Column 19 (23, 24, 25).

Eastern Region (G. Kelling)

Column 21 The sequence for Scandinavia is essentially that for the Oslo region of Norway and eastern Jamtland in Sweden. The late Precambrian Moelv and Sjoutalven tillites appear to be southerly products of the Varanger glaciation. The Middle Cambrian to Arenig black shale facies (Alum Shale) is widespread, but succeeding Ordovician carbonate sediments in the easterly platform regions pass westwards into thicker successions of deeper marine clastics, now seen in the parautochthonous tectonic slices of west-central Sweden. Late Silurian and Devonian redbeds are known only as a sedimentary cover to the Caledonian nappes further to the west (26-28).

Column 22 The succession for southern Britain is a composite of sequences in central England and South Wales (east of the Towy axis). Folding and localized low T and P metamorphism which occurred prior to the Early Cambrian is here assigned to the Avalonian "orogeny" and dating of the Longmyndian redbeds in Shropshire places an upper limit of c. 580-560 Ma on these events. Cambrian and Ordovician faunas are European, with the *Hirnantia* faunas developed in the Ashgill. Redbeds appear in the Wenlock and of South Wales and similar terrestrial sedimentation persists into the Devonian, the principal Caledonian deformation occurring here during the Middle Devonian, which is consequently poorly represented by sediments. However, marine carbonates and clastics of this age are well developed in south-west England as part of the Variscan orogen (29-32).

Columns 23, 24, 25 The successions are derived from the Avalon Peninsula (Newfoundland), eastern New England, and the area east of the King's Mountain Belt of southeastern U.S.A., respectively. Although not indicated on Columns 23 and 24, a late Precambrian (?Avalonian) folding event, without significant metamorphism but accompanied by granitic plutonism, is responsible for marked (but localized) discordances at the base of the Cambrian. The absence of later Ordovician and Lower Silurian rocks from these areas has been attributed to Taconic folding and uplift, although penetrative deformation and regional metamorphism of this age are rare. The Acadian event is manifested by folding, regional metamorphism, and granitoid plutonism, most clearly displayed in western Nova Scotia and southwestern New Brunswick. Early Palaeozoic faunas in the northern Appalachians are typically Atlantic, whereas mildly deformed Palaeozoic rocks penetrated by boreholes in southern Georgia and Alabama have yielded European faunas (15, 33-38).

Column 26 (P. E. Schenk only for this column) The Meguma Terrane may be underlain by a pre-Hadrynian basement complex. This is an important conclusion for trans-orogenic correlation. If true, this complex could be quasicontinental or even continental crust. As such, this crust could be part of the Avalon plate. Certainly the Meguma Terrane would better fit into the African - North American framework (59). The evidence is that zircons in a displaced gneiss within the Meguma have been dated by U/Pb with an upper intercept of 2350 Ma and a lower intercept of 1100 Ma. However, the Meguma becomes an orthogneiss in the area, and detrital zircons occur in relatively unmetamorphosed Meguma sediments; moreover, other detrital clasts give a rare-earth age of 1800 Ma (Clarke, pers. com.).

The pre-Acadian succession is a shoaling upward succession from deep-sea fan, quartz wacke turbidites; to black slates with slump structures; possible marine tillite; then shelf quartz arenites, shales, and sandstones, and minor carbonates. The oldest fossils in the complex are Tremadocian; the youngest are Seigenian, possibly Emsian (390 Ma). Post-folding granitic intrusives are mainly 380 to 360 Ma (Givetian to Famennian); however, another intrusive pulse at 320 to 300 Ma (Pennsylvanian) was associated with folding, faulting, metamorphism, and metallogenesis (Maritime Disturbance - Hercynian Orogeny). Post-orogenic sediments consist of fluvial to lacustrine, terrigenous material generally deposited in wrench-fault basins. These rocks range in age from Eifelian (387 Ma) to Early Jurassic, with many unconformities. The youngest lithified sedimentary rocks of the Meguma Terrane consists of conglomerates and marine clays of Cretaceous age (34, 40, 41).

This succession in the Meguma, both pre- and post-orogenic, is very similar to that of the Rabat-Tiflet Zone of northwestern Morocco (41). Unfortunately, it also is a displaced, suspected terrane. Trans-orogenic correlation of the Meguma Group is difficult because its source-area lay to the present east-southeast. This source-area had to supply a volume of sediment perhaps equivalent to the present Ganges-Brahmaputra deep-sea fan, and for a very long span of time. The simplest candidate is the Saharan Platform which shed sediment to the present west and northwest into shallow-water, (shelf?) environments throughout the early Paleozoic.

Columns 27, 28 The succession for NW France (Col. 28) is a compilation from the North Armorican Massif of Brittany. The post-Cadomian transgression here occurred in several stages, beginning in the Early Cambrian in the south and extending into the Arenig/Llanvirn further north. These scarcely metamorphosed Palaeozoic rocks pass southwards (across the South Armorican Shear Zone) into higher-grade metamorphics of uncertain age and geotectonic affinity (42).

Column 28 is broadly typical for the Central Iberian zone, although throughout much of this region there is stratigraphic continuity between (a) the late Precambrian greywackes, (b) the Archaeocyathid-bearing Early Cambrian carbonates and conglomerates, and (c) the major unconformity below the Arenig "Armorican" Quartzite. A further widespread hiatus is developed within the Lower Silurian (Llandovery). To the south, in the Ossa-Morena and South Portuguese zones, a thick Lower to Middle Cambrian carbonate-clastic sequence rests unconformably on late Proterozoic volcanics and sediments and is succeeded disconformably by Ordovician and Silurian shales and volcanics (43-45).

Columns 29, 30, 31 The succession shown in column 30 is a composite one, derived from both the eastern and western parts of the Meseta. In the eastern Meseta, the early Palaeozoic outcrops are considered to be largely allochthonous, in the form of nappes (and olistoliths) emplaced during Hercynian orogenesis. However the late-Precambrian-Cambrian transition is known from some autochthonous structures, although a sub-Middle Cambrian unconformity is present throughout much of the western Meseta. Ordovician and Silurian sequences are generally thin, fine-grained, and include minor hiatuses. Although depicted on the Table as Carboniferous in age, the Hercynian in much of the western Meseta (46-48).

The Coastal Block of Morocco, separated from the western Meseta by the Rabat-Western Atlas shear-zone, displays a thin, virtually undeformed sequence of uppermost Precambrian to Upper Devonian sediments and volcanics. Locally, molassic Devonian rests unconformably on mildly deformed Lower Palaeozoics as the sole indicated in Morocco of an Acadian-Caledonian event (summarized in 37).

In southern Morocco the Anti-Atlas region displays an epicratonic succession (Column 30) of Eocambrian and Palaeozoic sediments and thin volcanics, virtually undeformed until the late Carboniferous, although stratigraphic gaps occur extensively in the Upper Cambrian and (more locally) in the Lower Silurian.

Throughout Morocco the Early Palaeozoic faunas are typically Bohemian or "Atlantic" but American faunas appear in the Middle and Upper Devonian.

Column 31 is derived from the eastern (foreland) part of Mauretania, which in its epicratonic situation broadly corresponds to the Anti-Atlas succession. (The stratigraphic sequence of more westerly sectors of the orogen is poorly known, largely because of the highly deformed and metamorphosed condition of the rocks). The absence of much of the Middle and Upper Ordovician and the presence of a widespread discordance below the late Ashgill/Llandovery glacigenic sediments have been attributed to a "Taconic" or early Caledonian phase, but the main deformation (including nappe emplacement towards the east) and metamorphism occurred towards the end of the Devonian (49).

COMMENTS

Western and Central Regions (W. S. McKerrow)

The platform sequences (P on Table) rest on old basement, whereas the ocean basin sequences (B on Table) show on basement, and the sequences starting with deep water greywackes, shales, cherts or pillow basalts.

During much of the Lower Palaeozoic, North America lay close to the equator (50), and, in the absence of any local source of clastic sediments the platform was covered by carbonates. Two exceptions to this generalisation are seen in: 1. the presence of late Precambrian tillites in Greenland, Scotland and Ireland, though not in the Appalachians; 2. the widespread development of sandstones at the base of the Cambrian.

Neither of these events is connected with any orogenic episode. In the Late Precambrian, it would appear that only a small part of North America was within range of the widespread ice sheet. The development of sandstones in the basal Cambrian appears to be linked to an end-Precambrian regression, followed by a rapid transgression with sandstones marking the successive positions of the shoreline. Some other local occurrences of Cambrian clastics may be due to similar changes in sea level (51).

The majority of clastic input to the Caledonian/Appalachian Orogen can be related to mountain building. One exception is the Grampian Orogeny (500 to 510 Ma = Lower Ordovician), when the Dalradian rocks of Scotland and Ireland were folded and metamorphosed. The Fleur de Lys rocks of Newfoundland may also have been affected by the same event. Coarse sediments do not appear in southern Scotland until some 40 Ma later. These sediments do not appear to be too young to be derived from the Grampian Highlands. Is there some major strike-slip fault not yet recognised?

The Gander belt of Newfoundland (Col. 10) may have been on the east side of the Iapetus Ocean, like England (Col. 6), Wales (Col. 7) and south-east Ireland (Col. 8). These areas, like Avalon, appear to have been founded on a late Precambrian island arc. They do not contain Cambrian and Ordovician carbonate sequences as in North America, and their faunas are also distinct (52, 29).

In the Llandeilo, the Humberian Orogeny in Western Newfoundland (Col. 12) was probably due to collision of the Lushs Bight island arc, which pushed slope clastic succession westwards over the edge of the platform. Some of the Ordovician and Silurian rocks of central Newfoundland (Col. 11) may have been derived from the uplift at this time.

Similar island arc collisions (21) took place towards the end of the Caradoc in New Brunswick (Col. 13) and New York (Cols. 16, 17). This Taconic Orogeny resulted in a widespread development of clastic over the eastern part of the United States (Cols. 14, 15, 16, 18 and 19) and some adjacent parts of Canada, and the Taconic landmass continued to dominate the palaeogeography of this region throughout much of the Silurian (53).

On the eastern margin of North America (from Alabama to Greenland), only Scotland and Ireland (and possibly Newfoundland) contain Lower Palaeozoic calc-alkaline igneous rocks. The remainder of this margin remained passive throughout the late

Precambrian and early Palaeozoic. The explanation for the Early Ordovician Grampian Orogeny is still very uncertain. Granitoids were intruded in Scotland and north-west Ireland during the Ordovician, but they became more abundant during the Silurian and Early Devonian. At least the latter part of this igneous history (which also includes many Early Devonian andesites) is now known to be related to subduction (54).

The earliest evidence for closure of the Iapetus Ocean is in Greenland (Col. 1). At the end of the Early Silurian, the platform carbonate sequence is overlain by shales and then greywackes of Wenlock age, derived from the east. These clastic sediments are then over-ridden by nappes (containing slope and deep shelf late Precambrian and Lower Palaeozoic sediments) also from the east. As similar nappes (but derived from the west) occur in Norway, it seems likely that continental collision occurred during the Late Silurian, possibly earlier in the north of Norway than in the south. It is in the Ludlow that benthic ostracodes and fresh-water fish first cross from Scotland to southern Norway (29).

In the British Isles, the timing of continental collision is still uncertain. It must have occurred late than the uppermost Silurian Pridoli greywackes of the Lake District (Col. 6). But at the north-east end of the Midland Valley (Col. 4) a continuous Pridoli to Seigenian sequence may also have been deposited entirely prior to collision. It is only after the Seigenian that an unconformity is present everywhere within 200 km of the Iapetus suture in the British Isles. Earlier (latest Silurian) stratigraphic breaks are only present in a few places.

In the North Appalachians, stratigraphic evidence suggests that the Acadian Orogeny is of early Middle Devonian age (55). Continental collision is not likely to have been earlier than this time, for both fresh-water fish and benthic ostracodes remain distinct through much of the Early Devonian (29). After the Acadian Orogeny much of the region close to the suture remained uplifted and received little permanent sedimentary cover. The sedimentary effects of the Acadian Orogeny are best seen in the Middle and Upper Devonian Catskill Delta of western New York (Col. 16) and Pennsylvania.

There is no clear evidence for the Acadian Orogeny in the Southern Appalachians, though some coarse clastics suggest uplift within sediment-transport distance of Virginia (Col. 18). World plate reconstructions (50) suggest that the Acadian collision of Europe (north of New York) was distinctly earlier than the collision between Africa and America south of Virginia, which did not occur until the end of the Palaeozoic. That is, perhaps, another story.

In summary, during the Early Palaeozoic, the equator straddled North America, so that carbonates were deposited in platform shelf environments unless local uplift provided a source of clastics. Similarly, the basinal facies shales form the background sediment, but the effects of orogenies are not so local: turbidites can be transported for much greater distances from orogenic centres than is common with shelf clastic sediments. This preliminary scheme shows up the clastic wedges associated with some of the known Lower Palaeozoic orogenies, but there are still some clastic wedges (e.g. the Southern Uplands accretionary prism of Scotland) which are not clearly associated with thick clastic sequences. Perhaps these will be cleared up when we have determined how much strike slip movements have affected the Caledonian/Appalachian Orogen subsequent to sedimentation.

Eastern Region (G. Kelling)

With the exception of the Meguma (Colum 26) all the sequences listed in Figure 1 are of platform type and are dominated by shallow marine to terrestrial sediments throughout most of the Palaeozoic. However dark shales, formed in somewhat deeper, more restricted-circulation environments are widespread in the Middle to Upper Cambrian, the Lower to Middle Ordovician and the Lower Silurian (56). Where subsequent tectonic disruption is minimal (as in Southern Britain, Column 22) the platform succession can be traced northwestwards into more basinal Lower Palaeozoic sequences of deep-water aspect, associated with ophiolitic slices. These presumably represent the southern margins of the Iapetus ocean.

In view of its important role in trans-orogen correlation and reconstruction the Meguma succession is treated separately elsewhere (see Schenk, this volume).

Two distinct types of basement occur below these Palaeozoic successions: (i) truly cratonic basement, generally consisting of lower to mid-Proterozoic metamorphics succeeded by variable thicknesses of mildly or undeformed late Proterozoic sediments and volcanics (Scandinavia, Anti-Atlas and Mauretania); (ii) "young" basement, composed of variably deformed and block-faulted latest Proterozoic sediments and acid or bimodal volcanics, with some plutonics (Columns 22-29). The type (i) basement represents an extension of existing continental cores such as the Baltic Shield or the Saharan Shield. The more variable type (ii) basement can be regarded as fragmented continental margin or as discrete microcontinental blocks, spawned during late Precambrian orogenic events. The largest and most distinctive of these is the "Avalon terrain", characterised by thick late Precambrian sediments (with Ediacarids) and rift-type bimodal volcanics, succeeded by Cambrian shales yielding European or "Atlantic" faunas. This type of terrain occurs in eastern Canada, New England and southern Britain (57), and is also traceable by means of its geophysical signature from Newfoundland into northwest France and Iberia (58). The Mole Cotiere or Coastal Block of Morocco also displays Avalonian characteristics (59) and it is tempting to see the Avalonian mosaic as disrupted slivers of the pan-African tectogene, rifted and partially welded during the latest Precambrian. Strong support for the original homogeneity of the eastern margin basement is provided by the presence of late Proterozoic tillites and the Ediacarid faunas in both the cratonic and Avalonian terrains (Colums 21-23, 27, 31).

It appears, therefore that the present eastern margin of the Caledonides around the North Atlantic rests upon a heterogenous basement, essentially cratonic at its northern and southern extremities but more juvenile and microcontinental in the central parts of the orogen. While the generation of the continental crust of this central basement (Avalonian and Cadomian) can be assigned with some confidence to the late Precambrian (770-570 Ma), accretion of the component fragments to the stable North American and Eurafrican blocks took place at different times and probably in diverse fashions. The docking of these terrains may be responsible for some of the minor phases of deformation and the accompanying stratigraphic breaks and plutonism discernible both within the cover-sequences depicted in Figure 1 (for example the widespread Late Cambrian/Early Ordovician events) and also manifest in the volcano-sedimentary aprons forming the southern margins of the Iapetus Ocean (60). Moreover, the role and magnitude of major thrusts and transcurrent faults in assembling the "suspect terranes" (61) is only now beginning to be recognised (38, 62, 63).

Another important element in the early Palaeozoic evolution of the eastern margin of the Caledonides is the physical relationship between the Baltic area and remaining parts of Europe and North Africa. Stratigraphic and faunal comparisons strongly suggest Early Cambrian initiation of two contrasting regions, comprising southern Scandinavia, southern Britain, Canada, and the northern Appalachians (acado-baltic province) and the ibero-armorican/north African region (euro-tethyan province) (64). However, by the Early Ordovician (Arenig) trilobite and brachiopod assemblages and distinctive lithologies (Armorican quartzites and oolitic iron-ores) define a "Gondwanan" continental shelf and slope-apron, including southern Britain, Newfoundland, Nova Scotia, New England, Brittany, Iberia and northwest Africa, which contrasts markedly in faunal and lithological nature with the Baltic or Fenno-Scandian pericratonic platform (29). The marine realm separating these two regions appears to have been broadly coincident with the Caledonide structural zone which trends southeastwards from the southern North Sea, known as Tornquist's Line (65).

Whether this "Tornquist's Sea" was floored by true ocean crust remains unclear, but the Polish subsurface data reveal strongly deformed Early Ordovician basinal (graptolitic) sequences separating a northern platformal carbonate facies of Baltic type from a mixed clastic/carbonate succession (with oolitic ironstones) of "Gondwanan" affinity to the South (66). Important strike-slip movements along the Tornquist Line may have been indicated during the Ordovician (65, 67). Such movements may have been responsible for the virtual elimination of Tornquist's Sea as a faunal barrier by the later Ordovician (Caradoc-Ashgill) when trilobite and brachiopod faunas from Scandinavia, southern Britain, eastern Canada, and New England become virtually identical (29). The relationship of this short-lived Tornquist's Sea to the main Iapetus Ocean suggests an analogy with modern triple junctions and may explain the structural nature and frequent re-activation of the Tornquist Line (68).

During the Silurian and Early Devonian the increasing faunal and lithological integrity of the north European and North American successions reflects closure of Iapetus. Simultaneously the increasing faunal disparity between this northern region and the Ibero-Armorican and African domains has been cited as evidence for a latitudinal Rheic Ocean (16), and minor ophiolite remnants of this age (e.g. in southwest England and northwest Spain) may testify to episodic extension within this realm.

CONCLUSIONS (W. H. Poole)

Introduction

Five figures summarize the dominant lithologies of stylized segments within the orogen. One figure was drawn for each of late Precambrian (Vendian, younger than about 690-720 Ma) (Fig. 2), Cambrian (Fig. 3), Ordovician (Fig. 4), Silurian (Fig. 5), and Devonian (Fig. 6). Within each system, the orogen was separated into a western zone depicting the western margin of Iapetus ocean, a central zone crudely representing the volcanic island arc terrane, an eastern zone representing Avalon zone and bordering cratonal cover of northern France and northern Africa, and finally the apparently unique Meguma terrane of southern Nova Scotia, Canada. Difficulties were encountered in assigning some Cambro-Ordovician rocks to one zone or another near the boundaries and in assigning Siluro-Devonian rocks where Iapetus ocean is interpreted to have no longer existed. The solution is really no solution: we simply made a "best guess" so to avoid foundering the exercise. The reader must amend, add to or delete from the figures as he sees fit.

Late Precambrian, Vendian (Fig. 2)

Iapetus Ocean is believed by many geologists to have been formed up of a crystalline continent during late Precambrian time, although models of continental wandering derived mainly from paleomagnetic interpretations show an already existing, wide Precambrian ocean in the location of Iapetus.

Western zone bordered a craton of Grenvillian and older rocks. Flysch (s.l) of thick greywacke turbidite with local sandstone, volcanics, and carbonate were deposited on a depressed margin. Tillite and possible tillite occur near the base of the sequence in British Isles and North America. Central zone lacks assured Precambrian rocks. Eastern zone is well represented by felsic and mafic volcanics, flysch, some redbeds, sandstone in Scandinavia, and tillite. Northern African strata are more platformal. Ediacarid fauna may be found to be characteristic only of the eastern zone. Meguma zone lacks known Precambrian strata.

Tillite and possible tillite occur in the northeastern end of the western and eastern zones including France and northern Africa, suggesting that these areas may have been not far apart in a polar region. The possible tillite in southern United States falls outside this cluster.

Cambrian (Fig. 3)

Iapetus Ocean was clearly evident during Cambrian time.

Western zone continued to receive continental-terrace flysch, sandstone, ribbon (deep water?) limestone, local limestone olistostrome, and rarely volcanics. Archaeocyathid fauna and the characteristic olenellid tribobite fauna are common in the Lower Cambrian. Central zone contains shale, sandstone and greywacke plus ophiolite whose age may be Late Cambrian or Early Ordovician. The sedimentary rocks may be underlain by a basement common with the eastern zone. Eastern zone is characterized by shale, mainly black, and minor local limestone. Limestone and volcanics are more common in northern France and northern Africa. Paradoxidid and olenid trilobite faunas are characteristic of the eastern zone, along with archaeocyathid fauna. Meguma zone is represented by unfossiliferous quartzose greywacke turbidites which are probably in

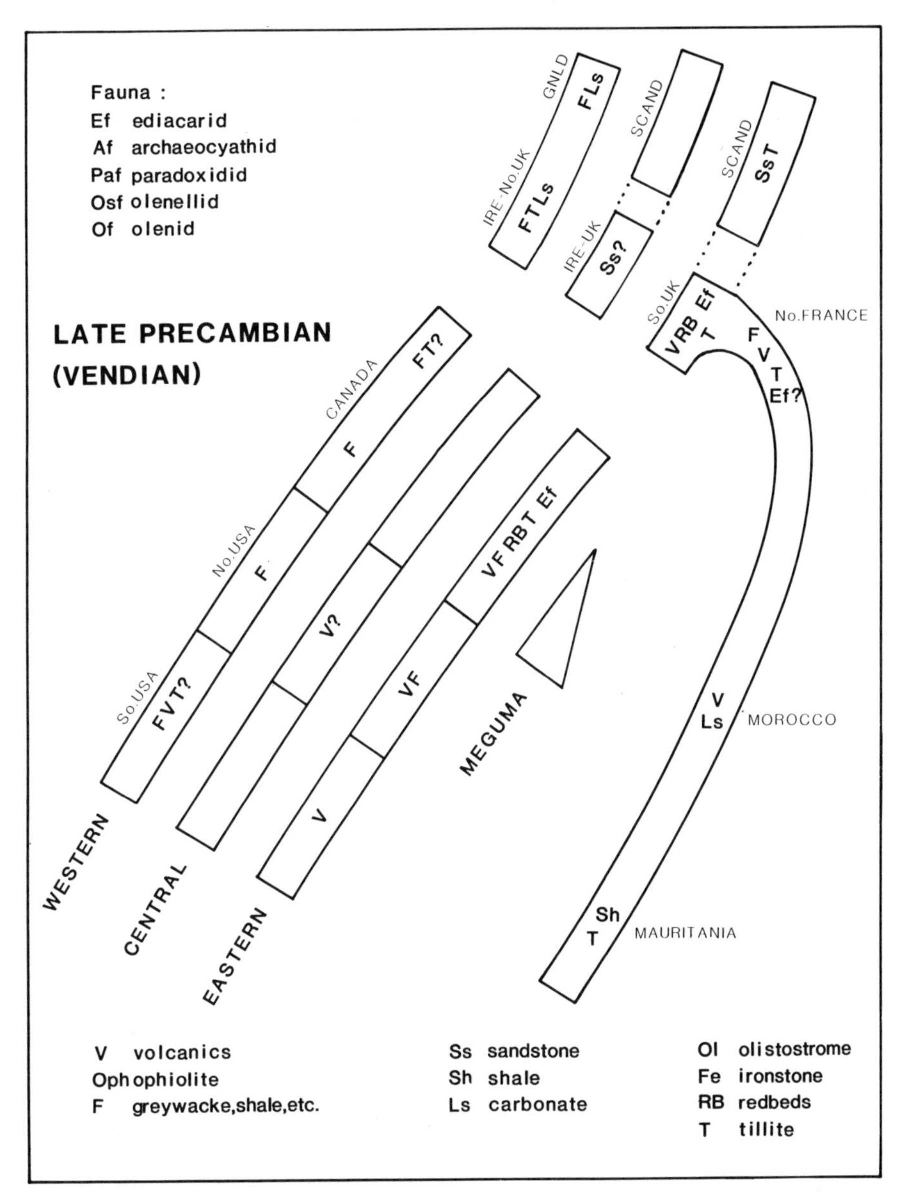

Figure 2 Dominant lithologies deposited during late Precambrian (Vendian) time in Caledonian belt and southeastern borderland. Western, central, eastern and Meguma zones plus borderland in northern France (Normandy and Brittany) and in northern Africa (Morocco and Mauritania) are stylized.

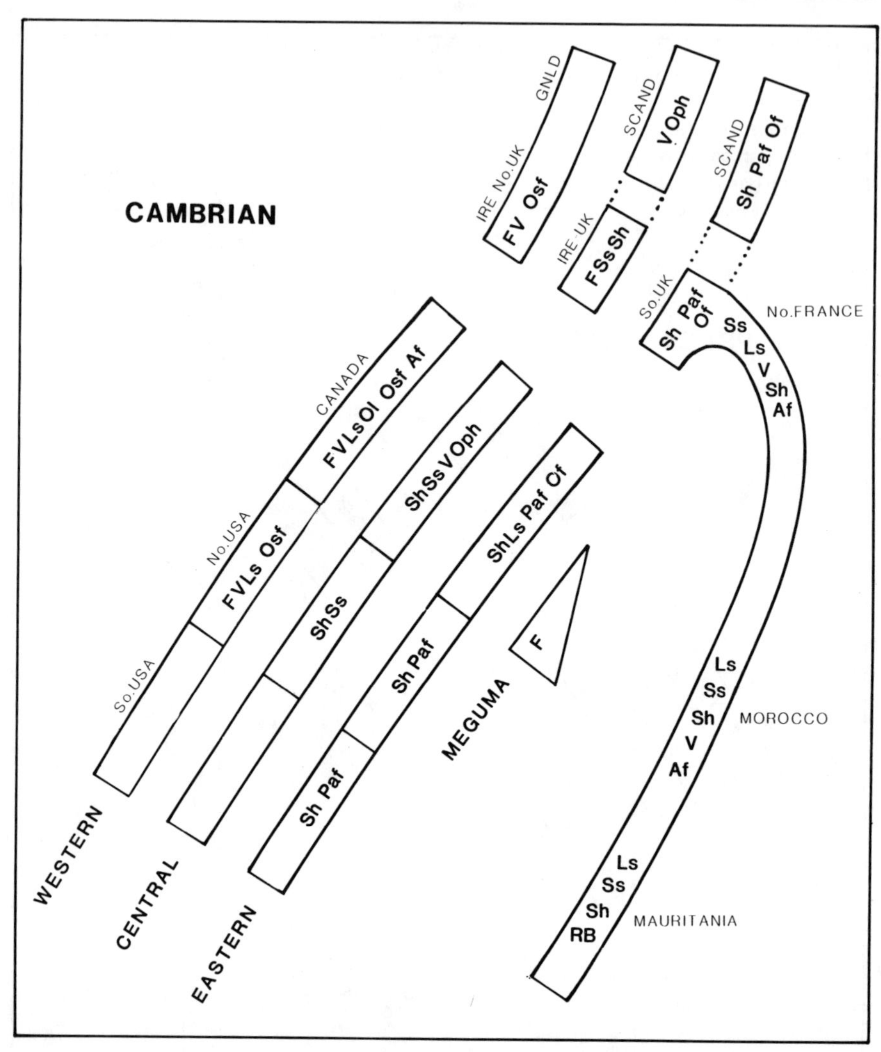

Figure 3 Dominant lithologies deposited during Cambrian time with characteristic faunas. Symbols as on figure 1.

large part of Late Cambrian or earlier age and which were derived from a southeastern source.

Sediments in the four zones during the Cambrian are distinctly different, as are the trilobite fauna in the western and eastern zones. Archaeocyathids in these two latter zones conceivably could be equally distinctive at the generic and specific levels. Ophiolites, notoriously difficult to date stratigraphically and paleontologically, seem to be no older than Late Cambrian.

Ordovician (Fig. 4)

Evidence of Iapetus Ocean is clear in Early Ordovician and much less so in Late Ordovician following subduction, volcanic island arc generation, and obduction in most parts of the orogen. Interpretations vary from place to place along the orogen.

Western zone continued the Cambrian depositional pattern. In North America, flysch derived from a southeastern source, probably from the encroaching allochthons, spread upon the depressed western zone and bordering carbonate platform during much of Ordovician. A similar process probably occurred in British Isles. Allochthons consist of continental terrace deposits and ophiolites from oceanic crust. Allochthons came to rest at different times: Caradoc in United States and mainland Canada; Llanvirn-Llandeilo in Newfoundland; and Caradoc (?) in the British Isles. Central zone is characterized by volcanic island-arc rocks (above ophiolite in North America and British Isles): felsic and mafic volcanics, greywacke, and some graptolitic black shale and chert and red ferruginous and manganiferous shale and chert. Volcanism occurred about the same time of obduction in the western zone. Greywacke, limestone, and volcanics were deposited during Late Ordovician. In Scandinavia, arc-type volcanics and associated sedimentary rocks were deposited presumably after Early Ordovician eastward thrusting (Finnmarkian) and do not appear to have been thrust during the later Ordovician. Eastern zone continued black shale deposition in Early Ordovician time, the only Ordovician strata in most of North America, while shale and limestone deposition with local volcanics continued throughout the Ordovician in British Isles and Scandinavia. Sandstone and ironstone are common to Newfoundland, France and northern Africa. Upper Ordovician tillite seems restricted to France and northern Africa and possibly the Meguma zone. Meguma zone deposition is restricted to black shale of Early Ordovician age, and diamictite presumably of Late Ordovician age.

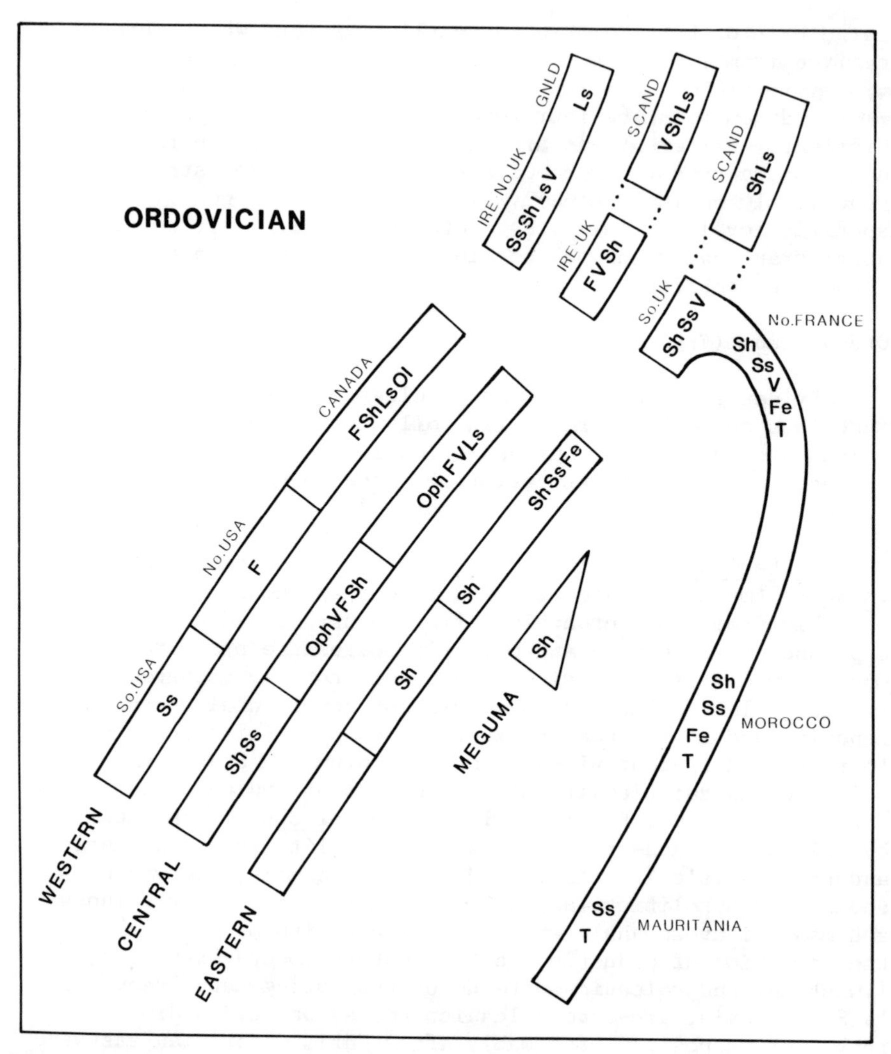

Figure 4 Dominant lithologies deposited during Ordovician time. Symbols as on figure 1.

Megafossil provincialism across Iapetus ocean, well established in the Cambrian, continued to Mid-Ordovician when more cosmopolitan forms became dominant, presumably as a result of ocean closure, changed ocean currents and achievable inter-zone migration. Plutonism (mainly subvolcanic?), metamorphism and deformation was common in the orogen although difficult to distinguish from effects of later orogenies.

Silurian (Fig. 5)

Interpretations differ on whether Iapetus Ocean continued to exist during the Silurian and Early Devonian. Some geologists attribute Siluro-Devonian volcanism, plutonism, metamorphism, and deformation to subduction, ocean-closure, and continental collision. In many parts of the Caledonide Orogen, the distinction between western, central and eastern zones established in the Cambro-Ordovician seems to be blurred and even lost during the Siluro-Devonian.

Western zone is characterized by sandstone, shale, limestone and local felsic and mafic volcanics in mainly shallow marine environments. Redbeds are known locally as well as ironstone in the United States. Central zone contains like strata as well as thick monotonous greywacke flysch commonly of deeper water aspect. Redbeds and subaerial volcanics are distinctive in Newfoundland. Eastern zone again is represented by similar sedimentary rocks, local volcanics and in the Late Silurian, redbeds as an early development of the Caledonian Orogeny. Eastward thrusting in Scandinavia appears to have begun in the Mid-Silurian. In France and northern Africa, black shale with sandstone and limestone were deposited beyond the immediate influence of the Caledonian orogeny. Meguma zone deposits are shale, quartz, sandstone, and felsic and mafic volcanics, all probably of Silurian age.

Devonian (Fig. 6)

The Caledonian-Acadian Orogeny during the Late Silurian and Devonian produced major changes in the types of deposits, not to mention deformation, metamorphism, and plutonism.

Western, Central and Eastern zones continued marine deposition of shale, sandstone and limestone in North America. At approximately the end of the Early Devonian, deposition changed to coarser sediments in nonmarine environments, here and there accompanied by felsic and mafic volcanics. Redbeds and local volcanics characterize the northeastern part of the

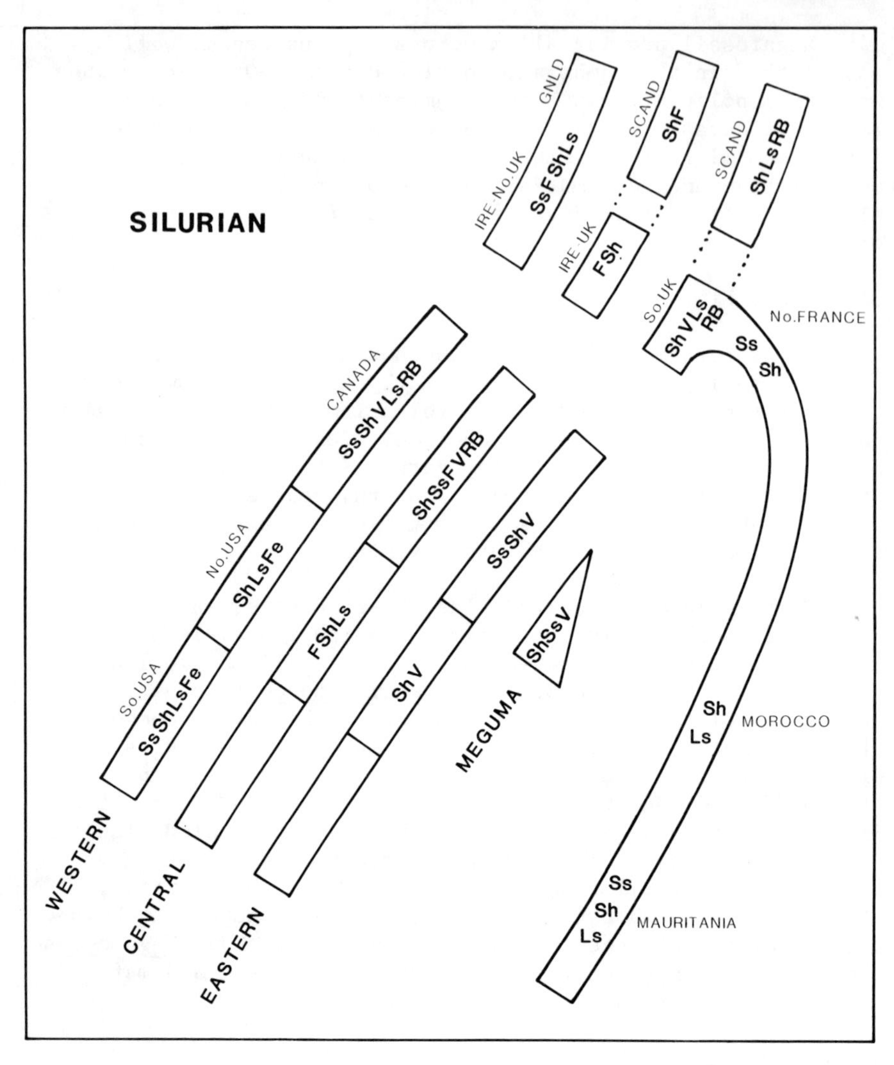

Figure 5 Dominant lithologies deposited during Silurian time. Symbols as on figure 1.

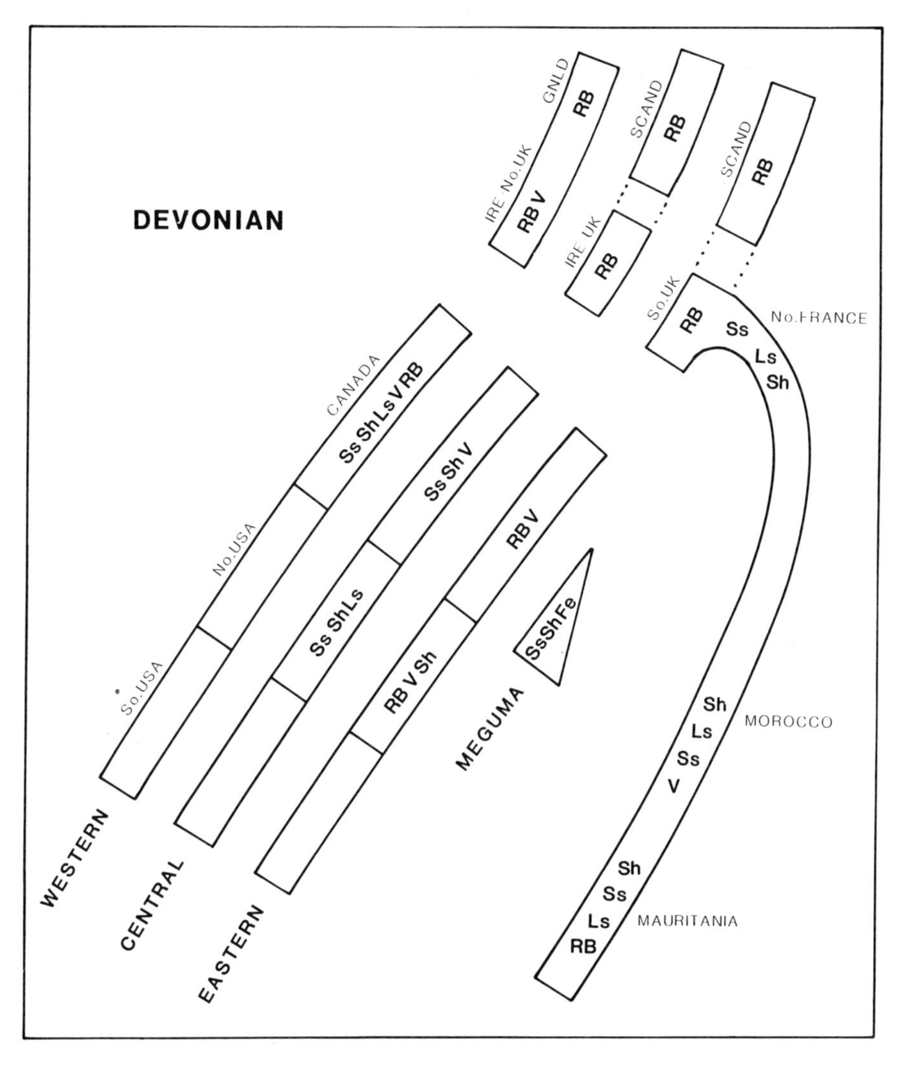

Figure 6 Dominant lithologies deposited during Devonian time. Symbols as on figure 1.

orogen in British Isles and Scandinavia. The Middle and Late Devonian in all zones are represented generally by interrupted sequences of nonmarine coarse redbeds and local volcanics. Southward into France and northern Africa deposition of sandstone, shale and limestone with local redbeds and volcanics continued the established, Silurian, depositional pattern. Meguma zone deposition is represented by only Lower Devonian sandstone, shale, and ironstone, in part non-marine.

Observations and Questions

Sedimentation, volcanism, and megafossil provinciality point convincingly to the presence of the Iapetus Ocean at least between late Proterozoic and mid-Ordovician. Did Iapetus exist in the early Late Proterozoic as well? Upper Proterozoic-Lower Cambrian ophiolites are rare if at all present in the geological record; could subduction have been so efficacious to have destroyed all such ophiolites? If Iapetus also existed during the Silurian, why are ophiolites of this age not abundantly recognized?

Subduction-generated Ordovician volcanics along the central zone are well accepted. The origin of the Late Proterozoic volcanics particularly in the eastern zone is quite controversial. Siluro-Devonian volcanics are attributed to operative subduction zones by many geologists.

If closure of Iapetus occurred in the Ordovician in at least northeastern Appalachians as held by many geologists, and given that no Siluro-Devonian ocean with ophiolites of this age existed in the same region, then can Siluro-Devonian volcanicsm, plutonism, metamorphism, and deformation be generated by subduction and ocean closure in this region? I would suppose not.

Did such processes close Iapetus during Siluro-Devonian in the British Isles? My prejudice suggests not. But what about Scandinavia and its incredibly far-travelled nappes? Perhaps such allochthoneity cannot everywhere and at all geological times be attributed to ocean closure and collision.

Plate tectonics has dominated our thinking for many years and continues to do so while we struggle to fit our geology to the theory and its variations. I wonder what the ruling theory will be in 30 years or more, and how much of present plate tectonic models will form part of it.

REFERENCES

1. Escher, A. and Watt, W. S. 1976. Geology of Greenland. Geol. Surv. Greenland, Copenhagen, 603 p.
2. Hurst, J. M. and McKerrow, W. S. 1981. Nature, 290, pp. 772-774.
3. Craig, G. Y. 1965. The geology of Scotland. Oliver and Boyd, Edinburgh 556 p.
4. Curry, G. B., Ingham, J. K., Bluck, B. J. and Williams, A. 1982. J. Geol. Soc. London, 139, pp. 451-454.
5. Harris, A. L., Baldwin, C. T., Bradbury, H. J., Johnson, H. D. and Smith, R. A. 1978. Ensialic basin sedimentation: the Dalradian Supergroup. In D. R. Bowes and B. E. Leake (Eds.) Crustal Evolution in northwestern Britain and adjacent regions, Geol. J. Spec. Issue 10, pp. 115-138.
6. Leggett, J. K., McKerrow, W. S. and Eales, M. H. 1979. J. Geol. Soc. London, 136, pp. 755-770.
7. Upton, B. G. J., Aspen, P., Graham, A. and Chapman, N. A. 1976. Nature, 260, pp. 517-518.
8. Moseley, F. (Ed.) 1978. The geology of the Lake District. Yorkshire Geol. Soc. Occasional Publ. 3, 284 p.
9. Bruck, P. M., Colthurst, J. R. J., Feely, M., Gardiner, P. R. R., Penney, S. R., Reeves, T. J., Shannon, P. M., Smith, D. G., and Vanguestaine, M. 1979. South-east Ireland: Lower Palaeozoic stratigraphy and depositional history. In A. L. Harris, C. H. Holland and B. E. Leake (Eds.) The Caledonides of the British Isles --- reviewed. Geol. Soc. London, Spec. Pub. 8, pp. 533-544.
10. Coward, M. P. and Siddans, A. W. B. 1979. The tectonic evolution of the Welsh Caledonides. In A. L. Harris, C. H. Holland and B. E. Leake (Eds.). The Caledonides of the British Isles --- reviewed. Geol. Soc. London, Spec. Pub. 8, 187-198.
11. Holland, C. H. (Ed.) 1981. A Geology of Ireland. Scottish Academic Press, Edinburgh 335 p.
12. Bird, J. M. and Dewey, J. F. 1971. Geol. Soc. America Bull., 81, pp. 1031-1060.
13. McKerrow, W. S. and Cocks, L. R. M. 1976. Can. J. Earth. Sci., 14, pp. 488-495.
14. Neuman, R. B. 1972. Brachiopods of Early Ordovican volcanic islands. Proc. 24th Int. Geol. Congr. Montreal, 7, pp. 297-303.
15. Williams, H. 1979. Can. J. Earth Sci., 16, pp. 792-807.
16. McKerrow, W. S. and Ziegler, A. M. 1971. J. Geol., 79, pp. 635-646.
17. St. Julien, P. and Hubert, C. 1975. Amer. J. Sci., 275A, pp. 337-362.
18. Skinner, R. 1974. Geol. Surv. Canada Mem. 371, 133 p.
19. Neuman, R. B. 1968. Paleogeographic implications of

Ordovician shelly fossils in the Magog Belt of the Appalachian region. In Zen, E-an, White, W. S., Hadley, J. B. and Thompson, J. B. (Eds.) Studies in Appalachian Geology: Northern and Maritme. Interscience, New York, pp. 35-48.

20. Osberg, P. H., Moench, R. H. and Warner, J. 1968. Stratigraphy of the Merrimack Synclinorium in west-central Maine. In Zen, E-an, White, W. S., Hadley, J. B. and Thompson, J. B. (Eds.). Studies in Appalachian Geology: Northern and Maritime, Interscience, New York, pp. 241-253.
21. Robinson, P. and Hall, L. M. 1980. Tectonic synthesis of southern New England. In Wones, D. R. (Ed.) The Caledonides in the U. S. A. V. P. I. and State Univ., Blacksburg, pp. 73-82.
22. Zen, E-an 1967. Time and space relationships of the Taconic autochthon and allochthon. Geol. Soc. America Spec. Paper 97, 107 p.
23. Glover, L. and Read, J. F. 1979. Guides to Field Trips 1-3 for Southeastern Section Meeting. Geol. Soc. America Blacksburg, Virginia. V. P. I. and State Univ., Blacksburg, 143 p.
24. Rogers, J. 1970. The Tecontics of the Appalachians. Wiley--Interscience, New York, 271 p.
25. Thomas, W. A. 1976. Evolution of Ouachita-Appalachian continental margin. J. Geol., 84, pp. 323-342.
26. Bjorlykke, K. 1978. The eastern marginal zone of the Caledonide Orogen in Norway; in I.G.C.P. Project 27, Caledonian-Appalachian Orogen of the North Atlantic Region, Geol. Surv. Canada, Paper 78-13, pp. 49-56.
27. Gee, D. G. 1978. The Swedish Caledonides - A Short Synthesis in I.G.C.P. Project 27, Caledonian-Appalachian Orogen of the North Atlantic Region. Geol. Surv. Canada, Paper 79-13, pp. 63-72.
28. Nicholson, R. 1974. The Scandinavian Caledonides in Nairn, A. E. M. and Stehli, F. G. (Editors), The Ocean Basins and Margins, Vol. 2, the North Atlantic. Plenum Press, New York and London, pp. 161-203.
29. Cocks, L. R. M. and Fortey, R. A. 1982. J. Geol. Soc. London, 139, pp. 467-480.
30. Kelling, G. 1978. The paratectonic Caledonides of mainland Britain in I.G.C.P. Project 27, Caledonian-Appalachian Orogen of the North Atlantic Region. Geol. Surv. Canada Paper 78-13, pp. 89-95.
31. Sturt, B. A., Soper, N. J., Bruck, P. M. and Dunning, F. M. 1980. Caledonian Europe, in Geologie de l'Europe (ed. J. Cogne and M. Slansky), Mem. B. R. G. M., No. 108, pp. 56-66.
32. Watson, J. and Dunning, F. W. 1979. Basement-cover rela-

tions in the British Caledonides in A. L. Harris, C. H. Holland and B. E. Leake (Eds)., The Caledonides of the British Isles - reviewed. Geol. Soc. London, pp. 67-92.

33. Hatcher, R. D. Jr. 1978. Synthesis of the Southern and Central Appalachians in I.G.C.P. Project 27, Caledonian-Appalachian Orogen of the North Atlantic, Geol. Surv. Canada Paper 78-13, pp. 149-157.
34. King, A. F. 1982 (Editor). I.G.C.P. Project 27 Field Guide for the Avalon and Meguma Zones, NATO Advanced Study Institute, Atlantic Canada (Memorial Univ., St. Johns, Newfoundland, Canada), 308 pp.
35. Murray, D. P. and Skehan, J. W. (SJ) 1979. A traverse across the eastern margin of the Appalachian-Caledonide Orogen, south-eastern New England in Skehan, J. W. and Osberg, P. H. (Editors), Geological Excursions in the North East Appalachians (IGCP Project 27). Weston Observatory, Boston College, Mass., pp. 1-36.
36. Osberg, P. H. 1978. Synthesis of the geology of the northeastern Appalachian, U. S. A., in I.G.C.P. Project 27, Caledonian-Appalachian Orogen of the North Atlantic Region. Geol. Surv. Canada Paper 78-13, pp. 137-148.
37. Schenk, P. 1978. Synthesis of the Canadian Appalachians in I.G.C.P. Project 27, Caledonian-Appalachian Orogen of the North Atlantic Region. Geol. Surv. Canada Paper 78-13, pp. 111-136
38. Williams, H. and Hatcher, R. D. Jr. 1982. Geology, 10, p. 530-536.
39. Harland, W. B., Cox, A. V., Llewellyn, P. G., Pickton, C. A. G., Smith, A. G., and Walters, R. 1982. A Geological Time Scale, Cambridge Earth Science Series, 131 p.
40. Reynolds, P. H., Zentilli, M., and Muecke, G. K. 1981. Can. J. Earth Sci. 18, pp. 386-394.
41. Schenk, P. E. 1982. The Meguma Zone of Nova Scotia -- a remnant of Western Europe, South America, or Africa? in Kerr, J. W. and Fergusson, A. J. (Eds.). Geology of the North Atlantic Borderlands. Can. Soc. Petroleum Geol. Mem. 7, pp. 119-148.
42. Autran, A. 1978. Synthese provisoire des evenements orogeniques caledoniens en France, in I.G.C.P. Project 27, Caledonide-Appalachian Orogen of the North Atlantic Region, Geol. Surv. Canada, Paper 78-13, pp. 159-176.
43. Guy-Tamain, A. L. 1978. L'evolution caledono-varisque des Hesperides, in I.G.C.P. Project 27, Caledonide-Appalachian Orogen of the North Atlantic Region, Geol. Surv. Canada, Paper 78-13, pp. 183-210.
44. Julivert, M., Martinez, F. J. and Ribeiro, A. 1980. The Iberian segment of the European Hercynian foldbelt, in Geologie de l'Europe (eds. J. Cogne and M. Slansky).

Mem. B.R.G.M. No. 108, pp. 132-158.
45. Matthews, S. C., Chauvel, J. J. and Robardet, M. 1980. Variscan geology of northwestern Europe, in Geologie de l'Europe (eds. J. Cogne and M. Slansky), Mem. B.R.G.M. No. 108, pp. 69-76.
46. Michard, A. 1976. Elements de geologie marocaine: Notes et Mem. Surv. geol. Maroc., 252, 408 pp.
47. Michard, A. 1978. Breve description du segment caledono-hercynien du Maroc, in I.G.C.P. Project 27, Caledonide-Appalachian Orogen of the North Atlantic Region. Geol. Surv. Canada Paper 78-13, pp. 213-230.
48. Michard, A. and Pique, A. 1980. The Variscan belt in Morocco : structure and developmental model, in The Caledonides in the U. S. A. (ed. D. R. Wones), Virginia Polytechnic Institute, Dept. Geol. Sci., Memoir 2, pp. 317-322.
49. Lecorche, J.-P and Sougy, J. 1978. Les Mauretanides, Afrique occidentale : Essai de synthese, in I.G.C.P. Project 27, Caledonide-Appalachian Orogen of the North Atlantic Region. Geol. Surv. Canada Paper 78-13, pp. 231-239.
50. Scotese, C. R., Bambach, R. K., Carton, C., Van der Voo, R., and Ziegler, A. M. 1979. J. Geol., 87, pp. 217-277.
51. Palmer, A. R. and James, N. P. 1980. The Hawke Bay event: a circum-Iapetus regression near the Lower Middle Cambrian boundary In Wones, D. R. (Ed.) The Caledonides in the U. S. A. V. P. I. and State Univ., Blacksburg, pp. 15-18.
52. McKerrow, W. S. and Cocks, L. R. M. 1976. Nature, 263, pp. 304-306.
53. Berry, W. B. N. and Boucot, A. J. 1970. Correlation of the North American Silurian rocks. Geol. Soc. America Spec. Paper 102, 289 p.
54. Thirwall, M. F. 1981. J. Geol. Soc. London, 138, pp. 123-138.
55. Boucot, A. J. 1968. Silurian and Devonian of the Northern Appalachians In Zen, E-an, White, W. S., Hadley, J. B. and Thompson, J. B. (Eds.) Studies in Appalachian Geology: Northern and Maritime. Interscience, New York, pp. 83-94.
56. Leggett, J. K. 1980. J. Geol. Soc. London, 137, pp. 139-156.
57. Rast, N. 1980. The Avalonian plate in the northern Appalachians and the Caledonides, in D. Wones (ed.). The Caledonides in the U. S. A., Virginia Polytechnic Inst., Dept. of Geol. Sci., Mem. No. 2. pp. 63-66.
58. Haworth, R. T. 1981. Geophysical expression of the Appalachian-Caledonide structures of the continental margins of the North Atlantic in J. W. Kerr and A. J. Ferguson (eds.). Geology of the North Atlantic Borderlands.

Canadian Soc. Petrol. Geol. Mem. No. 7, pp. 429-446.

59. Pique, A. 1981. Geology, 9, pp. 319-322.
60. Williams, H. 1978b. Geological development of the Northern Appalachians and its bearing upon the evolution of the British Isles, in D. R. Bowes and B. E. Leake (Eds.), Crustal Evolution in northwestern Britain, Geol. Jour. Spec. Issue No. 10, pp. 1-22.
61. Coney, P. J., Jones, D. L. and Monger, J. W. H. 1980. Nature, 288 pp. 329-333.
62. Van der Voo, R. and Scotese, C. 1981. Geology, 9, pp. 583-589.
63. Kent, D. V. 1982. J. Geophys. Res. 87, B10, pp. 8709-8716.
64. Babin, C., Cocks, L. R. M. and Walliser, O. H. 1980. Facies, faunas et paleogeographie ante-carbonifere de l'Europe, in Geologie de l'Europe. Mem. B. R. G. M. No. 108, (ed. J. Cogne and M. Slansky), pp. 191-201.
65. Brochwicz-Lewinski, W., Pozaryski, W. and Tomczyk, H., 1981. Strike-slip fault movements along the SW edge of the Eastern European Platform during the early Palaeozoic. Pozeglad Geol. (Varsovie), 29, pp. 385-397. (in Polish with English summary).
66. Tomczykova, E. and Tomczyk, H. 1970. The Ordovician, in J. Czaplicka (Ed.), Geology of Poland, Vol. 1, Stratigraphy, Part 1, pp. 177-236.
67. Pozaryski, W., Brochwicz-Lewinski, W. and Tomczyk, H. 1982. Sur let caractere heterochronique de la Ligne-Teisseyre-Tornquist, entre Europe centrale et orientale C. R. Acad. Sci. Paris, 295. pp. 691-696.
68. Ziegler, P. 1978. Geol. en. Mijnb., 57, pp. 589-626.

COMMENTS ON THE STRATIGRAPHIC/STRUCTURAL EVOLUTION OF SCANDINAVIA

(1) D. G. Gee, (2) B. A. Sturt

(1) Geological Survey of Sweden, S-75128, Uppsala, Sweden, (2) University of Bergen, 5104 Bergen, Norway

The Caledonian fold belt of western Scandinavia outcrops over a strike length of some 1800 km with a maximum transverse width of 30 km in the central and southern parts. The structure is characterized by a series of nappes, remarkable for their great areal extent in relation to their limited vertical thicknesses, that have all been displaced southeastwards onto the Baltoscandian Platform.

The evolution of the Scandinavian Caledonides involved two major orogenic cycles. The earlier, the Finnmarkian, occurred during Late Cambrian to Early Ordovician time, and the later, or Scandian (main Scandinavian) event had its climax during the latter part of the Silurian. Although both of these cycles involved thrusting, polyphasal folding, deep-seated metamorphism and syn-tectonic igneous activity, the major Scandian nappe transport dominates the overall tectonostratigraphic disposition of the belt. The nappes have been conveniently grouped into the Lower, Middle, Upper and Uppermost Allochthons. Within the various allochthons, relationships between transported "basement" and cover are complex. The Finnmarkian nappes of the northernmost segment contain sialic "basement" plinths, varying in age from Archaean through Karelian to possibly Grenvillian, which is overlain by platform sequences of late Proterozoic to Cambrian age. The Scandian nappes, on the other hand, contain not only these pre-Caledonian elements, but also Finnmarkian metamorphic rocks and oceanic crust as "basement" to Ordovician and Silurian sequences.

The Lower Allochthon is dominated by platform sequences of

P. E. Schenk (ed.),
Regional Trends in the Geology of the Appalachian-Caledonian-Hercynian-Mauritanide Orogen, 113–115.

latest Proterozoic and Early Palaeozoic age. In southernmost areas, these successions reach into the lowermost Devonian; in the north they extend only into the lowermost Ordovician (Tremadoc). Along much of the Caledonian front, these sedimentary rocks lie above gently west-dipping sole thrusts which overlie the autochthonous Precambrian crystalline basement with its thin sedimentary veneer. In the far north, the frontal nappes may have been emplaced during the Finnmarkian; further south the emplacement was of Scandian age. West of the Caledonian front, in the antiformal windows, the units in the Lower Allochthon contain substantial sheets of Precambrian rocks. The Middle Allochthon is dominated by Precambrian crystalline rocks, penetratively deformed during Caledonian reworking, together with late Proterozoic sandstones derived from the Baltoscandian miogeocline and transported distances of over 200 km onto the platform. A major tholeiitic dyke swarm occurs in the upper part of the Middle Allochthon and, in northern Norway, this unit is intruded by the diverse suite of the Seiland Igneous Province. The Upper Allochthon, derived distances of several hundred kilometres from the Caledonian eugeocline, contains oceanic crustal elements, unconformably capped by Ordovician and Silurian sediments. The Early Ordovician faunas are essentially of North American type. This complex also contains abundant island-arc volcanic rocks and volcaniclastic sediments. A complex of schists and gneisses, comprising the Uppermost Allochthon, overlies the Upper Allochthon in the central and northern parts of the foldbelt. It is in the Uppermost and Upper Allochthons that the major syn-orogenic granitic batholiths are located, as are gabbros and trondhjemites.

The entire nappe succession along with the underlying basement is folded by major folds with axes parallel to the length of the orogen. In the east, these are open in style; further west, they tighten and, in the westernmost parts of the belt, the basement and allochthonous cover are folded and refolded together and subject to metamorphism at least into amphibolite facies. Recent geochronological studies (Nd-Sm) of eclogites in the basement gneisses of southwestern Norway, suggest that very high pressure conditions existed beneath the nappe-pile during the Caledonian orogeny, as a result of the build-up of a tectonic overburden.

Uplift of the orogen and advance of the nappes during Middle and Late Silurian, was accompanied by deep erosion of the internal parts and by deposition in the early-Middle Devonian (possibly also in the latest Silurian) of thick continental sandstones in intermontane, fault-controlled basins. Similar continental sandstones occur in the Caledonian front in the south (Oslo) as a clastic wedge, derived from the advancing nappes to the northwest. Recent work, by one of us (B.A.S.) shows that major thrusting, allied with low-grade metamorphism, has affected many of the younger intermontane sequences. Correlation in the North Atlantic area suggests a Middle to Late Devonian age for this deformation.

The Finnmarkian orogenic event spanned the interval 530-490 Ma and involved the emplacement of thin-skinned nappes. It is dated by radiometric age-determinations on rocks of the syn-orogenic Seiland Igneous Pronvice and on cleaved shales in the autochthon/perauthochthon of northern Norway. The Finnmarkian thus broadly corresponds in time to the Grampian of Scotland, and involves thrusting, polyphasal folding, metamorphism ranging up into high amphibolite facies and a complex plutonic igneous history. There is increasing evidence that, during the Finnmarkian, obduction of ophiolites occurred on to a continental margin. Studies of the oceanic cap-rock sediments reveal that polyphasal orogenic deformation and metamorphism (at least into upper greenschist facies) of the ophiolities occurred before Middle Ordovician times, possibly pre-dating the deposition of Upper Arenig sediments. It is suggested that the Finnmarkian probably represents a major stage of arc-continent collision.

Post-Finnmarkian orogenic history in the Upper and Uppermost Allochthons was initially characterized by widespread uplift with deep erosion. Coarse Ordovician clastic sequences unconformably overlie ophiolites, early arc volcanic rocks, Finnmarkian metamorphic complexes, and Precambrian crystalline rocks. The Ordovician to Early Silurian was a period in which volcanic island-arcs dominated, together with sedimentation in back-arc and, possibly, fore-arc basins. The tectonic evolutions during this period was apparently conditioned by the arc, arc-trench gap, and marginal basin settings.

The major and most obvious thrust displacements in the fold belt occurred after the Llandovery and extended into earliest Devonian times. This, Scandian (main Scandinavian) orogenic event, involved deep-seated metamorphism, the intrusion of major granitic batholiths, and long-distance nappe transport. It was also responsible for the fragmentation and tectonic interleaving of ealier Finnmarkian elements into the Scandian nappe pile. Subsequent, large-scale thrusting in the middle-late Devonian (?) has far-reaching implications that need to be taken into account in future reconstructions of the orogenic evolution.

COMMENTS ON THE STRATIGRAPHIC EVOLUTION OF MOROCCO

M. Ben Said

Direction de la Geologie, Ministere de l'Energie et des Mines, Rabat, Moroc

INTRODUCTION

The Moroccan Palaeozoic displays a common evolution in sedimentology and biostratigraphy throughout Morocco from Cambrian to Early Devonian times: platform sedimentation in the north and northeast passes south into three basins:

a) the Tindouf Basin, south of the South Atlas Fault; (Col. 31, Poole et al., this volume, p. 75);
b) the Central Morocco Basin (Meseta and western High Atlas) (Col. 30);
c) the allochthonous Palaeozoic within the Alpine Rif chain.

The Tindouf Basin is a relatively quiet trough containing 6 to 11 km of Palaeozoic sediments ranging from Precambrian III to Stephanian in age. These are mainly in epicontinental facies with stronger subsidence indicated in southwestern Morocco (see Col. 31). They are affected mostly by mild tectonic movements during the late Hercynian Orogeny and subsequently during the Alpine events.

The Central Moroccan Trough has a general trend in sedimentary character resembling that of the Tindouf Basin within the Lower Palaeozoic and Lower Devonian. From the Middle Devonian the synsedimentary tectonism led to differentiation of the Basin, with development of a series of sub-basins and intervening highs. In the Rif region the Ghomaride nappes are formed by different series of the Palaeozoic.

P. E. Schenk (ed.),
Regional Trends in the Geology of the Appalachian-Caledonian-Hercynian-Mauritanide Orogen, 117–119.

CAMBRIAN

Dolomites and shales are characteristic for the Lower Cambrian, together with a thick series of pyroclastics and shales. Pyroclastic greywackes with Paradoxides in the intervening shale facies, accompanied by Conocoryphe sandstones, characterise the Acadian. The Upper Cambrian (including the Tremadocian) is generally absent in North Africa and Spain.

ORDOVICIAN

Sedimentation throughout Morocco is generally detrital, with shales, sandstones and quartzites, some of which are hard and continuous enough to build ridges such as the Jebel Bani in the Anti-Atlas.

The Ashgillian is marked by glacial and periglacial terrestrial and shallow marine deposits, including well-developed tillites.

SILURIAN

This system attains a maximum thickness of 1100m in the SW part of the Anti-Atlas but does not exceed 120m in the Meseta. However, graptolitic shale facies dominate in all areas. Continuous sections of the Silurian are rare in Central Morocco, because in this area the Silurian has formed a zone of decollement and sliding involved in dislocation of the Nappes in the eastern Meseta during the Variscan Orogeny. The generally transgressive nature of Early Silurian deposition is related to eustatic changes of the sea level following Ashgillian glaciation.

Carbonate sedimentation was resumed in this region during the Silurian, particularly in the upper Ludlow (Scyphorinites Limestones).

DEVONIAN

No major change of facies occurred during the Upper Silurian and Lochkovian. However prolific fossiliferous carbonates became widespread throughout Morocco during the Middle Devonian, and a magna-facies with some Bohemian influence occurs in the southwestern part of the Anti-Atlas (W of the Tindouf basin). In the Tafilalt and Maider areas the facies is also of Bohemian character in the platform facies and is accompanied by shallow-marine sediment, whereas in the Meseta there is a mixed carbonate/terrigenous facies association.

Basin differentiation is prominent during the Givetian and Frasnian, and is accompanied by a general spread of reef facies, throughout Morocco (Meseta, Tafilalt, and Smara regions).

The Meseta terrane appears to have been affected by the first bretonnic movements which has differentiated the Basin, with carbonates developed on subaquatic highs and flysch in the trough areas by Famennian time.

The Strunian was an age in which there was a general spread of sandstone and arkose facies, premonitory to the Carboniferous Period.

CARBONIFEROUS

In Morocco the early Carboniferous displays a distribution of sedimentary facies resembling that of the later Devonian. In the western Tindouf Basin detrital sediments (sandstones and shales) pass upwards into limestones (Ouarkziz) in the Upper Visean. The continental influence becomes more important from the Upper Namurian upwards. Redbeds are very common in the Westphalian and in the Stephanian both in the Tindouf and in Central Morocco. Localised major synsedimentary movements are known in the eastern Meseta, in the high Atlas, and Jebilet.

In the Atlas domain, no record of the Tournaisian is known. The most widespread transgression of the Moroccan Palaeozoic occurred during the Upper Visean. In Central Morocco these transgressive sediments rest upon different series ranging from the Precambrian up to the Upper Visean. This stage is typified by chaotic and olistostrome styles of sedimentation.

Cannibalistic reworking of basin-margin deposits in the Meseta is indicated by widespread Upper Visean conglomerates containing Upper Visean limestone clasts. There is no record of the Westphalian or Stephanian in the Atlas region but sedimentation continued during these stages to the NE (Oujda region). In Central Morocco important NE-SW trending folds and local metamorphism of early Westphalian age are followed by thick molassic redbeds of later Westphalian of Permian age.

THE MEGUMA TERRANE OF NOVA SCOTIA, CANADA - AN AID IN TRANS-ATLANTIC CORRELATION

P. E. Schenk

Dalhousie University
Halifax, Canada, B3H 3J5

ABSTRACT

GENERAL SETTING

The Appalachian Orogen is divisible into the ancient North American miogeocline to the west, and some 18 suspected terranes to the east (Fig. 1) (1). A terrane is an internally homogeneous geologic province with features that contrast sharply with those of nearby provinces. The most extensive terranes along the orogen are, from west to east, the Piedmont, Dunnage, Gander, Avalon, and Meguma. Two other large terranes occur east of the Avalon in the southeast; however, these are buried beneath very thick sediments of the Atlantic Coastal Plain (Fig. 1) (1). In general, these terranes are believed to have accreted against ancient North America in a west to east progression through time. Earlier and more western dockings may have been by head-on obduction and/or subduction; later and more eastern ones perhaps by oblique convergence or major transcurrent movements during latest accretion (1).

The Meguma Terrane is a good example of a terrane, at least until Carboniferous time. It is unlike the adjacent Avalon Terrane in metallogenesis (2), plutonism (3), tectonism (4,5,6), and sedimentology (7). Indeed the Meguma has been interpreted as a fragment of the margin of another continent, presumably now located across the Atlantic Ocean (8).

The great problem of terranes is their source. The derivation of the last one to arrive should be the easiest to solve,

P. E. Schenk (ed.),
Regional Trends in the Geology of the Appalachian-Caledonian-Hercynian-Mauritanide Orogen, 121–130.

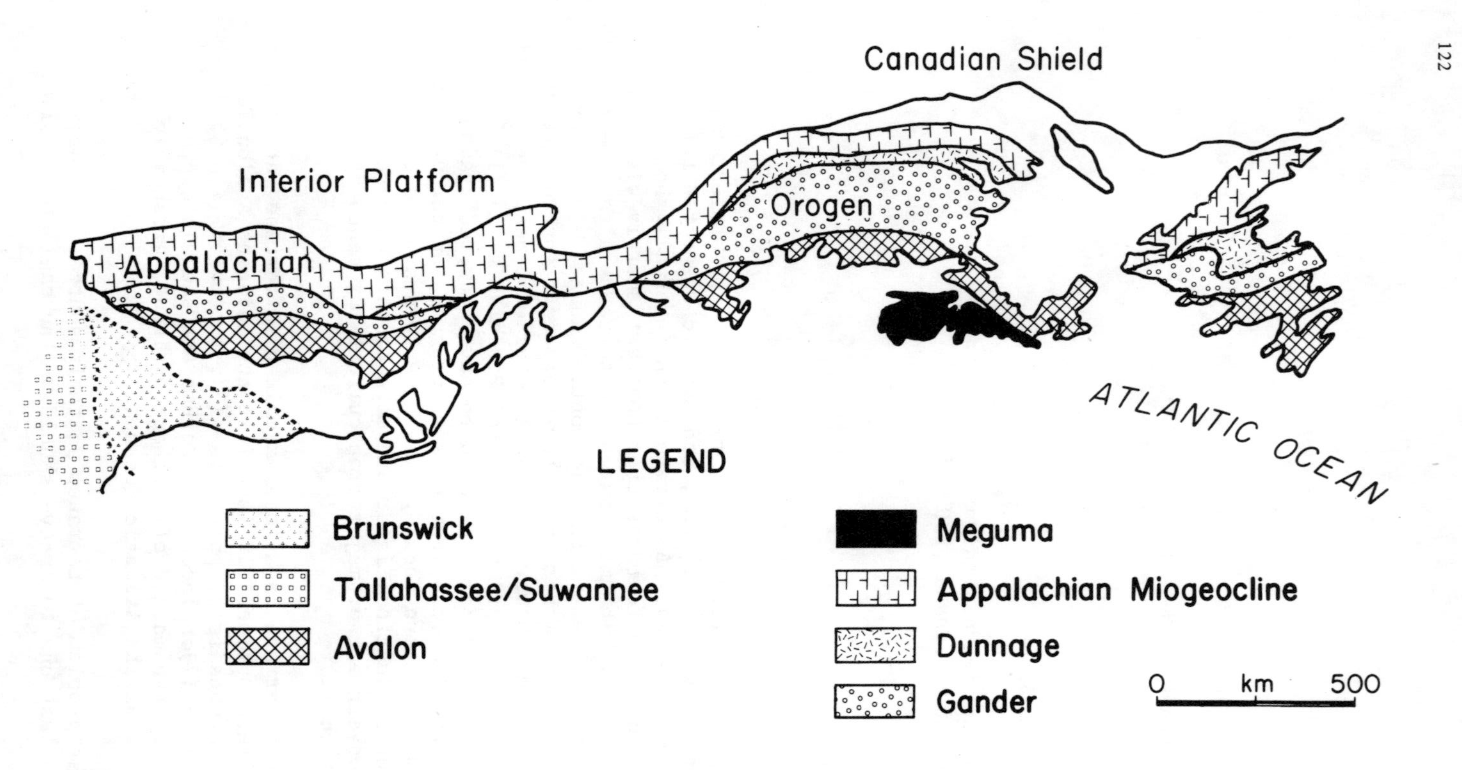

fig. 1. The seven largest suspected terranes of the Appalachian Orogen. Eleven others are much smaller and are not shown (1).

presumably by trans-Atlantic correlation. This is the Meguma as shown by: 1) it is the most eastern exposed terrane in the Orogen; and 2) it shows the youngest linkage (Carboniferous) of any terrane in the Northern Appalachians (1). Thus of all 18 suspected terranes in the Appalachians, the Meguma promises greatest success for the determination of its origin; indeed it itself is understandable only in the context of its departed source. Some suggestions for the identification of this source-continent are given below.

SIZE OF THE MEGUMA TERRANE

It occupies a very large segment of the Appalachian Orogen. The Terrane forms the Early to Middle Paleozoic bedrock of all of southern Nova Scotia (Fig. 2). Offshore, the Meguma is the foundation of both the Scotian Shelf to the south (9) and presumably the Bay of Fundy to the northwest. It may also underlie parts of the continental shelf both southeast of Cape Breton and also south of Newfoundland (10). If so, the total area would be approximately 200,000 km2 - almost the size of England and Scotland, or a third of France.

Consider only the presently exposed part of the Terrane and its continuation across the western part of the Scotian Shelf. This is an area of approximately 160,000 km2; however, the strata have been folded. Assuming that the folds approximate semicircles, and that there has been no rock flowage (so that foreshortening is approximately 36 percent) then the Meguma area becomes 210,000 km2. Assuming a flowage due to strain of 30 percent in a direction perpendicular to fold axes, the area becomes a square of 270,000 km2. The width of the restored Meguma Terrane becomes equal to the distance from Halifax to the St. Lawrence River - i.e. the entire width of the presently exposed Canadian Appalachian Orogen.

The volume of siliceous sedimentary rock in the Meguma Zone is also enormous. The thickest unit of the Terrane is the Meguma Group which is apparently at least 14 km in the thickness (7). Overlying strata of the "super-Meguma" range in total thickness from 2 to 5 km. If a thickness of 18 km was maintained across the 270,000 km2, almost 5 million km3 would be indicated. This is a block with an area equal to that of combined Spain and Portugal and a height of 5 km, or to the area of Morocco and a height of almost 7 km. The source rock of the Terrane must have been at least this size, and because of the very fine-grained nature of the Meguma rock, probably much larger.

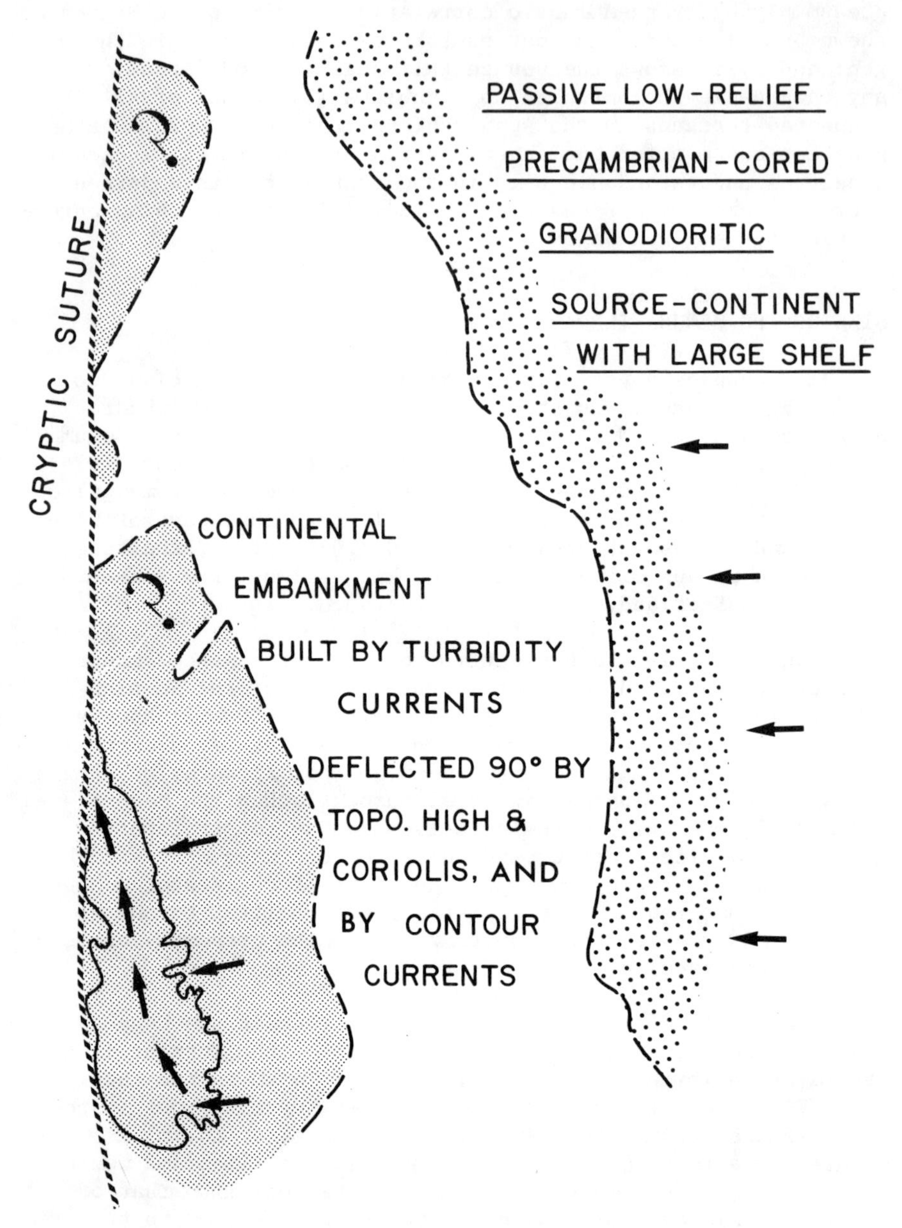

fig. 2. The Meguma Terrane showing provenance and dispersal pattern of the Meguma Group, and an interpretation of its presumed source area (8).

AGE OF THE MEGUMA TERRANE

Fossils are very scarce in the block until the Lower Devonian Torbrook Formation (Fig. 3). In the Meguma Group, graptolites, acritarchs, and chitonozoa all indicate a Tremadocian age, apparently regardless of stratigraphic position. This suggests either that lithostratigraphic divisions are very diachronous, or that indeed some 14 km of very fine-grained sedimentary rock was deposited within approximately 10 MA. If the latter is true, then this time was one of erosion in the source area. Because of the textural and compositional maturity of Meguma sediment, intermediary shallow-water, depositional basins are supposed (see below). Conceivably Tremadocian and even Upper Cambrian strata should be partially or entirely missing from such areas flanking the source (as is true in Morocco (11)).

The great scarcity of fossils until the Early Devonian is curious. One suggestion is that the environment was hostile to life because of the supposed low temperature of the ocean (12). Pebbly and bouldery mudstones at the top of the Group have been interpreted as marine tillites, possibly related to the Ashgillian glaciation of northwestern Africa (12, 13, and 14). This diamictite immediately underlies the first occurrence of quartz arenites which are interpreted to be of shallow neritic origin (12, 15). This glacioeustatic event must have been a worldwide event and should be a very useful chronozone in relatively unfossiliferous strata, such as the Meguma Group.

Radiometric ages have been determined in the Meguma Group. Grains of muscovite, apparently detrital and from the lower, Goldenville Formation, are Tremadocian to Arenigian (15). Metamorphism of the Group is Early Devonian in age (415 to 400 MA.) (17). Granitic plutons begin to intrude the entire Terrane at 408 MA and reach a climax after folding at 370 to 360 MA (18).

Taken in total, the Meguma Zone appears to be a conformable succussion extending in age from at least the Late Cambrian through the Early Devonian. Although lacunae undoubtedly exist and sedimentation was not "continuous", nevertheless notable unconformities are not definite. Thus the source area must be able to produce large amounts of detritus throughout this time. In part because of the nature of this detritus, its age, and volume, the source was probably a Precambrian Shield. Unpuplished study of rare earths in the Meguma Group indicate that the source rock was approximately 1.8 billion years old (Clarke, pers. com.). This continent should show, at least at the

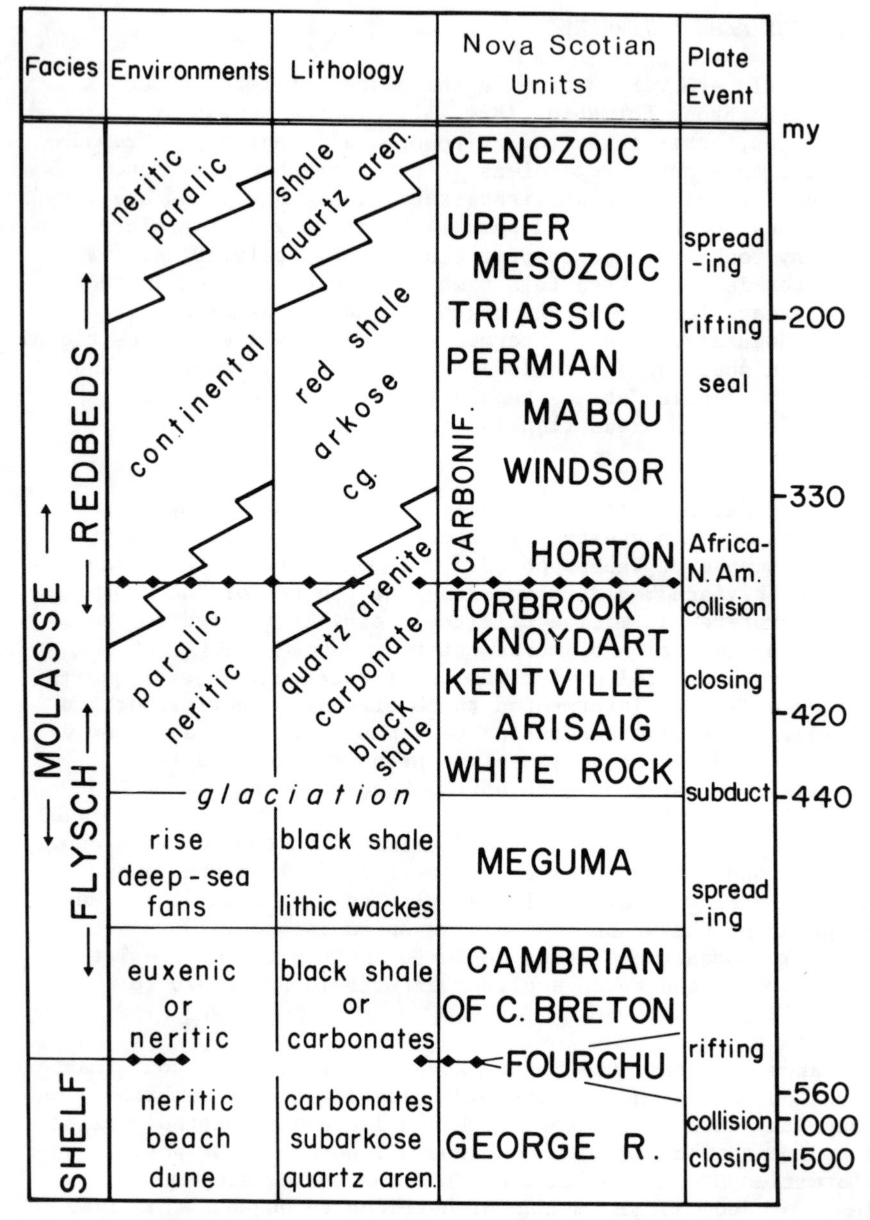

fig. 3. Stratigraphic column of Nova Scotia showing main units, and interpreted paleoenvironments, tectonic cycles, and general plate events (24).

latitude of the Meguma, uplift of its margin in the Early Devonian, followed by folding and intrusion. Moreover, this continent may still be rimmed with the remainder of Meguma-like succession which may be identified despite their possible involvment in subsequent episodes of orogeny and erosion.

LITHOLOGIES AND DEPOSITIONAL ENVIRONMENTS

The Meguma Group is distinctive in both lithology and succession; indeed, these are its greatest potential for trans-Atlantic correlation. It displays all of the characteristics of flysch (7) except that the succession is curiously inverted. Normally, "shaley" or "distal" flysch is overlain by "sandy" or "proximal" flysch as tectonism quickens in the mobile belt (19). The Meguma Group is an exception to this rule, and as such is somewhat unique.

The Group is also unusual in that it is sedimentologically mature in both texture and composition. This suggest extensive winnowing and sorting within sluggish streams, reworking and/or an extensive continental margin. Rare granules of gneiss, granites, and metasediments support a deeply eroded terrain, possibly of low relief (Fig. 2). Preliminary geochemical analysis also support a granodioritic source-land (20).

In general, the Terrane is interpreted as an upwardly shoaling complex (8, 21). The Goldenville Formation resembles modern channel deposits in the mid-fan areas of deep-water fan systems. The Halifax Formation resembles modern overbank turbidites, and in its upper parts, slope and outer shelf environments. The uppermost Halifax contains diamictite which may be glacial. Regional trends of both scalar and vectoral properties indicate that the source area lay to the present southeast. The overlying "super-Meguma" succession is inner shelf to paralic in character-the White Rock Formation is paralic with volcanics; the overlying Kentville Formation is neritic; and the capping Torbrook Formation is inner shelf to estuarine. In general, a continental embankment on a passive margin is envisioned; however, the volcanics in the White Rock Formation are important. At their maximum thickness of approximately 0.5 km, they consist of water-laid lapilli tuffs, lava and ash flows, and volcaniclastic conglomerates (15). The primary flows are a biomodal suite of alkaline to peralkaline metabasalts and metatrachytes (22, 23). They have been interpreted as products of an island arc (24), either a spreading ridge or subduction zone (23), and finally even a hot spot (25). Thus the significance of these volcanics remain in question.

SUMMARY

The source and origin of the Meguma Terrane should be the easiest to identify and explain of all the exposed Appalachian terranes because it is the latest to arrive. It is immense and is yet but a fragment of the original. The remainder, presumably still clinging to the source-continent, must be a major component of that landmass. It should be recognizable despite possible involvment in subsequent orogenies and cycles of erosion.

The source-area itself was a Precambrian Shield of meta-sedimentary/meta-igneous lithology. Winnowing and recycling of sediment on an intervening continental shelf achieved both textural and compositional maturity. This shelf may have been stripped during Tremadocian time. Thick shelf assemblages should be recorded both before and after this erosional vacuity.

Scarcity of fossils in the Meguma Group, coupled with possible glacial erratics in its uppermost part, suggest at least sub-polar conditions. In contrast, the abundance and diversification of fossils in Lower Devonian strata suggest a warming trend, conceivably due to the approach to ancestral North America which straddled the paleo-equator throughout Paleozoic time.

The Meguma Terrane records the growth of a continental embankment upon a passive margin. Silurian volcanism of disputable tectonic origin should be sought in the source-continent.

Several large landmasses are candidates for the source-area of the sediments of the Meguma terrane (8). Although problems certainly exist, northwestern Africa remains most likely.

REFERENCES

1. Williams, H. and Hatcher, R.D. 1982: Geology, 10, pp. 530-536.
2. Zentilli, M. 1977: Can. Instit. Mining Met., 70, p. 69.
3. Clarke, D.B., Barss, S.M., Donohoe, H.V. 1980: Granitoid and other plutonic rocks of Nova Scotia, in Caledonides in the USA, ed. by D.R. Wones, Virginia Polytech. Instit. State University, Mem. 2, pp. 107-116.
4. Webb, G.W. 1969: Amer, Assoc. Petroleum Geol. Mem. 12, pp. 754-786.
5. Eisbacher, G.H. 1969: Can. J. Earth Sci. 6, pp. 1095-1104.

6. Donohoe, H.V. and Wallace, P.I. 1969: Geol. Assoc. Canada Annual Meeting, Toronto, 2, p. 391.
7. Schenk, P.E. 1970: Regional variations of the flysch-like Meguma Group (Lower Paleozoic) of Nova Scotia compared to recent sedimentation off the Scotian Shelf, in Flysch sedimentology in North America, edited by J. Lajoie, Geol. Assoc. Canada Spec. Publ. 7, pp. 127-153.
8. Schenk, P.E. 1981: The Meguma Zone of Nova Scotia - a remnant of Western Europe, South America, or Africa? in Geology of the North Atlantic Borderlands, edited by J.M. Kerr and A.J. Ferguson, Can. Soc. Petroleum Geol. Mem. 7, pp. 119-148.
9. King, L.H. and MacLean, B. 1976: Geol. Survey Canada Paper 74-31, 31 p.
10. Lefort, J-P, and Haworth, R.T. 1981: Bull. Soc. geol. mineral. Bretagne, XIII, 2, pp. 103-116.
11. Michard, A. 1976: Elements de Geologie Marocaine, Notes et Mem 252, Serv. Geol. Maroc, p. 94.
12. Schenk, P.E. 1972: Can. J. Earth Sci. 9, pp. 95-107.
13. Schenk, P.E. and Lane, T.E. 1981: Early Paleozoic tillite of Nova Scotia, Canada, edited by M.J. Hambrey and W.B. Harland, Cambridge University Press, 6 p.
14. Schenk, P.E. in press: Evidence of Late Ordovician/Early Silurian marine glacial deposits in the Meguma Group of Nova Scotia (Nouveau Maroc?), in Paleoglaciations Ouest Africaines: caracterisations et evolution du phenomene glaciaire dans l'espace et le temps, edited by M. Deynoux, Universite de Poitiers.
15. Lane, T.E. 1980: The stratigraphy and sedimentology of the White Rock Formation (Silurian), Nova Scotia, Canada, unpubl. M.Sc. thesis, Dalhousie University, Halifax, Canada, 269 p.
16. Poole, W.H. 1971: Geol. Survey Canada Paper 71-A, pp. 9-11.
17. Reynolds, P.H. and Muecke, G.K. 1978: Earth Plan. Sci. Letters, 40, pp. 111-118.
18. Cormier, R.F. and Smith, T.E. 1973: Can. J. Earth Sci. 10, pp. 1202-1210.
19. Pettijohn, F.J. 1975: Sedimentary Rocks, 3rd edition, Harper and Row, 583 p.
20. Liew, M.Y.C. 1979: Geochemical studies of the Goldenville Formation at Taylor Head, Nova Scotia, unpubl. M.Sc. thesis, Dalhousie University, 204 p.
21. Schenk, P.E. and Lane, T.E. 1983: Pre-Acadian sedimentary rocks of the Meguma Zone, Nova Scotia - a passive continental margin juxtaposed against a volcanic island arc, Excursion 5B, Internat. Assoc. Sed. Congress, Hamilton, 85 p.

22. Muecke, G.K. and Sarkar, P.K. 1977: Geol. Soc. Amer. Northeast Meeting.
23. Sarkar, P.K. 1978: Petrology and geochemistry of the White Rock meta-volcanic suite, Yarmouth, Nova Scotia, unpubl. Ph.D. thesis, Dalhousie University, 350 p.
24. Taylor, F.C. 1967: Geological Survey Canada Mem. 349, 83 p.
25. Schenk, P.E. 1978: Geological Survey Canada Paper 78-13, pp. 111-136.

RELATIONSHIP BETWEEN PRECAMBRIAN AND LOWER PALEOZOIC ROCKS OF SOUTHEASTERN NEW ENGLAND AND OTHER NORTH ATLANTIC AVALONIAN TERRAINS

James W. Skehan, S.J., Weston Observatory, Department of Geology and Geophysics, Boston College, Weston, MA 02193, U. S. A. and
Nicholas Rast, Department of Geology, University of Kentucky, Lexington, KY 40506, U. S. A.

ABSTRACT

The eastern margin of the Appalachian orogen of eastern North America, as well as parts of western Europe and northwestern Africa, is comprised of Late Precambrian strata and igneous intrusions known collectively as the Avalonian terrain. This terrain is defined, in part, by Late Precambrian volcanic and sedimentary rocks, and by mafic to intermediate plutons cut by a series of granitic plutons dated from 595 to 650 Ma. In part, also, it is defined by Cambrian sedimentary cover bearing an Acado-Baltic trilobite fauna, and also Cambrian to Ordovician volcanics. The basement and cover rocks are cut by anorogenic alkaline plutons ranging from Ordovician to Devonian in age. In southeastern New England, all of the above mentioned rocks are present, except for Ordovician sedimentary rocks, but they are deeply eroded and are overlain by Upper Carboniferous fluvial basin sediments. The Late Precambrian rocks were deformed in the Avalonian orogeny; the Upper Carboniferous rocks were deformed during the polyphase Alleghanian-Variscan orogeny; and the Lower Paleozoic strata were mildly deformed at an as yet undetermined time between the above orogenic events, possibly during a Cadomian III orogenic episode. The Avalon Platform records Avalonian orogenic events of Late Precambrian age related to Cadomian II of western Europe and northwestern Africa. The mild deformation of the Lower Paleozoic rocks probably records an episode of orogenesis in nearby Gondwanaland related to deformational events of from Caledonian (or Taconian ?) to Post-Acadian time. This terrain also preserves a rich record of the Variscan-Alleghanian orogeny.

P. E. Schenk (ed.),
Regional Trends in the Geology of the Appalachian-Caledonian-Hercynian-Mauritanide Orogen, 131–162.

AVALONIAN TERRAIN

The eastern margin of the Appalachian-Caledonide orogen of North America, as well as related belts of Europe and Africa, now consists of a series of inliers and fault fragments considered as parts of the Avalonian terrain (Fig.1). This terrain was identified by Williams (1964) as the southeastern margin of a two-sided symmetrical mountain system with an ocean basin in between. Kay and Colbert (1965) described Cambrian strata of this belt extending from Sweden to Massachusetts through eastern Newfoundland as having a distinctive trilobite fauna representative of the Atlantic Coast. Lilly (1966), at the personally communicated suggestion of John Rodgers, was the first to apply the term Avalonian to the orogenic phase that affected eastern Newfoundland before the early phase of Paleozoic tectonism recognized in western Newfoundland.

The Avalonian terrain is defined, in part, by the characteristic presence of thick sequences of Late Precambrian, principally bimodal volcanic, volcaniclastic, and sedimentary rocks cut by numerous mafic to felsic plutons (Cameron & Naylor, 1976; Rast et al., 1976; Skehan and Murray, 1980a, 1980b). The latter form stocks and batholiths which typically cut all the older rocks. The Late Precambrian rocks are nonconformably or paraconformably overlain by fossiliferous Cambrian and Ordovician sedimentary rocks characterized by Acado-Baltic trilobite faunas (Hutchinson, 1952; Hayes and Howells, 1937; Theokritoff, 1968; Palmer, 1969). The Avalonian terrain is also characterized by somewhat diagnostic associations of intrusive rocks, including calc-alkaline granitic plutons and partially or completely migmatized mafic intrusives collectively referred to as gabbro-diorites (Wones, 1978).

These rocks, as well as volcanics, are intruded by numerous mafic dikes that are normally meta-diabases (Rast, 1979) but sometimes are strongly lineated amphibolites. The basement to the Avalon Platform, where exposed (New Brunswick-Nova Scotia), consists of older metasediments, gneisses, and granites yielding metamorphic ages of about 800+ Ma (Rast et al., 1976; Olszewski and Gaudette, 1982). The terrain in many of its parts shows structural, intrusive, and metamorphic effects of Late Precambrian age often referred to as the Avalonian orogeny.

A Cambro-Ordovician cover is identified in most of the larger fragments and inliers of the terrain in North America and Europe. Additionally the Avalonian terrain of New England contains Paleozoic alkaline granitic and mafic complexes that range in age from Ordovician to Devonian (Zartman and Marvin, 1971; Zartman, 1977; Barosh & Hermes, 1981; Zartman & Naylor, in press 1983). This implies that during this period of time the

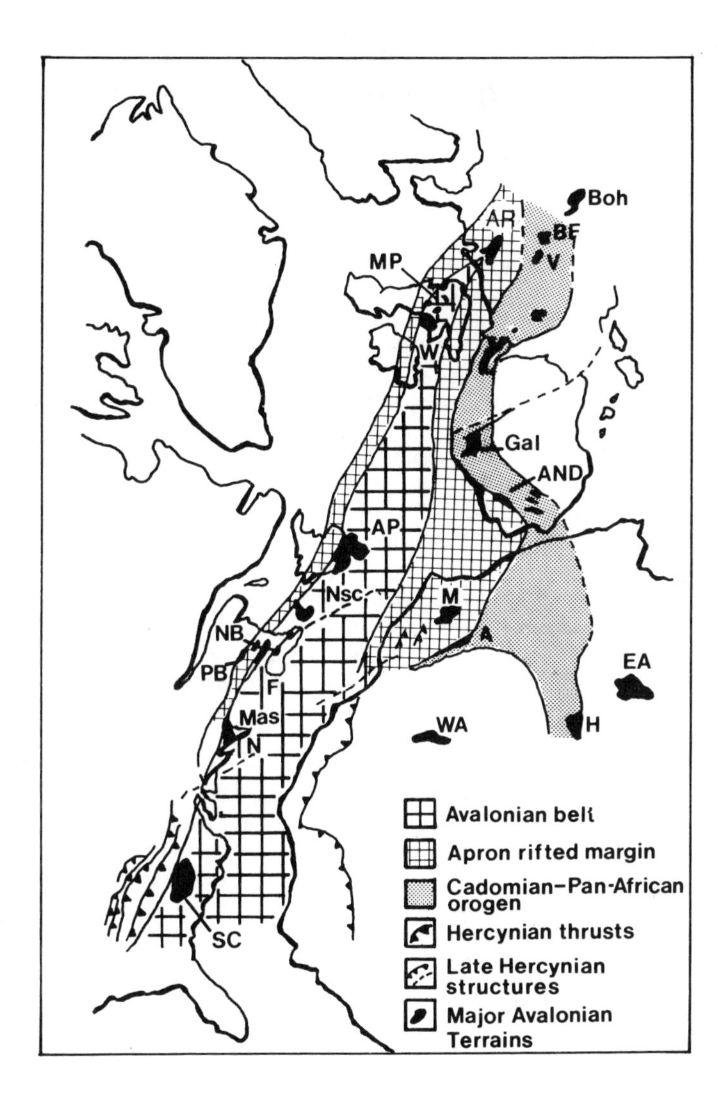

Figure 1. Sketch map of the Avalonian belt showing component terrains of the North Atlantic region: (clockwise) SC- South Carolina; Mas- Massachusetts; PB- Penobscot; NB- New Brunswick; NSc- Nova Scotia; AP- Avalon Peninsula, Newfoundland; W- Wales; MP- Midland Platform of the British Isles; AR- Ardennes; Gal- Galacia; V- Vosges; BF- Black Forest; Boh- Bohemian Massif; And- Andalusia; M- Moroccan Meseta; WA- West African Shield; A- Anti-Atlas; H- Hoggar; EA- East African block or shield; features closely related to specific terrains noted in text are N- Narragansett Basin and F- Fundy Basin (based on Rast, 1980, with modifications by Piqué, 1981, and by the authors).

terrain was a part of an anorogenic platform (Avalon Platform).

Upper Carboniferous (Pennsylvanian) coal basins, although not an integral part of the definition of Avalonian terrain, are present in this terrain and also in the Cadomian of western Europe (Fig. 1). In any case their flora, sediments, and deformation provide some controls on the age and relative position of Avalonian fragments prior to and during the Alleghanian-Variscan orogenic cycle (Lyons, 1971; Skehan and Murray, 1980a). Consequently, the relationship of the Avalonian fragments to deformational phases in the development of the Appalachian-Caledonide mountain system is an important dimension of the subject matter. Since in this paper references are made to widely dispersed European and American localities, a generally accepted stratigraphic scheme that we use in this paper follows as Figure 2. The identity of the the Avalonian Platform is indicated by a comparison of successions of sedimentary and volcanic rocks and time and character of plutonic rocks identified in North America and Europe (Fig. 2).

The Late Precambrian Avalonian succession has been first defined principally from Newfoundland, but lithologically it is so distinctive that analogs were found also in Nova Scotia, New Brunswick, New England, Southern Appalachians (Rodgers, 1964, 1967, 1968, 1969; Williams, Kennedy, and Neale, 1972, 1974; Rast et al., 1976; Skehan, 1969, 1973; Rast and Skehan, 1981a and 1981b; Samson et al., 1982), in the United Kingdom (Rast et al., 1976; Williams, 1978b). In the British Isles, where several Avalonian inliers have been identified in Wales and in the Midlands of England (Rast and Crimes, 1969), detailed correlations from one inlier to the other are very difficult, although attempts have been made by Bennison and Wright (1969).

In Newfoundland the deformed volcanic rocks of the Harbour Main and Love Cove Groups are said to be conformably overlain by, and near their tops, interbedded with the Conception and Connecting Point Groups (Williams and King, 1976). King (1980) indicates that the Harbour Main Group appear to be the oldest of the exposed rocks in the Avalon Zone and locally is overlain by the Conception Group with local unconformity. Anderson (1972) suggested an age range of 800-600 Ma for the Conception Group while Stukas (1977 cited by Strong, 1979) gave $^{40}Ar/^{39}Ar$ ages for Harbour Main volcanics as old as 1500 Ma (see also Stukas and Reynolds, 1976). The undeformed Belle Bay Volcanics yield an age date 660 ± 20 thus suggesting that the deformation of the older units noted above antedates the Belle Bay (Strong, 1979). However, since rhyolites of the Love Cove have been dated isotopically (U-Pb zircon) at 590 ± 30 (Dallmeyer et al., 1981 quoted by O'Brien and O'Driscoll, 1982) either the Group's stratigraphic position as essentially coeval with the Harbour

Main Group or the identification of the unit sampled for age dating is in question. Thus in Figure 2 we have adopted Williams and King (1976) stratigraphic interpretation until this problem can be resolved. The Marystown Group (Fig. 2) disconformably underlies fossiliferous Cambrian rocks, and has been dated by Dallmeyer (1980 cited by O'Brien and King, 1982) at 608 ± 25 Ma.

These data and implications suggest a need for an extension of the Newfoundland succession into other parts of the Avalonian terrain. In Nova Scotia, as in Newfoundland, a possible gneissic basement to the Avalonian Platform has been recorded (Keppie, 1980, 1982) and a complex sequence of sediments and volcanics overlying the basement gneisses, and major intrusives have been recognized (Keppie and Schenk, 1982). A similar basement gneiss recorded from New Brunswick, the Brookville Gneiss, and the Greenhead Group are older than 800 Ma and probably younger than 1200 Ma (Olszewski and Gaudette, 1982). Here the metasediments and metavolcanics of the Avalonian succession are intruded by later granites and granodiorites (Rast et al., 1978; Fyffe et al., 1981).

AVALONIAN TERRAIN OF NEW ENGLAND

A generalized map of southeastern New England (Fig. 3) shows the Avalonian terrain in relation to the adjacent major component parts of the Appalachian-Caledonide orogen of this region. In the southeastern New England area, no obvious older basement has been recognized (Zen, 1981). Avalonian rocks, however, (Rast et al., 1976) as such were claimed in this area (Rodgers, 1967, 1968; Palmer, 1969; Skehan, 1969) shortly after Lilly (1966) applied the term, Avalonian, in southeastern Newfoundland. Therefore in this paper, we emphasize the succession of deformed intrusive rocks and other structural deformations in this area with a view to establishing a better documented relation with the other areas of Avalonian terrain in Maritime Canada, the British Isles, and France.

It is generally agreed that the Avalonian terrain of southeastern New England includes all of the rocks that lie east and southeast of the Bloody Bluff fault zone. However, we propose additionally that the rocks of the Putnam-Nashoba belt (Skehan and Abu-moustafa, 1976; Bell and Alvord, 1976; Barosh et al., 1977) west of the Bloody Bluff and east of the Clinton-Newbury fault zone, are probably part of the Avalonian terrain as well, and are probably in large part of Late Proterozoic age based on arguments related to polymetamorphism, intensity of polydeformation, and polyintrusion history (Skehan and Murray, 1980b). These commonly have been considered as Paleozoic (Skehan and Abu-moustafa, 1976; Robinson and Hall, 1980; Hall and

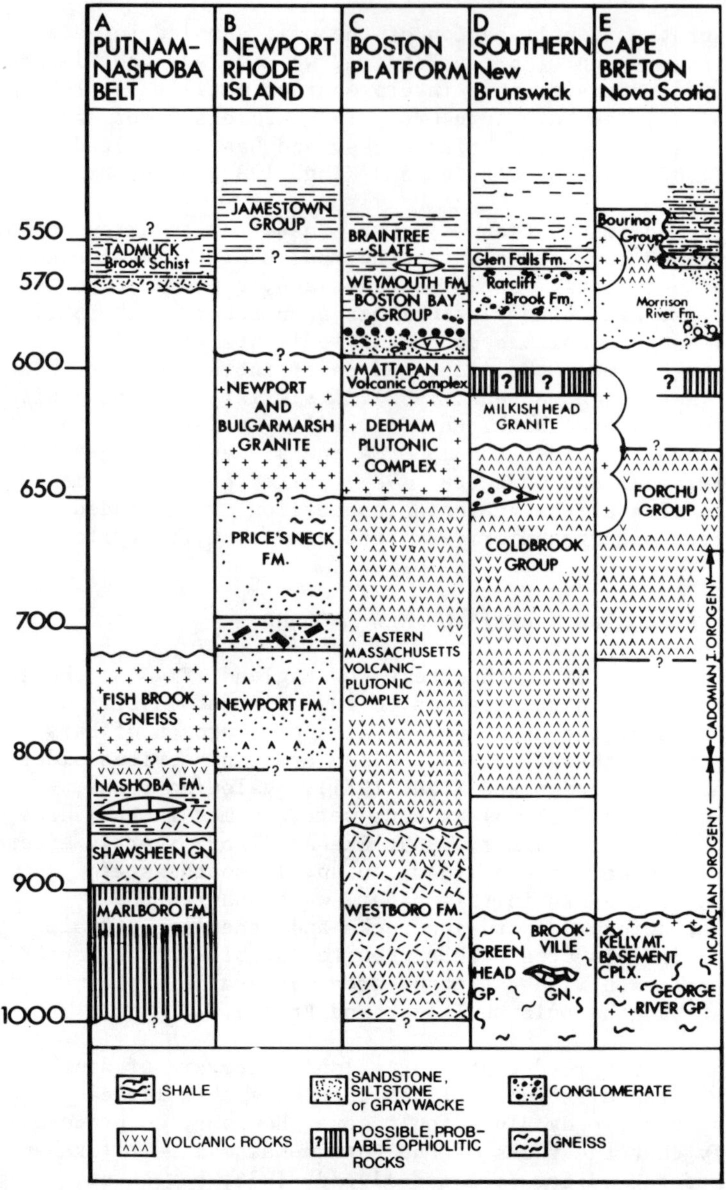

Figure 2. Stratigraphic settings of Proterozoic to Cambrian sequences of the North Atlantic region. Age control for the Fish Brook Gneiss of Putnam-Nashoba Belt is from Olszewski (1980) and relative position by concurrence of Olszewski (pers. comm. 1982) with Skehan that it is probably plutonic and not volcanic rock. Newport section is based on Rast and Skehan, (1981a, 1981b); that for Boston, on data from sources cited in text; Eastern Massachusetts volcanic- plutonic complex (informal

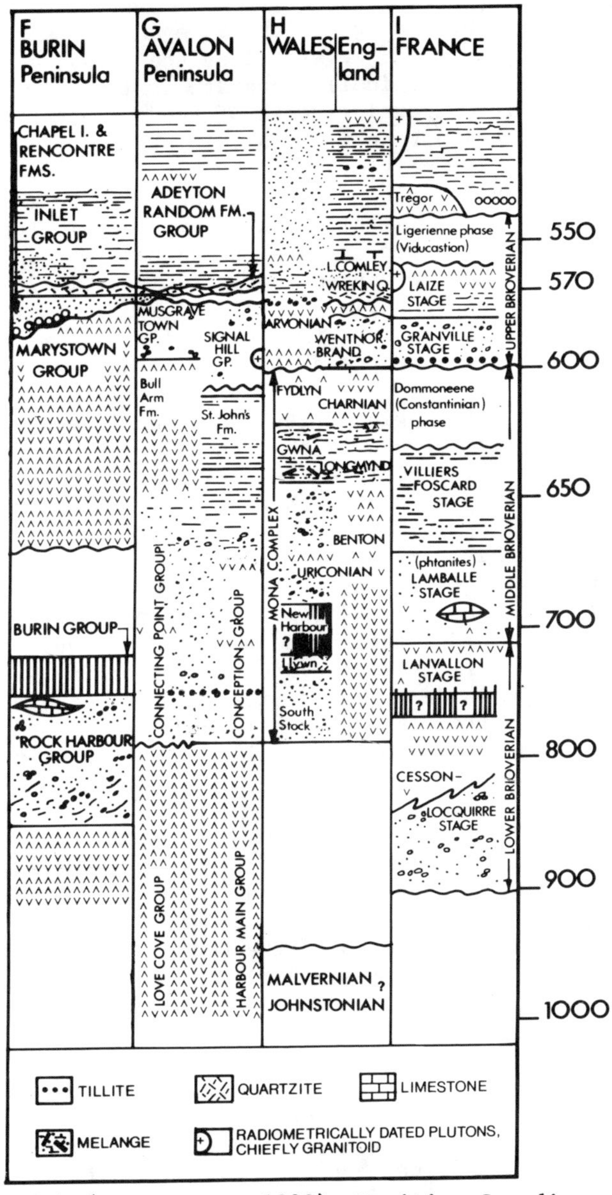

term) by Goldsmith (pers. comm. 1983); Maritime Canadian and British sections are based mainly on data reviewed by Rast et al., (1976) and Strong, (1979) with contributions by Keppie, (1980, 1982); Keppie and Schenk, (1982); Anderson, (1981); and King, (1980, 1982). Sources cited by Rast et al., (1976) and Strong (1979), for remaining sections, as well as Anderson (1981- Cambrian of Britain) and Adams (1976- L. Brioverian of France), are referenced below in text.

Robinson, 1982) but the age of these important units is as yet only partially resolved. Nevertheless the rocks of the Putnam-Nashoba belt are tectonically separate from the Avalonian of the Boston Block. The intrusive rocks of the Avalonian terrain of the Boston Block (Zartman & Naylor, in press 1983) consist of two groups of dominantly granitic rocks. The first, dominantly a calc-alkaline suite, yields ages in the range of about 600-650 Ma and the second, an alkaline suite provides isotopic data, the most diagnostic of which appears to indicate a range from Late Ordovician to Middle Devonian (Kovach et al., 1977; Hermes et al., 1981b; Zartman & Naylor, in press 1983). In at least the southern part of this block the crystalline basement with later intrusives is more widespread than the metasedimentary strata.

A significant discriminant between volcanic and volcaniclastic sequences of this terrain is that those which are cut by the 600-650 Ma granitic complexes belong to one of the older Proterozoic groups of rocks, whereas those not cut by them belong to the Infracambrian. Thus in the Boston area proper, as well as in southeastern Rhode Island, only the relatively lightly deformed granitoids of the Dedham plutonic complex and rocks cut by the Dedham underlie the strongly deformed but only lightly metamorphosed metasedimentary and metavolcanic strata of the Mattapan Volcanic Complex and the Boston Bay Group. However, in southeastern Rhode Island similar plutons, the Newport and Bulgarmarsh granites (Figs. 2 and 3), cut across the Newport and related sequences. In addition, all of these plutons cut across older, dominantly quartzitic and mafic volcanic and plutonic complexes (Fig. 2).

In western Rhode Island, and adjacent parts of Massachusetts and Connecticut, on the other hand, granitic plutons, such as the Ponaganset, the Sterling, the Esmond, and the Milford, essentially age equivalents of the Dedham (Hermes et al., 1981b; Zartman and Naylor, in press 1983), are strongly deformed and foliated together with the metasediments and metavolcanics that they intrude.

There is evidence that Cambrian strata outside the Boston Basin lie unconformably on Proterozoic metasediments and/ or nonconformably on 595-650 Ma plutons. These strata around Boston are very weakly deformed and at Hoppin Hill, Massachusetts (Fig. 3) Dowse (1950) and Anstey (1979) described fossiliferous Lower Cambrian sediments resting nonconformably on coarse granitic rocks presumed to represent a regolith of the Dedham (514 ± 17 Ma, Fairbairn et al., 1967). Elsewhere near the type locality typical ages for the Dedham and related plutons are 591 ± 28 Ma (Fairbairn et al., 1967) and 630 Ma (Zartman quoted by Nelson, 1975).

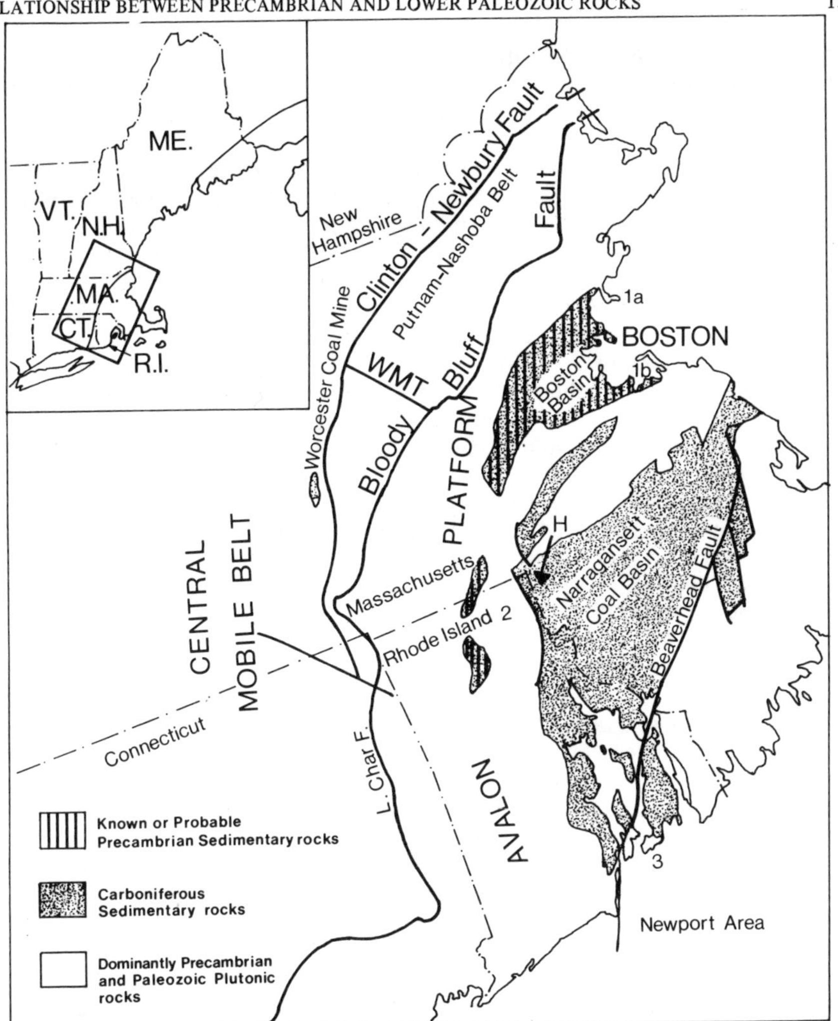

Figure 3. Generalized map of southeastern New England showing major divisions of the Avalonian or Boston Platform with respect to the Putnam-Nashoba and the Central New England Mobile Belt and their bounding faults. H is Hoppin Hill locality in which fossiliferous Lower Cambrian, resting on basal quartzite, nonconformably overlies the Hoppin Hill Granite, thought to be the equivalent of the Dedham plutonic complex. Numbers refer to localities described in the text.

Age determinations at present group most of these granitic bodies (Smith, 1978; Smith & Giletti, 1978; Hermes et al., 1981b; Zartman & Naylor, in press 1983), based on Rb-Sr whole rock isochrons and U-Pb concordia of zircons, in the range 595 to 650 Ma. Few isotopic determinations have been made on mafic plutons, many of which are cut by Late Proterozoic granitic intrusions, but one K-Ar determination on hornblende from a gabbroic pegmatite west of Boston (Fig. 3) yielded a date of 886 ± 22 Ma (Zartman & Naylor, in press 1983).

The widespread intrusion of the above dominantly calc-alkaline plutons in southeastern New England, and the preceding deformational episode recognized in Newport, R.I., we attribute to the Avalonian orogeny (Rast, 1980), broadly equivalent to the Monian of the British Isles (Wood, 1974) and possibly also to the Cadomian II of Europe and Africa (Bishop et al., 1975). In North America the Avalonian orogeny was claimed in Newfoundland (Lilly, 1966; Rodgers, 1967; Hughes, 1970) on the basis of an unconformity toward the top of the Precambrian succession. Although Dallmeyer (in press 1982) advances arguments against the occurrence of the Avalonian orogeny in Newfoundland, Anderson's (1981) study of the relationship of the Random Formation to older rocks (Fig. 2) supports its existence. The Avalonian orogeny has been recognized elsewhere (Rast et al., 1976), and may even have affected parts of the southern Appalachians, where it is referred to as the Virgilina phase (Glover and Sinha, 1973; Rast, 1980). Thus it seems reasonable to conclude that the Newport and related granitic batholiths of 595 to 650 Ma were emplaced late in the Avalonian-Cadomian II orogeny.

LATE PROTEROZOIC STRATIFIED ROCKS OF SOUTHEASTERN NEW ENGLAND

Late Proterozoic sedimentary, volcanic-plutonic, and volcaniclastic rocks of the Avalonian terrain in southeastern New England appear to reflect two groups. Boston and northward they consist largely of: 1) a bimodal volcanic-plutonic complex and 2) metasedimentary and metavolcanic strata. The latter are dominantly represented near the base of the exposed section by quartzites and gneisses in the form of unnamed stratified xenoliths and pendants probably going up conformably into quartzites, mafic metavolcanic rocks and schists of the Westboro Formation (Fig. 2, and Bell and Alvord, 1976). The Middlesex Fells Volcanic Complex, dominantly of mafic metavolcanic rocks, is reported by Bell and Alvord to overlie the Westboro unconformably. The Eastern Massachusetts volcanic-plutonic complex consists of the Middlesex Fells Volcanic Complex as well as associated gabbro, diorite, and syenite (Goldsmith, pers. comm. 1983). All of these units are cut by the Dedham plutonic complex and associated granitic plutons of 595 to 650 Ma age.

The lower portion of this succession of rocks (Fig. 2) is probably the equivalent of the well known Blackstone Group of Rhode Island, with its Quinnville Quartzite belonging to the quartzitic basal part and the Hunting Hill Volcanics (Quinn, 1971) corresponding to the basal Middlesex Fells Volcanic Complex. Similarly the Plainfield Quartzite of western Rhode

Island and eastern Connecticut is probably the equivalent of the lower part of the Blackstone and of the Westboro. In western Rhode Island and eastern Connecticut, distributed around the periphery of, and even within the Rhode Island dome or "batholith", the metasediments are metamorphosed carbonates, orthoquartzites, argillites, and pillow lavas (Quinn, 1971; Dreier & Mosher, 1981), extensively engulfed by Late Precambrian plutons which are the approximate age equivalent of the Dedham plutonic complex, and the Newport and Bulgarmarsh granites (Hermes et al., 1981b; Zartman and Naylor, in press 1983).

Quinn (1971) states that the oldest layered rocks in Rhode Island are gneisses (Absalona, Woonasquatucket, and Nipsachuck Formations) which Skehan (in press 1983) suggests should be called Harmony group. On the basis of regional stratigraphic arguments we tentatively consider the dominantly felsic volcanic, volcaniclastic, and sedimentary succession of the Harmony group to be the equivalent of the volcanic portion of the Eastern Massachusetts volcanic-plutonic complex (Middlesex Fells Volcanic Complex). Both of the authors have noted that lithologically certain parts of the Eastern Massachusetts volcanic-plutonic complex have counterparts in the Coldbrook Group of New Brunswick, and one (JWS) observed that the Harmony group has counterparts in the Marystown Group of the Burin Peninsula. Lastly, the highly feldspathic metapelites, metavolcaniclastics, and metavolcanics found near Newport, Rhode Island (Rast and Skehan, 1981a, 1981b) probably are a dominantly clastic facies of the Eastern Massachusetts volcanic-plutonic complex and of the Harmony group.

CAMBRIAN SEDIMENTARY ROCKS OF SOUTHEASTERN NEW ENGLAND

Fossiliferous Lower and/or Middle Cambrian rocks have been found in four localities (Figs. 2 and 3): 1) in the northern part of the Boston Basin (Nahant- Locality 1a); 2) in the eastern and southeastern part of the same basin (Quincy, Weymouth and Braintree- Locality 1b); 3) at Hoppin Hill, North Attleboro, Massachusetts, an inlier in northwesternmost Narragansett Basin east of Locality 2; and 4) the recently discovered extensive Middle Cambrian deposits of Jamestown, Rhode Island, the western part of Locality 3, Figure 3 (Skehan et al., 1978).

In the Avalonian terrain of southeastern New England there are no proven stratified rocks of Upper Cambrian or Ordovician age, although Ordovician carbonate rocks have been claimed for a long time near the Blue Hills, south of the Boston Basin (Fig. 3, and Foerste in Shaler et al., 1899; Billings, 1982). The only stratified rocks of Siluro-Devonian age comprise the Newbury Volcanics, north of Boston, which carry a brachiopod, pelecypod, and ostracode fauna. Studies of ostracode zonation in coastal Maine by Berdan and Martinsson (Shride, 1976) and correlations with the Newbury led them to conclude that the paleontologic material most narrowly diagnostic for the Newbury may be wholly of Pridoli age. Because the leperditiids may range into the Gedinnian, the possibility that higher strata of the Newbury may be as young as Early Devonian cannot be ruled out (Shride, 1976). There are, however, no other known Devonian or Mississippian sedimentary rocks in the Avalon Platform of southeastern New England.

The fluvial sedimentary rocks of Pennsylvanian age make up the approximately 3000 m thick sequence of Narragansett and Norfolk basins, and rest unconformably on Cambrian-Precambrian formations. The almost total absence of Upper Cambrian through Lower Devonian stratified rocks in the Avalon Platform is in sharp contrast with their widespread development in the Mobile Belt of the rest of southern New England.

ALKALINE PLUTONIC ROCKS

A distinctive suite of peralkalic and alkalic plutonic rocks of Ordovician to Devonian age (Zartman, 1977; Kovach et al., 1977; Hermes et al., 1981b; Zartman & Naylor, in press 1983) are widely represented over the Avalonian Platform of southeastern New England including the Bay of Maine (Hermes et al., 1978). The main examples are the Quincy Granite, the Blue Hills Granite Porphyry, and the Rattlesnake Pluton of Sharon (Lyons and Krueger, 1976), all south of the Boston Basin (Fig. 3); the Cape Ann Granite, the Peabody Granite, and Wenham Monzonite located north of Boston; the Cowesett Granite and related rocks of the East Greenwich Group, and part of the Scituate Granite Gneiss of western Rhode Island. Although shear zones in places cut these rocks there is little evidence of pervasive deformation in them (Hermes et al., 1981b).

UPPER CARBONIFEROUS (PENNSYLVANIAN) ROCKS

Bailey (1928) noted the convergence of the folded Appalachians of Late Paleozoic age and the earlier deformed belt of crystalline rocks of Caledonian age. He noted that Pennsylvanian rocks of the folded Appalachians reappeared on the eastern margin of the deformed Caledonian rocks in North America. More specifically these Pennsylvanian rocks of the eastern margin are represented by the Narragansett (Fig. 3) and adjacent Norfolk Basins. The Narragansett Basin represents the most deformed part of the Upper Paleozoic Appalachian belt in New England and possibly in the Appalachian belt as a whole. The stratigraphic succession consists of from 3000 to 6000 m of fluvial Upper Carboniferous sedimentary rocks ranging, on the basis of macroflora, from Westphalian B or C to Stephanian A or even younger (Lyons, 1979).

The rocks of the Narragansett Basin have been highly deformed by several episodes of deformation (Skehan et al., 1976; Murray and Skehan, 1979; Skehan and Murray, 1980a; and Burks et al., 1981), affected by Barrovian metamorphism up to the K-spar sillimanite zone, and cut by thrust faults. A voluminous literature on the Narragansett and Norfolk Basins is cited in Cameron, 1979; and Skehan et al., 1979, 1982. The deformed and metamorphosed succession was intruded by the Narragansett Pier and the Westerly Granites whose isotopic age dates are 276 and 247 Ma respectively (Kocis et al., 1977, 1978; Hermes et al., 1981a; Zartman & Naylor, in press 1983). Internally, the Carboniferous rocks show numerous dislocations including low lying thrust faults. Some of these (e.g. Jamestown thrust in Locality 3) are cut by faults, such as the Beaverhead fault (Fig. 3), which is interpreted as a late, possibly Mesozoic, transcurrent fault.

Arguments have been developed by Skehan and Murray (1980b) to show that the deformation features of the Narragansett and Norfolk Basins bear a strong geometrical relationship to the Bloody Bluff and to the Clinton-Newbury fault zones and to other faults in southeastern New England of similar orientation. The inference is that these westward-dipping thrusts and related features were active during the Alleghanian-Variscan orogenic cycle but may have originated earlier (Skehan and Murray, 1980a). McMaster et al., (1980) developed arguments which interpret the Narragansett Basin as a series of rhombohedral horst and graben blocks, developed between en echelon zones of left lateral, northeast striking faults, to form effectively a pull-apart basin.

GEOLOGIC RELATIONSHIPS IN NEW ENGLAND

In terms of present day findings, a sequence of metasedimentary and metavolcanic rocks of Avalonian affinities can be recognized in several parts of southeastern New England (Fig. 3) including: 1) Boston Basin, 2) northern and eastern Rhode Island, and 3) Newport and adjacent areas of Narragansett Bay.

1. Boston Basin. In the Boston Basin, Late Precambrian layered rocks, until recently considered as Carboniferous, have been extensively recognized on the basis of: 1) their isotopic age (Kaye & Zartman, 1980; Zartman & Naylor, in press 1983); 2) microfossils (Lenk et al., 1982); 3) their state of deformation that is much stronger than in the numerous small fault fragments of the locally recognized Cambrian strata; 4) regional lithostratigraphic comparisons (Skehan, 1978); and 5) intrusive

contacts with dated granites. At present the Mattapan Volcanic Complex and the Boston Bay Group are accepted generally as Precambrian (Billings, 1982).

The apparently oldest rocks in this area are granites (Dedham plutonic complex) which according to Billings (1976a, 1976b; 1979a, 1979b) are nonconformably overlain by the Mattapan Volcanic Complex. Thus between the Late Precambrian plutonic rocks and the fossiliferous Cambrian rocks of the Boston area there is a succession of dated rocks, the Mattapan Volcanic Complex (602 ± 3 Ma, Kaye and Zartman, 1980) overlain by the Boston Bay Group of Late Precambrian Vendian age (Lenk et al., 1982). The latter successions, referred to the Infracambrian, may pass conformably up into fossiliferous Lower Cambrian rocks.

The Mattapan Volcanic Complex of the Boston area consists of hard dense white, pink, and red rhyolites commonly mapped locally as felsite. Dark green to light green "melaphyres" or altered basaltic and andesitic volcanics, are also present but are composed largely of secondary minerals (Billings, 1979a). Pyroclastic rocks include crystal tuff, lapilli tuff, breccia, and lahars (Nelson, 1975). These extrusives appear to have arrived at the surface as flows, ashfalls, and ashflows, but many vent fillings and dikes in older basement rocks are also found and are related to the extrusion of these complexes.

Billings (1929) called attention to detrital boulders of Dedham-type granitic rock in the basal part of the Mattapan Volcanic Complex leading him to conclude that the volcanics are distinctly younger than the granite. LaForge (1932) recognized that the volcanic complex is separated from the Dedham intrusive by a pronounced unconformity, which we interpret as having formed during the uplift and erosion following the Avalonian orogeny. Billings (1929) interpreted the relationship of the Mattapan Complex to the overlying Roxbury Conglomerate of the Boston Bay Group as an angular unconformity. It seems equally probable, however, that the angular relationship may be a local and penecontemporaneous structure related to positioning of volcanic rocks over irregular topography, rather than being of orogenic significance. There are no known deformational structures in the Mattapan that are not present in the Boston Bay Group.

Kaye and Zartman (1980) report U-Th-Pb analytical results on zircons from samples of rhyolite in the Mattapan Volcanic Complex and conclude that the U-Pb data provide evidence of a concordia intercept age of 602 ± 3 Ma.

Because of the importance of the Infracambrian succession of rocks in several Avalonian terrains, Table 1 has been provided as a summary of these volcanic and sedimentary units. The Boston Bay Group (Table 1) is divided into two formations, the Roxbury Conglomerate below and the Cambridge Argillite above.

The Roxbury Conglomerate is an assemblage of conglomerate, sandstone, arkose, shale, volcanic ashes, flows, lahars (Brighton Volcanic Member), and a diamictite (Squantum Tillite Member). But the Boston Bay Group also includes other conglomerates and shallow water sediments with, in places, current cross bedding. Most of the rocks are present throughout the formation except for an important horizon of diamictite, known as the Squantum "tillite" which is discontinuous. The Squantum is considered by some to be a glacial or reworked glacial deposit (Rehmer and Hepburn, 1974; Bailey et al., 1976; and Wolfe, 1976). A voluminous literature on both sides of the controversy as to whether the Squantum is a glacial deposit or not is referenced in Bailey et al., (1976). The Brighton Volcanics form lens-shaped or irregular bodies within the Roxbury Conglomerate. We interpret the composition of this sequence making up the Boston Bay Group as the product of erosion of a volcanic-plutonic source terrain.

The Cambridge Argillite consists of gray to green argillite with beds ranging in thickness from 1 mm to 1 m, the light gray layers being silt or fine sand and the dark gray, being clay or fine silt. Within the Cambridge Argillite locally, some 850 m above the Squantum Member, lies the Milton Quartzite (Billings, 1929, 1976a), which is a white sericite quartzite 150 m thick. If the Cambridge should be transitional into the Cambrian, as Kaye (1980) contends, the stratigraphic position of the Milton Quartzite becomes important with respect to the Cambrian-Precambrian boundary. If it is basal Cambrian and conformable with the underlying argillites within the Boston Basin, and nonconformable outside the Basin, a situation would prevail similar to that in the Avalon Peninsula (Anderson, 1981) relative to the Random Formation and underlying units.

Billings (1976a, 1976b, 1979a, and 1979b) has long advocated a Pennsylvanian age for this group but in the light of fossil evidence (Lenk et al., 1982) now agrees that the age of these rocks is Late Precambrian. Because he maps the Blue Hills fault as separating the fossiliferous Cambrian rocks (Fig. 3, Locality 1b) from the Boston Bay Group to the north, Billings (pers. comm., 1981) does not accept Kaye's conclusions that the Boston Bay Group passes conformably up into Cambrian rocks. Kaye (Kaye and Zartman, 1980 and pers. comm., 1981), on the other hand, maintains that recent mapping and examination of cores (Kaye, 1980) shows that rocks of the Boston Bay Group grade upward without a break into fossiliferous Lower Cambrian rocks in some localities. In any case it now seems certain that the Boston Bay

TABLE 1

Description and lithology of the sedimentary formations and members of the Boston Bay Group (adapted from Rehmer and Roy, 1976, p. 72)

	Brookline Member	Dorchester Member	Squantum "Tillite" Member	Cambridge Argillite
Thickness	150-1300 m; thins rapidly to the northern margin of the basin	180-500 m; generally approximately 300 m	20-180 m	Estimates have varied widely; minimums of 600-1200; from tunnel data, must be 2300 m min. and may exceed 5500 m
Lithology	43-60% cong., 20-55% argillite, 2-20% sandstone. Matrix: fine to medium feldspathic sand Clasts: well rounded generally 2.5-8 cm with average 10 cm; locally up to 30 cm. Mostly quartzite, granite, felsite, lesser melaphyre and argillite; basal clasts coarser and typically of underlying formation. Interbedded volcanics.	60% argillite, 25% sandstone, 15% conglomerate, 1% tuff. Fine to medium feldspathic sand; quartz grains rounded. Pebble clasts: mostly quartzite, some granite, lacks clasts of argillite, av. max. pebble size of 14 cm.	Diamictite, 50-63% matrix of silt and clay-sized, locally sandy Pebble clasts, subrounded to angular; rare striated, some faceted. Mostly of quartzite and granite, also felsite, argillite. Av. 8-15 cm but range to 0.25-1 m commonly; some to 6 m. Some large angular argillite fragments bent and deformed.	Fine-grained, mostly argillaceous (qtz.-sericite-chlorite); some siltstone and tuff; typically 90% argillite, 10% feldspathic sandstone; slightly calcareous

Features of current deposition	Ripple marks and current bedding (lamination and cross-bedding). Ripples generally oscillation type.	Cross-bedding common; ripple marks		Common pinch and swell bedding in association with small-scale cross-bedding; oscillation and interference ripple marks (no current ripples) scour marks
Features of gravity deposition	Graded bedding		"dropstones" slump structures and contorted zones in the asociated argillites	Graded bedding; slump structures and contorted zones; load-casts
Bedding	Unbedded to obscure bedding; sand and shale partings and lenses; well-stratified in only a few places.	Absent or poor in the sandstones; argillite has well-stratified beds and lenses	Obscure or unstratified; associated thin sandstone and argillite layers.	Rhythmic banding comprising about half of formation; beds generally 1-8 cm; pinch and swell in beds 0.5 cm thick
Other				Color mostly gray to the north; 60% reddish to purplish and 40% gray or greenish to the south.

Sources: Billings (1929, 1976), Billings and Rahm (1966), Billings and Tierney (1964), Caldwell (1962), Dott (1961), LaForge (1932), Rahm (1962), and Sayles (1914) - all referred to in Rehmer and Roy (1976).

Group must be Late Proterozoic but younger than 602 Ma, the age of the underlying Mattapan Volcanic Complex. The microflora of Vendian age found in the Cambridge Argillite and the absence of diagnostic acritarchs of Early Cambrian age in a sediment which contains well-preserved Bavlinella appears to limit the analyzed part of that unit to the late Proterozoic (Lenk et al., 1982). The question of the age of the uppermost part of the Boston Bay Group awaits further documentation.

The Cambrian strata of Nahant (Fig. 3, Locality 1a) are weakly deformed Lower Cambrian carbonates and quartzites that have been preserved in a Devonian gabbro-diabase complex (Zartman and Naylor, in press 1983). The Weymouth and Braintree carbonates and argillites (Fig. 3, Locality 1b) have yielded respectively Lower and Middle Cambrian faunas and are also weakly deformed. At Hoppin Hill somewhat more strongly deformed Lower Cambrian pebbly quartzites, resting nonconformably on Hoppin Hill Granite, are overlain by maroon and green slates which enclose thin beds of fossil-bearing pink carbonates (Anstey, 1979). Alkaline intrusions in the Blue Hills dated as Late Ordovician or Early Silurian age (Zartman, 1977) invade and preserve some of the Cambrian strata of the Boston area (Emerson, 1917; Billings, 1982).

2. Northern and Eastern Rhode Island. The Blackstone Group of Rhode Island is about 5000 to 6500 m thick (Quinn, 1971). Another group, referred to as the "older gneisses of northwest Rhode Island", and called the Harmony group by Skehan (in press 1983), consists of a 2500 to 3000 m thick sequence of dominantly volcanic and volcaniclastic rocks in northwestern Rhode Island. The protoliths of the Harmony group are interpreted by Quinn (1971) as thick accumulations of graywacke, feldspathic sands, and ash flows mainly of felsic composition. These are, in part at least, cut by Late Precambrian plutonic rocks. All of these stratified rocks occur in isolated outcrops separated by Late Proterozoic and later plutons (Barosh and Hermes, 1981; Hermes et al., 1981b; Zartman and Naylor, in press 1983).

The Blackstone Group metasediments and volcanics have been correlated with the Marlboro Formation of nearby Massachusetts (Quinn & Moore, 1968), and tentatively correlated by Bell and Alvord (1976) with the Westboro Formation, the Middlesex Fells Volcanic Complex, and some older unnamed gneissose and quartzitic strata. Preliminary observations on the Blackstone Group by Rast and Skehan suggest that the pillow basalts are at the base going up into marble, quartzite, schist and back into pillow basalts. As such it is completely dissimilar from the layered rocks of the Boston Bay Group or the Mattapan Volcanic Complex. We concur generally with the above correlations of Bell and Alvord (1976) within the Avalonian Block of southeastern New

England. Moreover, this succession is possibly equivalent to the Greenhead Group and Brookville Gneiss of New Brunswick, and the Kelly Mountain complex and George River Group of Cape Breton Island. Quinn and Moore's (1968) suggestion that the Blackstone Group may be the correlative of the Marlboro Formation awaits testing as the Marlboro Formation is in the Putnam-Nashoba Belt which is separated by a major thrust zone (Figs. 2 and 3; Wintsch, 1978).

3. Newport, Rhode Island Area. Southern Rhode Island contains a bedded succession of Late Proterozoic rocks (Newport formation overlain by Price's Neck formation). The Newport formation, consists of a thick sequence of graded feldspar-rich tuffaceous sandstones (turbidites), siltstones, slates, and a melange of distorted graywacke and chlorite-quartz slate containing blocks of quartzite, dolomite, gabbro, and serpentinite. The upper part, the Price's Neck formation, consists of purple volcaniclastic turbidites and tuffs cut by a stockwork of felsic and mafic dikes (Rast & Skehan, 1981a, 1981b; Webster pers. comm. 1982). At least the layered rocks and Newport granite show evidence of polyphase deformation. In addition, there are localities where other limited metasedimentary rocks are exposed, as for example, on the Sakonnet River, Tiverton, Rhode Island (Fig. 3, northeast of Locality 3), but at present they cannot be related easily to any of the above-mentioned groups, although in places they resemble argillaceous parts of the Boston Bay Group.

The evidence presented above suggests that the oldest metasediments are probably the Westboro Formation and the Eastern Massachusetts volcanic-plutonic complex (Fig. 2). Moreover, their probable equivalents to the west, the Blackstone Group and the Harmony group (informal name), have been deformed together with granitoids that invade them. The Newport and Price's Neck formations may be younger than the oldest units noted above and may be lateral equivalents of the Eastern Massachusetts volcanic-plutonic complex and the Harmony group. These formations are invaded by undeformed granite over much of the eastern part of the Avalon Platform (Fig. 3). The Boston Bay succession, and possibly also those of the North Scituate and Woonsocket Basins (Fig. 3, Locality 2 near the Rhode Island-Massachusetts state line)(Skehan & Murray, 1979), is probably the youngest of the preserved Late Proterozoic rocks.

STRUCTURAL RELATIONSHIPS IN NEW ENGLAND

So far, the most complete evidence is available only from Newport, Rhode Island. The Newport granite, (Figs. 2 and 3, Locality 3) whose isotopic age date of 595 ± 12 Ma (Smith, 1978; Smith and Giletti, 1978) demonstrates that it was intruded in Late Proterozoic time, shows certain critical relationships relative to rocks which it cuts. Here the Newport granite cuts the S_1 cleavage of metasedimentary rocks which it intrudes, but the granite and the older sedimentary rocks are both cut by a later S_2 cleavage (Rast and Skehan, 1981a). Also, thin sections from the Price's Neck formation contain fragments of rocks having a strong metamorphic fabric (Skehan & Webster, pers. comm. 1982).

Thus we conclude that the Late Precambrian Newport sedimentary succession had undergone a deformational episode prior to the intrusion of the Newport granite, but then both were affected together later by the S_2 cleavage. The age of this structure is debatable, since polyphase deformation features also can be seen in nearby Cambrian and Carboniferous strata.

In Jamestown, Rhode Island, trilobite-bearing Middle Cambrian phyllites (Skehan et al., 1981) show extensive polyphase deformation including the development of isoclinal first folds (F_1), refolded by recumbent open second folds (F_2) accompanied by a strong penetrative cleavage, which, in many places, is affected by later kinks (F_3).

Additionally tectonic slides or fold-faults, as they were called by Bailey (1910) who first described them from the Highlands of Scotland, have been identified (Skehan et al., 1981). Some of these slides are pre- to syn-F_1; all are pre-F_2 and commonly have a breccia developed at the sharp contact. Commonly there is a well developed angular fault relationship between beds at the contact. On the Beavertail Peninsula (Fig. 3, Locality 3 and southeast of the Beaverhead fault), following the widespread tectonic sliding event, the Cambrian rocks have been folded into an upright major asymmetrical syncline and were flattened during the development of the dominant S_2 cleavage.

The relationships among F_1, F_2, and F_3 of the Late Proterozoic Newport sequence and that of the Cambrian of Beavertail present considerable difficulties of interpretation. The F_2 folds and associated S_2 cleavages in both sequences are subparallel, but it is difficult to to equate them because the preceding F_1 structures cannot be equivalent since one is manifestly Precambrian and the other post-Lower Cambrian. At present it seems that the early deformation of the Cambrian strata has to be attributed to some Lower or Middle Paleozoic (Caledonian-Acadian) or possibly to a Late Paleozoic Alleghanian event. It can be shown, however, that the latter correlation probably does not apply.

In Mackeral Cove (Fig. 3, Locality 3), the Cambrian strata are overthrust by a Precambrian granite and overlain nonconformably by Pennsylvanian Pondville Conglomerate. Since the Pondville is in direct fault contact with the highly deformed Cambrian phyllites, the overthrust block probably represents the overturned, faulted limb of a granite-cored anticline. On approach to the thrust plane the Cambrian phyllites develop a strong steep crenulation (S_3) cleavage that deforms the dominant S_2 cleavage of the Cambrian rocks. Thus the crenulation cleavage is interpreted as Alleghanian in age since it occurs in Pennsylvanian beds where it is referred to as Alleghanian S_2 by Burks et al., (1981). These relationships demonstrate that S_2 and S_3 in the Cambrian rocks appear to correspond with S_1 and S_2 respectively of the Alleghanian orogenic cycle. Therefore, at least the tectonic slides and the F_1 structure in the Cambrian phyllites are pre-Alleghanian but later structures are probably of Alleghanian-Variscan age.

Therefore, in the area effects of Precambrian, Lower to Middle Paleozoic, and Alleghanian orogenies are recognized. The surmised relations of the various phases of these deformations are very similar to the situation encountered in southern New Brunswick, where all three orogenic events are known (Rast, 1980; Rast, in press 1983).

CONCLUSIONS

If one assumes that the undeformed to weakly deformed granites intruding layered Avalonian rocks of the Boston Platform are generally contemporaneous then it can be held that there is the following sequence of depositional and intrusive events which serve as a basis of proposed international correlations below and in Figure 2:

1) Older gneissose and quartzitic units followed by quartzitic, calcareous, and shaly sediments, and dominantly mafic volcanics (Westboro Formation and lower part of Blackstone)

We enlarge upon Rodgers' (1967) and Naylor's (1975) correlations to propose that these rocks are correlatives of the Green Head Group and the Brookville Gneiss of New Brunswick; of the Kelly Mountain and George River Groups of Cape Breton Island; of the Harbour Main Group of the Avalon Peninsula; of the Malvernian and Johnstonian complexes of the British Isles which are regarded by Shackleton (1975) as younger than 1400 Ma but not younger than Uriconian; and of the Lower Brioverian of Brittany and Normandy, and possibly the uppermost part of the Pentevrian Complex (Bishop et al., 1975).

2) Mafic to felsic volcanics, volcaniclastics, and gabbro, diorite, and syenite plutons (Middlesex Fells Volcanic

Complex, Eastern Massachusetts volcanic-plutonic complex, upper part of Blackstone Group, Harmony group volcanic and volcaniclastic rocks, Newport and Price's Neck volcaniclastic graded sandstones, pelites, and volcanic and plutonic stockwork)

In this case also, we enlarge on Rodgers' (1967) and Naylor's (1975) suggestions that southeastern New England has correlatives in the Maritime Provinces. More specifically we propose that these packages of rocks (Fig. 2) are correlatives of the Coldbrook, the Forschu, Georgeville, the Marystown and possibly the Love Cove Groups; of the Mona Complex and equivalent rocks in the British Isles; correlations with Brittany and Normandy are more tentative but it seems certain (Bishop et al., 1975) that the Brioverian extends upward to at least 660 ± 25 Ma.

3) Intrusion of undeformed and weakly deformed calc-alkaline granites of the Dedham plutonic complex, the Newport, Bulgarmarsh, and the more highly deformed Ponaganset, Ten Rod, Hope Valley Alaskite, and Scituate Granite Gneiss

We propose that these plutonic complexes are the equivalent of the Milkish Head Granite of New Brunswick; of the Swift Current Granite (O'Brien and O'Driscoll, 1982); of the Cape Roger Mountain Granite of the Burin Penisnula; of the Holy Rood Granite of the Avalon Peninsula; of the Coedna Granite of Anglesey, North Wales (Shackleton, 1975); and of the Mancellian granites of Normandy (Adams, 1967 cited in Bishop et al., 1975) and a wide range of other plutons in Brittany (Adams, 1976).

4) Mattapan Volcanics and fossiliferous Boston Bay Group

We propose that these Infracambrian rocks, dominantly volcanic near the base; coarse clastic, commonly volcaniclastic and containing volcanics, and with a glacially derived(?) "tillite" in the middle; grading up into fine argillite are the correlative (Fig. 2) of at least the lower Ratcliff Brook Formation; of the Morrison River and part of the Bourinot Group; of the Upper Marystown Group, the Chapel Island and Rencontre Formations; of the Musgravetown and Signal Hill Groups of the Avalon Peninsula; of the Arvonian Volcanics and the Wentnor and Brand beds of Wales and England; and possibly the lower part of the Upper Brioverian including the tillite near its base (Cogné, 1972; Adams, 1967; Auvray, 1975; Peucat and Cogné, 1974; and Roach et al., 1972).

5) Fossiliferous Cambrian sedimentary rocks

Although Lower and Middle Cambrian rocks are represented in the Avalon Platform of southeastern New England, in only Locality 1b (Fig. 3) are both found. The Tadmuck Brook Schist has been referred to the pre-Silurian (Bell and Alvord, 1976; and to the Silurian- Devonian (Skehan and Abu-moustafa, 1976); to the Silurian?, Ordovician, or Proterozoic Z (Zen, 1981). On the basis of data in Skehan and Abu-moustafa and on regional litho-stratigraphic considerations we refer the Tadmuck Brook to a position just above a major unconformity which is probably either the Cambrian-Proterozoic boundary or within the Late Proterozoic.

We follow Walcott (1890), Howell et al., (1936), Palmer (1969), and many others (see Theokritoff, 1968 and Skehan, 1973) in correlating the fossiliferous Cambrian of southeastern New England with sequences containing a comparable Acado-Baltic fauna in New Brunswick, Cape Breton Island, and in Newfoundland. Sequences in the Burin and Avalon Peninsulas, the Inlet and the Adeyton Groups respectively, are nonconformable on the lowermost Cambrian Random Formation (Anderson, 1981), which in places goes down gradationally into the Chapel Island and Rencontre Formations (Fig. 2) and elsewhere rests unconformably on the older rocks of the Infracambrian. Anderson also correlates the Random with the lowermost Wrekin Quartzite of the English Midlands which rests unconformably on Late Proterozoic rocks. Palmer (1969), Seilacher and Crimes (1969) note that Cambrian and Tremadocian sequences of the Acado-Baltic Province in central England, northern Spain, and Portugal (Crimes, pers. comm.) have remarkable similarities to those in eastern North America.

6) Alkaline intrusions ranging from Late Ordovician to Middle Devonian, probably in two major series of pulses

7) Fossiliferous Siluro-Devonian sedimentary and volcanic rocks of the Newbury Volcanics (small fault bounded blocks in northernmost Avalonian Platform, Fig. 3)

8) Fossiliferous Carboniferous fluvial sedimentary rocks

In this sequence, therefore, the events recorded in New England can be related to Maritime Canada, and even the British Isles, Normandy and Brittany, where the existence of Late Precambrian, early to mid-Paleozoic, and late Paleozoic orogenic events succeeded by Mesozoic faulting is well recorded and is widely known.

Acknowledgements. We thank Robert E. Zartman, Richard S. Naylor, and R. David Dallmeyer for furnishing unpublished data. We profited from discussions with them, and with T. P. Crimes, F. Doré, H. E. Gaudette, Lynn Glover III, J. Christopher Hepburn, O. Don Hermes, Rudolph Hon, Clifford Kaye, J. P. Lefort, J. Le Gall, R. Neuman, W. J. Olszewski Jr., S. Samson, D. T. Secor Jr., D. F. Strong, H. R. Williams and many others who participated in symposia and field trips of the International Geological Correlation Program (IGCP), Project 27, and other conferences.

We are grateful also to Frances Ahearn who drafted the illustrations, to Patricia C. Tassia, James P. McCaffrey, S.J., Dorothy M. Sheehan, Marylou Coyle, Greta E. Gill, Rebecca Meacham, Peter J. Canning, Mary L. Gannon and John J. Felock who assisted in the preparation of various revisions of the manuscript. We acknowledge National Science Foundation (NSF) Grants Nos. EAR-77-14429, EAR-82-00047, and U. S. Geological Survey Contract No. P089071 for partial support of field studies on which a portion of this compilation is based.

REFERENCES

Adams, C. J. D. 1967: K/Ar ages from the basement complex of the Channel Islands (United Kingdom) and the adjacent French mainland. Earth Planet. Sci. Lett., 2, 52-56.

Adams, C. J. D. 1976: Geochronology of the Channel Islands and adjacent French mainland, J. Geol. Soc. Lond., 132, 233-250.

Anderson, M. M. 1972: A possible time span for the late Precambrian of the Avalon Peninsula, southeast Newfoundland in light of fossils, tillites and rock units within the succession, Can. J. Earth Sci., 9, 1710-1726.

Anderson, M. M. 1981: The Random Formation of southeastern Newfoundland: a discussion aimed at establishing its age and relationship to bounding formations, Amer. Jour. Sci. 281, 807-830.

Anstey, R. L. 1979: Stratigraphy and depositional environment of Early Cambrian Hoppin Slate of southern New England and its Acado-Baltic Fauna, Northeastern Geology 1, (1), 9-17.

Auvray, B. 1975: Relations entre plutonisme acide et volcanisme ignimbritique: exemple des manifestations magmatiques cambriennes du Nord de la Bretagne. Pétrologie, 1, 125-137.

Bailey, E. B. 1910: Recumbent Folds in the Schists of the Scottish Highlands. Quart. Jour. Geol. Soc. London 66, 586-618.

Bailey, E. B. 1928: The Ancient Mountain Systems of Europe and America, Scottish Geog. Mag. 44, (5), 703-711.

Bailey, R. H., Newman, W. A., & Genes, A. 1976: Geology of Squantum "Tillite", pp. 92-106 in Cameron, B. (ed.), Geology of Southeastern New England, New England Intercollegiate Geol. Conf., 68th Ann. Mtg., Princeton Press, Princeton, N. J.

Barosh, P. J., & Hermes, O. D. 1981: General Structural Setting of Rhode Island and Tectonic History of southeastern New England, pp. 1-16 in Boothroyd, O. D. and J. C. (eds.), Guidebook to Geologic Field Studies in Rhode Island and Adjacent Areas, New England Intercollegiate Geol. Conf., 73rd Ann. Mtg., U. of R. I., Kingston, R. I.

Barosh, P. J., Fahey, R. J., & Pease, M. H. Jr. 1977: Preliminary Compilation of the bedrock geology of the land area of Boston 2 degree sheet, Massachusetts, Connecticut, Rhode Island, Maine, and New Hampshire, in Barosh, P. J. (ed.), New England Seismotectonic Study Report, Weston Observatory, Boston College.

Bell, K. G., & Alvord, D. C. 1976: Pre-Silurian stratigraphy of northeastern Massachusetts, pp. 179-216 in Page, L. R. (ed.), Contributions to the stratigraphy of New England, Geol. Soc. Am. Mem. 148.

Bennison, G. M., & Wright, A. E. 1969: The Geological History of the British Isles, St. Martins Press, N.Y.

Billings, M. P., 1929: Structural geology of the eastern part of the Boston Basin, Am. J. Sci., 5th ser., 18, 97-137.

Billings, M. P. 1976a: Bedrock geology of the Boston Basin, pp. 28-45 in Cameron, B. (ed.), Geology of Southeastern New England, New England Intercollegiate Geol. Conf., 68th Ann. Mtg., Princeton Press, Princeton, N. J.

Billings, M. P. 1976b: Geology of the Boston Basin, pp. 5-30 in Lyons, P. C. & Brownlow, A. H. (eds.), Studies in New England Geology, Geol. Soc. Am. Mem. 146.

Billings, M. P., 1979a: Bedrock geology of the Boston Basin, pp. 47-64 in Cameron, B. (ed.), Carboniferous Basins in southeastern New England, Field Guidebook for Trip No. 5, Ninth Internat. Cong. of Carb. Strat. and Geol., Amer. Geol. Inst., Falls Church, VA.

Billings, M. P. 1979b: Boston Basin, Massachusetts, pp. A15-A20 in Skehan, S.J., J. W., Murray, D. P., Hepburn, J. C., Billings, M. P., Lyons, P. C., & Doyle, R. O. (eds.), The Mississippian and Pennsylvanian (Carboniferous) Systems in the United States - Massachusetts, Rhode Island, and Maine, U. S. Geol. Survey Prof. Pap. 1110 A-L.

Billings, M. P. 1982: Ordovician cauldron subsidence of the Blue Hills Complex, eastern Massachusetts, Geol. Soc. Am. Bull. 93, 909-920.

Bishop, A. C., Roach, R. A., & Adams, C. J. D. 1975: Precambrian rocks within the Hercynides, pp. 102-107 in Harris, A. L., Shackleton, R. M., Watson, J., Harland, W. B., & Moorbath, S., (eds.), A Correlation of Precambrian rocks in the British Isles, Spec. Rept. 6, Geol. Soc. London.

Burks, R. J., Mosher S., & Murray, D. P. 1981: Alleghenian Deformation and Metamorphism of Southern Narragansett Basin, pp. 265-275 in Hermes, O. D., & Boothroyd, J. C. (eds.), Guidebook to Geologic Field Studies in Rhode Island and

Adjacent Areas, New England Intercollegiate Geol. Conf., 73rd Ann. Mtg., U. of R. I., Kingston, R. I.

Cameron, B. (ed.) 1979: Carboniferous Basins in Southeastern New England Field Guidebook for Trip No. 5, Ninth Internat. Cong. of Carb. Strat. and Geol., Am. Geol. Inst., Falls Church, Va.

Cameron, B., & Naylor, R. S. 1976: General Geology of southeastern New England, pp. 13-27 in Cameron, B. (ed.), Geology of Southeastern New England, New England Intercollegiate Geol. Conf., 68th Ann. Mtg., Princeton Press, Princeton, N. J.

Cogné, J. 1972: Le Brioverien et le cycle orogénique cadomien dans le cadre des orogènes fini-précambriens, Notes Mém. Serv. Géol. Maroc., 236, 193-213.

Dowse, A. M. 1950: New evidence on the Cambrian contact at Hoppin Hill, North Attleboro, Massachusetts: Am. J. Sci., 248, 95-99.

Dreier, R. B., & Mosher, S. 1981: The Blackstone Series: Evidence for an Avalonian Plate margin in northern Rhode Island, pp. 93-102 in Hermes, O. D. & Boothroyd, J. C. (eds.), Guidebook to Field Studies in Rhode Island and adjacent areas, New England Intercollegiate Geol. Conf., 73rd Ann. Mtg., U. of R. I., Kingston, R. I.

Emerson, B. K. 1917: Geology of Massachusetts and Rhode Island, U. S. Geol. Survey. Bull. 597.

Fairbairn, H. W., Moorbath, S., Ramo, A., Pinson, W. H. & Hurley, P. M. 1967: Rb-Sr age of the granitic rocks of southeastern Massachusetts and the age of the Lower Cambrian at Hoppin Hill, Earth and Planet. Sci. Lett., (2), 321-328.

Fyffe, L.R., Pajari, G.E., and Cherry, M.E., 1981, The Acadian plutonic rocks of New Brunswick, Maritime Seds. and Atlantic Geol., 17, 23-36.

Glover, Lynn III, & Sinha, A. K. 1973: The Virgilina deformation, a Late Precambrian to early Cambrian(?) orogenic event in the central Piedmont of Virginia and North Carolina, Am. J. Sci., 273-A, 234-251.

Hall, L. M. & Robinson, P. 1982: Stratigraphic-Tectonic Subdivisions of southern New England, pp. 15-41 in St-Julien, P. & Beland, J. (eds.), Major Structural Zones and Faults of the Northern Appalachians, Geol. Assoc. of Can. Spec. Pap. 24.

Hayes, A. O., & Howell, B. F. 1937: Geology of Saint John, New Brunswick. Geol. Soc. Am. Spec. Pap. 5.

Hermes, O. D., Ballard, R. D., & Banks, P. O. 1978: Upper Ordovician peralkalic granites from the Gulf of Maine, Geol. Soc. Am. Abs. with Prog., 11, 436.

Hermes, O. D., Barosh, P. J., & Smith, P. V. 1981a: Contact Relationships of the Late Paleozoic Narragansett Pier Granite and Country Rock, pp. 125-152 in Hermes, O. D., & Boothroyd, J. C. (eds.), Guidebook to Field Studies in Rhode Island and adjacent areas, New England Intercollegiate Geol. Conf., 73rd Ann. Mtg., U. of R. I., Kingston, R. I.

Hermes, O. D., Gromet, L. P., & Zartman, R. E. 1981b: Zircon geochronology and petrology of plutonic rocks in Rhode Island, pp. 315-338 in Hermes, O. D. & Boothroyd, J. C. (eds.), Guidebook to Field Studies in Rhode Island and adjacent areas, New England Intercollegiate Geol. Conf., 73rd Ann. Mtg., U. of R. I., Kingston, R. I.

Howell, B. F., Shimer, H. W., & Lord, G. S. 1936: New Cambrian Paradoxides fauna from eastern Massachusetts, Geol. Soc. Amer. Proc. for 1935, 385.

Hughes, C. J. 1970: The late Precambrian Avalonian orogeny in the Avalon Peninsula, SE. Newfoundland, Am. J. Sci., 269, 183-190.

Hutchinson, R. D. 1952: The Stratigraphy and Trilobite Faunas of the Cambrian Sedimentary Rocks of Cape Breton Island, Nova Scotia, Geol. Surv. Can. Mem. 263.

Kay, M., & Colbert, E. H. 1965: Stratigraphy and Life History, John Wiley and Sons, New York, 736 p.

Kaye, C. A. 1980: Bedrock Geologic Map of the Boston North, Boston South, and Newton Quadrangles, U. S. Geol. Survey Map MF-1241, 2 maps, scale 1:24,000.

Kaye, C. A., & Zartman, R. E. 1980: A Late Proterozoic Z to Cambrian Age for the Stratified Rocks of the Boston Basin, Massachusetts, U. S. A., pp. 257-261 in Wones, D. R. (ed.), Proceedings "The Caledonides in the USA", Internat. Geol. Correl. Prog. (IGCP), Project 27: Caledonide Orogen, VA Polytec. Inst. and State U. Mem. 2.

Keppie, J. D. 1980: An Introduction to the Geology of Nova Scotia, pp. 107-117 in Rast, N. (ed.), Field Trip Guidebook, Trip 3: The Northern Appalachian Geotraverse: Quebec-New Brunswick-Nova Scotia, Geol. Assoc. of Canada & Min. Assoc. of Canada.

Keppie, J. D. 1982: Geology and Tectonics of Nova Scotia, pp. 125-152 in King, A. F. (ed.), Guidebook for Avalon and Meguma Zones, Internat. Geol. Correl. Prog. (IGCP), Project 27: Caledonide Orogen, Dept. of Earth Sciences, Memorial U. of Newfoundland.

Keppie, J. D., & Schenk, P. E. 1982: Geology and Tectonics of Nova Scotia, pp. 159-187 in King, A. F. (ed.), Guidebook for Avalon and Meguma Zones, Internat. Geol. Correl. Prog. (IGCP), Project 27: Caledonide Orogen, Dept. of Earth Sciences, Memorial U. of Newfoundland.

King, A. F. 1980: The Birth of the Caledonides: Late Precambrian Rocks of the Avalon Peninsula, Newfoundland, and their Correlatives in the Appalachian-Orogen, pp. 3-8 in Wones, D. R. (ed.), Proceedings "The Caledonides in the U. S. A.", Internat. Geol. Correl. Prog. (IGCP), Project 27: Caledonide Orogen, VA Polytec. Inst. and State U. Mem. 2.

King, A. F. (ed.) 1982: Guidebook for Avalon and Meguma Zones, Internat. Geol. Correl. Prog. (IGCP), Project 27: Caledonide Orogen, Dept. of Earth Sciences, Memorial U.of Newfoundland.

Kocis, D. E., Hermes, O. D., Cain, J. A., & Murray, D. P. 1977: Reevaluation of Late Paleozoic igneous activity and accompanying contact metamorphism in southeastern New England. Geol. Soc. Am. Abs. with Progs., 9, (3), 286-287.

Kocis, D. E., Hermes, O. D., & Cain, J. A. 1978: Petrologic comparison of the pink and white facies of the Narragansett Pier Granite, Rhode Island, Geol. Soc. Am. Abs. with Prog., 10, (7), 71.

Kovach, A., Hurley, P. M., Fairbairn, H. W. 1977: Rb-Sr whole rock age determination of the Dedham Granodiorite, eastern Massachusetts, Am. J. Sci., 277, 905-912.

LaForge, L. 1932: Geology of the Boston area Massachusetts, U. S. Geol. Survey Bull. 839.

Lenk, C., Strother, P. K., Kaye, C. A., & Barghoorn, E. S. 1982: Precambrian Age of the Boston Basin: New Evidence from microfossils, Science, 216, 619-620.

Lilly, H. D. 1966: Late Precambrian and Appalachian tectonics in the light of submarine exploration of the Great Bank of Newfoundland and in the Gulf of St. Lawrence: preliminary views, Am. J. Sci., 264, 569-574.

Lyons, P. C. 1971: Correlation of the Pennsylvanian of New England and the Carboniferous of New Brunswick and Nova Scotia, Geol. Soc. Am. Abs. with Prog., 3, (1), 43-44.

Lyons, P. C. 1979: Biostratigraphy in the Pennsylvanian of Massachusetts and Rhode Island, pp. A20-A24 in Skehan, S.J., J. W., Murray, D. P., Hepburn, J. C., Billings, M. P., Lyons, P. C., & Doyle, R. G. (eds.), The Mississippian and Pennsylvanian (Carboniferous) Systems in the United States-Massachusetts, Rhode Island, and Maine, U. S. Geol. Survey Prof. Paper 1110 A-L.

Lyons, P. C., & Krueger, H. W. 1976: Petrology, chemistry, and age of the Rattlesnake pluton and implications for other alkalic granitic plutons of southern New England, pp. 71-102 in Lyons P. C. & Brownlow, A. H. (eds.), Studies in New England Geology, Geol. Soc. Am. Mem. 146.

McMaster, R. L., deBoer, J., & Collins, B. P. 1980: Tectonic development of southern Narragansett Bay and offshore Rhode Island, Geology 8, 496-500.

Murray, D. P., & Skehan, S.J., J. W. 1979: A Traverse across the eastern margin of the Appalachian-Caledonide orogen, southeastern New England, pp. 1-35 in Skehan, S.J., J. W., & Osberg, P. H. (eds.), The Caledonides in the U. S. A.: Geological Excursions in the Northeast Appalachians, Internat. Geol. Correl. Prog. (IGCP), Project 27: Caledonide Orogen, Weston Observatory, Weston, MA.

Naylor, R. S. 1975: Age Provinces in the Northern Appalachians, Ann. Rev. Earth Planet. Sci. 3, 387-400.

Nelson, A. E. 1975: Bedrock geologic map of the Natick quadrangle, Middlesex and Norfolk Counties, Massachusetts, U. S. Geol. Survey Geol. Quad Map GQ 1208.

O'Brien, S. & King, A. F. 1982: The Avalon Zone in Newfoundland, pp. 1-27 in King, A. F. (ed.), Guidebook for Avalon and Meguma Zones, Internat. Geol. Correl. Prog. (IGCP), Project 27: Caledonide Orogen, Dept. of Earth Sciences, Memorial U. of Newfoundland.

O'Brien, S. & O'Driscoll, C. 1982: Southwestern Avalon Zone: Burin Peninsula, pp. 99-110 in King, A. F. (ed.), Guidebook for Avalon and Meguma Zones, Internat. Geol. Correl. Prog. (IGCP), Project 27: Caledonide Orogen, Dept. of Earth Sciences, Memorial U. of Newfoundland.

Olszewski, Jr., W. J. 1980: The geochronology of some stratified metamorphic rocks in northeastern Massachusetts, Can. J. Earth Sci., 17, 1407-1416.

Olszewski, Jr., W. J. & Gaudette, H. E. 1982: Age of the Brookville Gneiss and associated rocks, southeastern New Brunswick, Can. J. Earth Sci., 19, 2158-2166.

Palmer, A. K. 1969: Cambrian Trilobite Distributions in North America and their Bearing on Cambrian Paleogeography of Newfoundland, pp. 139-144 in Kay, M. (ed.), North Atlantic Geology and Continental Drift, Mem. 12, Am. Assn. of Pet. Geol.

Peucat, J. J., & Cogné, J. 1974: Les schistes cristallins de la Baie d'Audierne (Sud Finistère): un jalon intermédiaire dans la socle antécambrien entre la Méséta Ibérique et les régions sud-armoricaines, C. R. Acad. Sci. Fr. Ser. D, 278, 1809-1812.

Piqué, A. 1981: Northwestern Africa and the Avalon plate: Relations during the late Precambrian and late Paleozoic time, Geology, 9, 319-322.

Quinn, A. W. 1971: Bedrock geology of Rhode Island, U. S. Geol. Survey Bull. 1295.

Quinn, A. W., & Moore, G. E. 1968: Sedimentation, tectonism, and plutonism in the Narragansett Basin region, pp. 269-280 in Zen, E-an, White, W. S., Hadley, J. B., & Thompson, Jr., J. B. (eds.), Studies of Appalachian Geology, Northern and Maritime, New York Interscience Publishers, John Wiley and Sons, Inc., N. Y.

Rast, N. 1979: Precambrian meta-diabases of southern New Brunswick-the opening of the Iapetus Ocean?, Tectonophys. 59, 127-137.

Rast, N. 1980: The Avalonian plate in the northern Appalachians and Caledonides, pp. 63-66 in Wones, D. R. (ed.), Proceedings "The Caledonides in the U. S. A.", Internat. Geol. Correl. Prog. (IGCP), Project 27: Caledonide Orogen, VA Polytec. Inst. and State U. Mem. 2.

Rast, N., & Crimes T. P. 1969: Caledonian Orogenic Episodes in the British Isles and Northwestern France and their Tectonic and Chronological Interpretation, Tectonophys. 7, 277-307.

Rast, N., & Skehan, S.J., J. W. 1981a: The geology of the Precambrian rocks of Newport and Middletown, Rhode Island, pp. 67-92 in Hermes, O. D. & Boothroyd, J. C. (eds.), Guidebook to

Geologic Field Studies in Rhode Island and Adjacent Areas, New England Intercollegiate Geol. Conf., 73rd Ann. Mtg., U. of

Shaler, N. S., Woodworth, J. B., & Foerste, A. F. 1899: Geology of the Narragansett Basin, U. S. Geol. Survey Mon. 33, 402 pp.

Shride, A. F. 1976: Stratigraphy and correlation of the Newbury Volcanic Complex, northeastern Massachusetts, pp. 147-178 in Page, L. R. (ed.), Contributions to the Stratigraphy of New England, Geol. Soc. Am. Mem. 148.

Skehan, S.J., J. W. 1969: Tectonic framework of southern New England and eastern New York, pp. 793-814 in Kay, M. (ed.), North Atlantic Geology and continental drift, Mem. 12, Am. Assn. Pet. Geol.

Skehan, S.J., J. W. 1973: Subduction zone between Paleo-American and Paleo-African plates in New England, Geofisca International, 13, 291-308.

Skehan, S.J., J. W. 1978: Puddingstone, Drumlins, and Ancient Volcanoes: A Geologic Field Guide along Historic Trails of Greater Boston, WesStone Press, Dedham, MA.

Skehan, S.J., J. W., & Abu-moustafa, A. A. 1976: Stratigraphic Analysis of Rocks Exposed in the Wachusett-Marlborough Tunnel, east-central Massachusetts, pp. 217-140 in Page, L. R. (ed.), Contributions to the Stratigraphy of New England, Geol. Soc. Am. Mem. 148.

Skehan, S.J., J. W., Murray, D. P. 1979: Woonsocket and North Scituate Basins of Massachusetts and Rhode Island, pp. A14-A15 in Skehan, S.J., J. W., Murray, D. P., Hepburn, J. C., Billings, M. P., Lyons, P. C. & Doyle, R. G. (eds.), The Mississippian and Pennsylvanian (Carboniferous) Systems in the United States - Massachusetts, Rhode Island, and Maine, U. S. Geol. Survey Prof. Pap. 1110 A-L.

Skehan, S.J., J. W., & Murray, D. P. 1980a: A Model for the Evolution of the Eastern Margin (EM) of the Northern Appalachians, pp. 67-72 in Wones, D. R. (ed.), Proceedings "The Caledonides in the U.S.A.", Internat. Geol. Correl. Prog. (IGCP), Project 27: Caledonide Orogen, VA Polytech. Inst. & State U. Mem. 2.

Skehan, S.J., J. W., & Murray, D. P. 1980b: Geologic Profiles across southeastern New England, Tectonophys. 69, 285-319.

Skehan, S.J., J. W., Murray, D. P., Belt, E. S., Hermes, O. D., Rast, N., & Dewey, J. F. 1976: Alleghenian deformation, sedimentation, and metamorphism in southeastern Massachusetts and Rhode Island, pp. 447-471 in Cameron, B. (ed.), Geology of Southeastern New England, New England Intercollegiate Geol. Conf., 68th Ann. Mtg. Princeton Press, Princeton, N. J.

Skehan, S.J., J. W., Murray, D. P., Palmer, A. R., Smith, A. T., & Belt, E. S. 1978: Significance of fossiliferous Middle Cambrian rocks of Rhode Island to the history of Avalonian microcontinent, Geology 6, 694-698.

Skehan, S.J., J. W., Murray, D. P., & Raben, J. D. 1979: Field Guide to the Narragansett Basin, southeastern Massachusetts

and Rhode Island, pp. 138-155 in Cameron, B. (ed.), Carboniferous Basins of Southeastern New England, Field Guidebook for Trip No. 5; Ninth Internat. Cong. of Carb. Strat. and Geol., Am. Geol. Inst., Falls Church, VA.
Skehan, S.J., J. W., Rast, N. & Logue, D. F. 1981: The geology of the Cambrian rocks of Conanicut Island, Jamestown, Rhode Island, pp. 237-264 in Hermes O. D., & Boothroyd, J. C. (eds.), Guidebook to Geologic Field Studies in Rhode Island and Adjacent Areas, New England Intercollegiate Geol. Conf., 73rd Ann. Mtg., U. of R. I., Kingston, R. I.
Skehan, S.J., J. W., Murray, D. P., Raben, J. D., & Chase, Jr., H. B. 1982: Exploration and Exploitation of the Narragansett coal basin, pp. 381-399 in Farquhar, O. C. (ed.), Geotechnology in Massachusetts: Proceedings of a Conference in March, 1980, Graduate School, U. of MA.
Smith, B. M. 1978: The geology and Rb-Sr whole-rock age of granite rock of Aquidneck and Conanicut Islands, Rhode Island; M. Sc. Thesis, Brown U., Providence, R. I.
Smith, B. M., & Giletti, B. J. 1978: Rb-Sr whole-rock study of the deformed porphyritic granitic rocks of Aquidneck and Conanicut Islands, Rhode Island, Geol. Soc. Am. Abs. with Progs., 10, (2), 86.
Strong, D. F. 1979: Proterozoic Tectonics Of Northwestern Gondwanaland: New Evidence from Eastern Newfoundland, Tectonophys. 54, pp. 81-101.
Stukas, V. J. 1977: Plagioclase Release Patterns: A High Resolution $^{40}Ar/^{39}Ar$ Study. M. Sc. Thesis, Dalhousie U., NS, Canada.
Stukas, V. J. & Reynolds, P. R. 1976: $^{40}Ar/^{39}Ar$ dating of terrestrial materials: a review of recent studies with particular emphasis on plagioclase release patterns, Prog. with Abs., 1. GAC-MAC Ann. Mtg.
Theokritoff, G. 1968: Cambrian Biogeography and Biostratigraphy in New England, pp. 1-22 in Zen, E-an, White, W. S., Hadley, J. B., & Thompson, J. B. Jr. (eds.), Studies in Appalachian Geology, Northern and Maritime, New York Interscience Publishers, John Wiley & Sons, Inc., N. Y.
Walcott, C. D. 1890: The fauna of the Lower Cambrian Olenellus zone, U. S. Geol. Survey 10th Ann. Rept., 509-763.
Williams, H. R. 1964: The Appalachians in northeastern Newfoundland-A two-sided symmetrical system, Am. J. Sci., 262, 1137-1159.
Williams, H. R. (compiler) 1978a: Tectonic-Lithofacies map of the Appalachian Orogen, Memorial U. of Newfoundland, Map No. 1.
Williams, H. R. 1978b: Geological Development of the Northern Appalachians: its bearing on the evolution of the British Isles, pp. 1-22 in Bowes, D. R., & Leake, B. E. (eds.), Crustal evolution in northwestern Britain and adjacent regions, Seal House Press, Liverpool, England.

Williams, H. R. 1979: Appalachian Orogen in Canada, Can. J. Earth Sci., 16, 792-807.
Williams, H. R., & King, A. F. 1976: Southern Avalon Peninsula, Newfoundland: Trepassy map-area, Geol. Surv. Can., Pap., 76-1A, 179-182.
Williams, H. R., Kennedy, M. J., & Neale, E. R. W. 1972: The Appalachian Structural Province, pp. 181-261 in Price, R. A., & Douglas, R. J. W. (eds.), Variations in tectonic styles in Canada. Geol. Assoc. of Can. Spec. Pap. 11.
Williams, H. R., Kennedy, M. J., & Neale, E. R. W. 1974: The northeastward termination of the Appalachian Orogen, pp. 79-123 in Nairn A. E. M. & Stehli F. G. (eds.), The ocean basins and margins, 2 vols., Plenum Press, New York.
Wintsch, R. P., & Hudson, M. R. 1978: Southeastward thrusting in eastern Connecticut, Geol. Soc. Am. Abs. with Progs, 10, (2), 91.
Wolfe, C. W. 1976: Geology of Squaw Head, Squantum, Massachusetts, pp. 107-116 in Cameron, B. (ed.), Geology of Southeastern New England, New England Intercollegiate Geol. Conf., 68th Ann. Mtg., Princeton Press, Princeton, N. J.
Wones, D. R. 1978: Plutonic Rocks of eastern Massachusetts. Geol. Soc. Am., Abs. with Prog. 10, (2), 91.
Wood, D. S. 1974: Ophiolites, melanges, blueschists, and ignimbrites: early Caledonian subduction in Wales?, pp. 334-344 in Dott, R. H., & Shaver, R. H. (eds.), Modern and Ancient Geosynclinal Sedimentation, S.E.P.M. Spec. Publ. 19.
Zartman, R. E. 1977: Geochronology of some alkalic rock provinces in eastern and central United States, Ann. Rev. Earth and Planet. Sci., 5, 257-286.
Zartman, R. E., & Marvin, R. F. 1971: Radiometric age (Late Ordovician) of the Quincy, Cape Ann, and Peabody granites from eastern Massachusetts, Geol. Soc. Am. Bull. 82, 937-958.
Zen, E-an, (ed.), 1981: Bedrock Geologic Map of Massachusetts, U. S. Geol. Survy Open-File Report 81-1327, 39 pp. and 3 map sheets.

VOLCANISM AND PLUTONISM IN THE CALEDONIDES OF SCANDINAVIA

M.B. Stephens[1], H. Furnes[2], B. Robins[3], and B.A. Sturt[4].

[1]Geological Survey of Sweden (S.G.U.), Stockholm, Sweden, [2]Geologisk Institut, Aud A. Allegt. 41, 5014 Bergen, Norway, [3]University of Bergen, 5014 Bergen University, Bergen Norway, [4]University of Bergen, 5014 Bergen University, Bergen, Norway.

Caledonian igneous rocks are concentrated in the upper part of the Middle Allochthon and in the higher nappes. Within these tectonically bounded stratigraphic packages there are preserved relics of pre-Iapetus rifting, Iapetus opening with oceanic crust, Iapetus closing with subduction-related arc volcanics and even late extentional magmatism developed on newly sutured crust. Different environments are characteristic of successively higher groups of nappes:-1) Rifting related dolerites and amphibolites occur in the Middle (Särv Nappe and equivalents) and Upper (Seve Nappes and equivalents) Allochthons respectively, intruded or extruded in an ensialic environment. 2) Arc-related (probably ensimatic) volcanic and sub-volcanic rocks occur within the Upper Allochthon (Köli Nappes and equivalents). 3) Fragmentary ophiolite slices and associated within-plate basalts occur in the Upper Allochthon (including the Støren Nappe) and possibly even the Uppermost Allochthon.

Arc-related and continental within-plate volcanic-subvolcanic complexes also occur in the same tectonic units as the ophiolite slices.

Synotogenic intrusive magmatism not apparently associated with volcanism is also seen in these nappes. Basic, ultrabasic and alkaline intrusive complexes are found in the Middle Allochthon and Uppermost Allochthon. Granitoids are concentrated in the higher nappes of the Upper Allochthon and Uppermost Allochthon.

P. E. Schenk (ed.),
Regional Trends in the Geology of the Appalachian-Caledonian-Hercynian-Mauritanide Orogen, 163–165.

Rifting in an ensialic environment related to the opening of Iapetus appears to have occurred after deposition of late Precambrian glaciogenic sediments, while most, if not all of the ophiolite fragments are pre-late-Ordovician in age and two of them (Vassfjellet, Støren) are inferred to be pre-mid Arenig. However there is some uncertainty as to what extent the ophiolite fragments represent remnants of Iapetus, or younger smaller marginal basins which opened during the protracted phase of plate convergence and closure of Iapetus. Although there is evidence for igneous activity in the Cambrian and Ordovician related to subduction processes, it is apparent that closure of Iapetus was complex and occurred over a period perhaps as long as c. 100 m.y. During the period of plate convergence, different phases of magmatic activity related to subduction processes occurred possibly at the same time as (Støren Nappe) or more confidently were followed by (Lower and Middle Köli Nappes) magmatism related to an extensional tectonic regime.

Evidence for magmatic evolutionary trends is only present within either a single magmatic province or within limited groups of related nappes. These include: 1) a change upward from low-K tholeiitic basalts to alkaline basalts in two of the ophiolite slices (Karmøy and Leka). 2) a development from early tholeiitic magmas through localised high-K calc-alkaline magmas to predominantly alkaline magmas in the Seiland Province (Cambrian-early Ordovician). 3) an evolution in both the Lower and Middle Köli Nappes from transitional basalts through bimodal sequences of low-K dacite-rhyodacite and arc-related tholeiitic basalt and basaltic andesite to rifting-related predominantly tholeiitic basaltic rocks (Ordovician - early Silurian). 4) a change from low-K, arc-related tholeiitic basalt and basaltic andesite through trondhjemite to granitic rocks in the Upper Köli Nappes and equivalents (Ordovician - Silurian). These granites are spatially associated with gabbro intrusions and show higher K_2O contents and more variable initial 87Sr/86Sr ratios than the trondhjemites. 5) an evolution in the Støren Nappe and equivalents from predominantly low-K tholeiitic basalts in ophiolite slices to medium/high-K calc-alkaline basalt-rhyolite sequences and either contemporaneous or later plutonic complexes of pre-dominantly quartz diorite to granite (pre-mid Arenig to Silurian).

The latter is complicated by the eruption of rifting-related tholeiitic and alkaline basalts in the Ordovician and early Silurian, several of which have been interpreted as forming parts of ophiolites. The concentration of synorogenic granitic rocks of mid to late Ordovician and Silurian age in the highest nappes, in spatial association with synorogenic gabbros and showing variable initial $^{87}Sr/^{86}Sr$ ratios is thought to reflect increased heat flow during the Ordovician and Silurian in westernmost continental environments.

The complex magmatic evolution postulated for the Ordovician and Silurian igneous rocks in the Upper Allochthon may be related to the presence of one or more relict oceanic segments and/or the opening of new basins during this period, thus allowing different areas to develop separate magmatic evolutionary trends. Elimination of oceanic segments during Silurian-Devonian continent-continent collision subsequently telescoped together these different sequences.

VOLCANISM AND PLUTONISM IN BRITAIN AND IRELAND

E. H. Francis[1], P. S. Kennan[2], and C. J. Stillman[3]

Dept. of Earth Sciences, University of Leeds, Leeds, LS2 9JT, United Kingdom[1], Geology Dept., University College, Belfield, Dublin 4, Ireland[2], Geology Dept. Trinity College, Dublin, Ireland[3]

VOLCANISM IN BRITAIN AND IRELAND (C. J. Stillman and E. H. Francis)

PROTEROZOIC TO ORDOVICIAN

North and south of the Iapetus Suture, the volcanic rocks show significant differences. On the Laurentian (or North American) plate north of the suture, Dalradian sediments and volcanics accumulated in late Proterozoic to Cambrian times and were deformed by the Cambro-Ordovician Grampian orogenic event. Submarine tholeiitic basalts, such as the Tayvallich lavas and tuffs occur in the Upper Dalradian and are associated with a tholeiitic dyke suite cutting the lower parts of the succession. They are thought to have been emplaced during a period of crustal extension associated with the opening of the Iapetus Ocean. Evidence of early ocean crust may be seen in the Highland Border Series which contains clasts of ophiolitic material apparently derived from the south in Arenig times. When extension gave way to Grampian compression, synorogenic magmatism, seen now in Connemara, produced a suite of ultrabasic to acid bodies with volcanic arc geochemistry thought to be the roots of an early Ordovician arc. Similar root zones may occur in the Tyrone Igneous Complex and perhaps the Aberdeenshire Gabbros.

Marginal now to this continental mass, though perhaps themselves exotic terranes, are a number of areas of early Ordovician island-arc volcanics. The Lough Nafooey volcanics, now situated just to the north of the Connemara migmatites, are submarine island-arc tholeiites which later become calc-alkaline; these were erupted on the flanks of the developing

P. E. Schenk (ed.),
Regional Trends in the Geology of the Appalachian-Caledonian-Hercynian-Mauritanide Orogen, 167–171.

South Mayo trough in Tremadoc to Arenig times. Younger sequences continuing up to the Llanvirn are bimodal basic and acid. Similar bimodal volcanics and intrusives are found at Charlestown. The Longford-Down - Southern Uplands zone, an accretionary prism stacked onto the Laurentian continental margin by obduction from the southeast, contains volcanic components. These include the pre-Arenig marginal basin ocean crust seen in the Ballantrae ophiolite, obducted between 490 - 470 m.y. ago, and other components derived from oceanic volcanos and ocean floor volcanism, which are seen in the tectonic slices of the northern edge of the Southern Uplands. Similar occurrences are seen in Ireland south of Belfast Lough, at Strokestown at the western end of the Longford Down massif, and in South Connemara. All are apparently the products of marginal basin spreading-axis volcanism, but emplaced above a northward subduction zone.

South of the Suture there is apparently a break between the Proterozoic and the Cambrian, when Caledonide sequences began. The late Proterozoic largely comprises a variety of igneous rocks produced in a major period of island arc - continental margin activity prior to c. 600 ± 50 m.y., which was uplifted and eroded prior to the Cambrian to form an ensialic floor to the Lower Palaeozoic of southern Britain. The Mona Complex ophiolitic rocks in Anglesey show ocean floor tholeiitic characters; the Arvonian of North Wales and the Pebidian of South Wales are calc-alkaline volcanic arc suites, as are the Uriconian of Shropshire and the Charnian of south Leicester. Only the Warren House volcanics of Malvern are tholeiitic.

The succeeding Lower Palaeozoic volcanics show a geochemical and petrographic zonation indicating a southeastward dipping subduction zone beneath this Proterozoic crust, with the trench somewhere north-west of the present position of the Iapetus Suture. In the northern Lake District the Arenig-Llanvirn Eycott Volcanics are basalt and basaltic andesites of outer island-arc transitional tholeiite chemistry. Their correlatives along strike in north Leinster, Ireland, are younger, Caradoc to Ashgill, and include calc-alkaline types. South of these the main Lake District volcanics are the Borrowdale Volcanics of Llandeilo to Caradoc age; these are mainly andesites, calc alkaline in composition and possibly extend southwestward along strike to the extensive volcanic belts of south-eastern Ireland. Here the Llanvirn-Llandeilo volcanism is essentially basaltic, calc-alkaline, and is followed by extensive acid volcanism in Caradoc times. This is apparently subduction-related volcanism erupting in arcs situated on the margins of the southern Britain microcontinental plate. In Wales the Ordovician volcanism took place in a major extensional basin developed actually on the Proterozoic continental crust. Tremadoc activity was subduction-related and calc-alkaline, but was rapidly succeeded by tholeiitic magmas emplaced in an environment of back-arc extension, which were then joined by the acid products of crustal fusion to give the characteristic bimodal association which culminated in the Caradocian activity of Snowdonia.

Silurian to Devonian

The Silurian period was one of limited and sporadic volcanism when the major phase of subduction-related eruptive activity seems to have died down. South of the Suture there is a Llandovery suite of alkali basalts and rhyolites at Skomer, S. Wales; further east Llandovery basalts are known from Tortworh and andesites from the Mendips. In the far west of Ireland on the Dingle peninsula and Inishvikillaun island Wenlock rhyolites and tholeiitic andesites occur. North of the Suture Llandovery to Wenlock acid pyroclastics in Mayo and Clare Island are regarded as crustal remelts associated with development of the margin of the South Mayo Trough. In the east Llandovery bentonites in the Birkhill Shales of the Southern Uplands and their correlatives in the Ards Peninsula of N. Ireland represent widespread but volumetrically minor pyroclastic eruption.

The Early Devonian provides an entirely new scene. Large volumes of calc-alkaline magma are emplaced in and on the crust, of which the "Newer Granites" are a major component. Extrusive activity provided the extensive andesitic piles of the Larne Plateau and Ben Nevis, and the volcanics related to the Midland Valley, in the Ochils, Sidlaws, Pentlands, and Clyde Valley, and in Ireland the Curlew Mountains and at Cushendall. Spatial variation in the chemistry (a progressive increase in Sr, Ba, K, light REE, P_2O_5, and in La/Y ratios) along a NW traverse from the Midland Valley to the Lorne Plateau is consistent with eruption above a NW subduction zone which would coincide with the current model for the continental margin between Scotland and Scandinavia, however it does provide a number of paradoxes:

1) The Iapetus is believed to have closed by this time; the volcanism post-dates the deformation of the Cambro-Silurian successions and interfingers with the intermontane continental molasse sediments being locally stripped off.
2) The Cheviot lavas (like the granites of the Lake District and beneath the Pennines) do not fit chemically into the pattern of the volcanics further north. They may be slightly younger and the mechanism for their production is unclear.

The Middle Devonian is absent, with a major unconformity between Lower and Upper Devonian in N. England and Scotland.

The Upper Devonian to Permian provides a number of separate scenaries. In the Shetlands there are Upper Devonian alkali basalts. In southern Scotland and north to central England there are within-plate continental rift-type alkaline lavas, mainly basalts. In Ireland there are two distinct suites; Upper Old Red Sandstone acid tuffs, felsites and dolerite intrusions form the Lough Guitane volcanics south of Killarney erupted into a fluviatile continental environment. In the Lower Carboniferous epicontinental seas further north, the Limerick Volcanics are erupted as submarine piles which sometimes built up to subaerial vents. These were dominantly basaltic eruptions of lavas and tuffs. Occasional isolated island volcanos are known elsewhere in the Irish midlands. In SW England Late Devonian and Early Carboniferous pillow lavas appear chemically to suggest a continental intra-plate environment; both tholeiitic and alkali olivine basalts are represented, they are preceded by an early trough formation phase via continental rifting, which can be discerned in earlier Lower Devonian basalts, but this does not appear to have developed to the stage of an opening ocean. No further volcanic activity is seen until after the emplacement of the Cornubian granite batholith, with some associated calc-alkali acid lavas. There followed the highly potassic graben-related basalts and lamprophyres of the Exeter Volcanic Series, in the early Permian.

PLUTONIC ROCKS IN IRELAND (P. S. Kennan)

Large granitic intrusions dominate any map of Irish plutonic rocks. The great majority have one major facet in common, they post-date the Iapetus closure. Their origin as calc-alkaline intrusions of considerable volume is thus not easy to relate in a simple way to common plate-subduction models. Any proposed genetic model must take account of basement rocks that are different on either side of the suture. To the north, the basement is probably very old, Lewisian in part, and granulitic. To the south the oldest basement is probably late Precambrian.

The Donegal granites are the best known. They are, in common with the other intrusions "Newer" (c. 410 Ma.) granites. They are remarkable in the range of intrusion mechanisms displayed, i.e. replacement (Thorr), diapiric with appinite satellites (Ardara), cauldron subsidence (Rosses) and wedging in a shear zone (Main Donegal). The Connemara intrusions show a similar range of emplacement styles but differ in an extended history of emplacement. The earliest elements in the Connemara magmatites are thought to be derivatives from the major basic/ultrabasic intrusions (c. 510 Ma), the last are probably Devonian. The Ox Mountain Granodiorite (486 Ma) is the oldest granitic intrusion of the Caledonian cycle in Ireland; it just post-dates the peak of the regional metamorphism and its structural setting is a matter of debate. The Longford Down Massif intrusions bear comparison with those of the Southern Uplands, where recent isotopic and petrographic studies have suggested the involvement of a) the partial melting of a variety of source rocks, and b) differentiation during ascent. The same might be said of other Irish granites and certainly of the Leinster Granite - an amalgam of separate zoned plutons emplaced with a suite of appinites into a low-grade envelope of slates and volcanic rocks. These plutons of megacrystic granite and their envelope rocks represent an obvious continuation of the Gander Zone of Newfoundland.

Isotopic studies indicate that the Irish granites originated in the upper mantle or lower juvenile crust. Some at least are certainly S-type granites derived, possibly, from not much older volcanogenic sediments.

The basic/ultrabasic intrusions of Connemara seem to be the dismembered parts of a major intrusion of indeterminate shape. It is tempting to speculate that they are, with the basic rocks of Tyrone, ophiolitic. Sheeted dolerite dykes close to the Leinster Granite might, with associated volcanic rocks and cherts, be viewed similarly.

VOLCANISM AND PLUTONISM IN THE APPALACHIAN OROGEN

L. R. Fyffe[1], D. Rankin[2], W. B. Size[3], and D. R. Wones[3]

New Brunswick Geological Surveys Branch, PO Box 6000 Fredericton, N.B.[1], US Geological Survey, Mail Stop 926-A, National Centre, Reston, VA, USA[2], Emory Univ. Dept. of Geology, Atlanta, Georgia, USA[3], Dept. of Geological Sciences, Blacksburg, VA, USA[4]

INTRODUCTION

The following four papers are summaries of national reviews submitted to the NATO Advanced Study Institute Colloquium held on August 12-17, 1982 at Fredericton, New Brunswick.

VOLCANISM IN THE APPALACHIANS OF THE UNITED STATES (D. Rankin)

Volcanism in the Appalachians can be summarised in terms of its tectono-magmatic environment.

Proterozoic North America

A. Volcanicity associated with continental rifting related to the opening of the Iapetus Ocean.

1. Strongly bimodal volcanics preserved along the Blue-Green Long axis. In general tholeiitic basaltic lavas are found along both sides of allochthonous basement massifs and record a shore line, being subaerial to the NW and submarine, interlayered with shale and greywacke (now schist and gneiss) to the SE. Rhyolites are only seen at embayments, e.g. Mt. Rogers, South Mountain, Sutton Mountains. Best known examples are at Mt. Rogers and comprise lavas and welded tuffs which are peralkaline, (with riebeckite and fluorite; high Nb and REE). Models proposed for the volcanism refer to a gross Appalachian structure related to Iapetus rifting with development of embayments and promentories; the

P. E. Schenk (ed.),
Regional Trends in the Geology of the Appalachian-Caledonian-Hercynian-Mauritanide Orogen, 173–185.

location of the volcanics being controlled by hot spots and triple junctions, according to Rankin, or to transform faults, according to Rodgers & Thomas. A discordant zircon age from 5 rhyolite samples gives 810 m.y. $^{39}Ar/^{40}Ar$ ages suggest that the Blue Ridge rocks are still deeply buried at this time. These rocks have traditionally been called Cambrian, but it is more likely that they are of latest Proterozoic age.

2. Thin continental basalt lavas at the base of the Cambrian, interlayered with alluvial sediments, on the NW flank of the Blue Ridge, near the Mt. Rogers Embayment.

3. Sparse submarine basalts in Taconic klippen, from displaced continental slope.

B. Basalts in ophiolites: N. Vermont - Belvedere Mountain
Baltimore, Maryland.

Exotic Terranes

Volcanism here is self-contained and not clearly related to collisions with North America. Several Terranes may be present, but currently most are grouped in "Avalonia".

A. Chain Lakes Terrane: (c. 1500 m.y. basement) in NW Maine, adjacent to Quebec. Volcanics comprise the Boyle Mountain ophiolite and associated basalts of the Jim Pond Formation (of possible Cambrian age).

B. Avalonia:

1. Ellsworth Terrane of SE Maine: highly deformed, laminated quartzofeldspathic phyllites and schists rich in chlorite. Marine volcaniclastic rocks of Cambro-Ordovician age.

2. Avalon platform of Rhode Island and SE Massachusetts; the Blackstone Group is an older complex of quartzite, marble, mafic volcanics (greenstone to amphibolite) and lesser felsic volcanics; comparable to the Greenhead, George River and Burin groups. It is intruded by granites of 620 to 600 my age, and is thought to be overlain nonconformably by the felsic volcanic rocks (602 $\pm$ 3 m.y.) that form the base of a section extending up to the Lower Cambrian Weymouth Formation.

3. Nashoba Terrane of adjacent but inland Eastern

Massachusetts; this is very similar to the Blackstone Group and includes a thick amphibolite unit, the Marlboro Formation, which is probably mafic volcanic in origin. Robinson interprets these as Middle Ordovician volcanics.

4. Maryland and Virginia: The James Run - Chopawomsuc volcanics are of Cambrian age. They are largely mafic but have some felsic members and the suite is again bimodal. Pavlides has suggested that they represent an island arc with a back-arc basin lying to the west and ocean to the east; the volcanics being erupted above a west-dipping subduction zone.

5. Carolina volcanic slate belt: a very large province of volcanic, volcaniclastic and epiclastic rocks which extend from Southern Virginia to Georgia. The rocks are largely fragmental, but there are some lava flows; compositionally they are mixed, with a relatively small percentage of intermediate composition, and a large volume of felsic types. They are interpreted as the products of an ensialic island arc, of late Proterozoic to Middle Cambrian age. An ediacara fauna is known from near Durham, N.C. and there are two localities with (?) Middle Cambrian Paradoxides. They probably extend from c. 650 to 550 m.y. in age, though some poorly documented zircon ages of 740 m.y. are known from the northern end of the belt. There is no known basement. Higher grade and more intensely intruded rocks of the Charlotte belt to the west are also considered to be part of the same volcanic province, as also (according to Wright Horton) are the volcanic rocks of the Kings Mountain belt.

6. Inner Piedmont: North Carolina, South Carolina, Georgia: it is uncertain as to whether this assemblage is North American or Exotic. It contains a mixture of quartzo-feldspathic gneisses, pelitic schists, amphibolites, derived from mafic volcanic rocks and magmatites. The age could be late Proterozoic; no fossils are known, but the sequence is definitely pre-Taconic.

Volcanism related to Taconic collisions

Three groups of volcanic rocks fall into this category, all in New England. None have been identifed south of this.

A. The volcanics in the Bronson Hill Anticlinorium; NW Maine to Long Island Sound. Despite the extremely complex geology, with amphibolite facies metamorphism, nappe structure, and mantled gneiss domes, these volcanics emplaced within and beneath black, sulphide-rich slates have been uncontroversially defined as the products of a Middle Ordovician island arc above an eastward-dipping subduction zone. According to Robinson and Shumaker, the main volcanic unit, the Ammonoosuc Volcanics, comprises two units: a Lower Unit of pillow basalts, andesites and low-potash dacites, very like the Tonga Arc volcanics, and an Upper Unit of rhyolite, which is now peraluminous, and presumably from a different source. The volcanic units thin both eastward and westward in Massachusetts, a distribution which further supports the idea that they belong to a discrete island arc.

B. The Hawley Ascott-Weedan Belt, of Connecticut-Massachusetts-Vermont-Quebec. This is also of Middle Ordovician age and comprises mafic volcanics. They may belong to a separate arc to the west of the Bronson Hill arc, but this interpretation is not certain.

C. The Ordovician volcanic rocks of Northern Maine. These have been studied by Hine who concludes that some at least are basalts and andesites of an island are.

Palaeozoic post-Taconic Volcanism

A. The Talladega Belt, of Alabama and Georgia. This consists largely of clastic sedimentary rocks of low metamorphic grade in a thrust slice overriding the Valley-and-Ridge belt and overridden by the Blue Ridge Terrane. The age is somewhere between Late Proterozoic and Lower Devonian. The volcanic component, the Hillabee Chlorite Schist, is structurally the highest unit, overlying the Lower Devonian Jemison Chert, though its stratigraphic position is uncertain. The rocks are believed to comprise mafic volcanics (quartz tholeiites) and quartz keratophyres. They are unique in the Appalachians.

B. The Tioga bentonite of Ohio, Virginia and Pennysylvania; of Middle Devonian age the source for this ash is unknown. With these two exceptions, there is no other Siluro-Devonian volcanism south of New England. In New England the Siluro-Devonian volcanism is concentrated along previous volcanic belts. Some Devonian volcanics are clearly related to Acadian compressional events; Silurian volcanics are not so clearly related but are found in the same general belts.

C. The Piscataquis Volcanic Belt of north-central Maine and related rocks found along strike, including the mafic Standing Pond Volcanics in Vermont. The Silurian volcanics are mainly mafic; by Early Devonian there is mixed volcanism in NE Maine, culminating in the rhyolite volcanism of the Piscataquis belt itself, of Siegenian age. These are peraluminous rhyolite lavas and welded tuffs, some with garnet phenocrysts. They have high Sc contents, and appear to be co-magmatic with gabbroic and granitic plutons. The whole assemblage may well represent an island arc.

D. The Coastal Volcanic Belt of Maine-Massachusetts. This is a Siluro-Devonian belt of fragmental volcanics and lavas of bimodal composition, interbedded with marine sediments.

GENERAL SUMMARY OF PLUTONIC ROCKS IN THE APPALACHIANS OF THE UNITED STATES - A GENERAL SUMMARY (D. R. Wones)

Distribution: Plutons occur along the full length of the orogen and across its width, with the exception of its western edge. The pattern of distribution is highly variable and does not coincide with the deformation.

Size, shape and spacing: The plutons are in general not large, few have a radius greater than 10 km and the largest is c. 2400 km^2 in area, yet they are numerous enough to occupy almost half of the area of the orogen. Except in New Hampshire, there are few coalescing plutons and in general there is a spacing of 10 to 30 km between them. They are generally ovoid to circular in shape, with the exception of Cambro-Ordovician plutons in the Blue Ridge and the Devonian in Bronson Hill, which are elongated parallel to the axis of the regional trend.

Distribution with time: Late Precambrian plutons are small in the west and large in the east. Cambrian plutons occur in the Slate Belt and in west North Carolina. Ordovician plutons are confined to linear belts in the Bronson Hills anticlinorium, the Blue Ridge and the Charlotte Belt. Silurian plutons are found in the Nashoba Belt, and Devonian plutons are widespread in New England and occur in a narrow zone in the Piedmont, Carboniferous plutons are widespread in the southern Appalachians.

Composition: They are dominated by granite; linear belts of tonalite and granodiorite are found in a) the Blue Ridge and its southern extension; b) the Peraluminous granites predominate; they usually precede calcic plutons and are taken to imply the melting of immature sediments. Gabbro bodies are also present in most times and places.

MAGMATIC ROCKS OF THE SOUTHERN APPALACHIANS OF THE UNITED STATES (W. B. Size)

Magmatic rocks of the southernmost exposed portion of the Appalachian Orogen can be divided into two principal groups. The most extensive consists of a series of granitic plutons extending along the Inner Piedmont of Alabama, Georgia and South Carolina. In Georgia these include the syn-tectonic Ben Hill, Palmetto and Austell granites and late- to post-tectonic Stone Mountain, Panola and Elberton granites. The least deformed granites in this group range in age from 320-355 m.y. while more tectonized granites are about 375 m.y., which may bracket a period of deformation in the Inner Piedmont.

A second group of magmatic rocks occurs north of the Brevard cata-clastic zone, along the Ashland-Wedowee belt and Blue Ridge. These intrusives have a compositional range from quartz-diorite to tonalite and trondhjemite. In the Northern Alabama Piedmont these rocks are restricted to one structural zone or suspect terrane - the Tallapoosa Block. The Elkahatchee Quartz Diorite is one of the oldest (490 ± 26 m.y.) and largest (ca. 1000 km^2) plutons within this zone. Further northeastward along this zone there are over 20 younger (360 - 324 m.y.) and less deformed intrusions collectively called the Almond trondhjemites. Along this zone in Georgia and North Carolina trondhjemite plutons (Villa Rica and Whiteside) and more highly tectonized. Initial $^{87}Sr/^{86}Sr$ ratios for the older, more deformed granitic rocks is relatively low (0.7024-0.7052) and higher for some of the younger granites (ca 0.7250) indicating a change in involvement of continental crustal rocks towards magma generation in this region.

UPPER PALEOZOIC PLUTONS ROCKS OF THE CANADIAN APPALACHIANS (L. R. Fyffe)

Introduction

Upper Paleozoic (Devonian to Carboniferous) plutonic rocks underlie an area of some 400,000 km^2 in the Atlantic Provinces of Canada (New Brunswick, Nova Scotia, Prince Edward Island, and Newfoundland). Emplacement of these plutons, which are the most voluminous in the Appalachians, took place during the waning stages of, and after cessation of Late Paleozoic deformation. The bulk of the plutons are granitic in composition in marked contrast to the abundance of tonalite and granodiorite in the Mesozoic of western North America.

Descriptions of Plutons

Upper Paleozoic plutonic rocks in New Brunswick occur in two belts underlying an area of 5,700 km^2. The Southern Belt is located on the northern flank of an east-northeast-trending belt of Siluro-Devonian volcanics that separates Precambrian rocks and their platformal cover from a eugeoclinal Cambro-Ordovician sequence to the north. The Axial Belt (1) crosses the centre of the province along the southeast flank of a northeast-trending belt of Siluro-Devonian volcanics that separates the Cambro-Ordovician eugeoclinal sequence from Upper Ordovician limestone to the northwest.

Early Devonian mafic intrusions occur in both the Southern and Axial belts (2) but occupy a relatively small area of only 500 km^2. Elongate mafic complexes composed of gabbro, norite, troctolite and peridotite were emplaced along major north-northeast- and north-northwest-trending faults within and along the boundaries of the Siluro-Devonian volcanic belts. Mineral layering, where present, is generally steep suggesting tilting of the plutons after emplacement (3). Small equant subvolcanic stocks of gabbro intrude the volcanics along the northwestern flank of the Axial Belt.

Bodies of hornblende-bearing granodiorite and monzogranite, forming a total area of 400 km^2, are associated with the mafic complexes or occur as stocks containing numerous dykes of diabase suggesting the presence of mafic rocks which they intrude possess a cataclastic foliation related to movement on faults that controlled their emplacement.

Rb-Sr whole rock ages of 403 $\pm$ 20 Ma (4, 5), 409 $\pm$ 20 Ma (6), and 370 $\pm$ 30 Ma (7) have been obtained from the felsic members of mafic complexes in southern, central, and northern New Brunswick, respectively.

Over half of the total area of Upper Paleozoic plutonic rocks in New Brunswick comprises granite of Early to Middle Devonian age. The western portion of the only batholith in the Southern Belt is composed of equigranular biotite granite with a Rb-Sr age of 406 $\pm$ 7 Ma (5). The granite is apparently intruded by an oval pluton of megacrystic biotite monzogranite containing feldspar phenocrysts between 4 and 5 cm long (8). Mildly peralkaline syenogranite, forming an apophysis off the eastern end of the batholith, has been dated at 383 $\pm$ 6 Ma (5).

A deeper level stock of biotite-muscovite granite to the northwest of the batholith intrudes Cambro-Ordovician sedimentary rocks containing staurolite, garnet, and andalusite up to 500 m from the contact (8).

Megacrystic to equigranular biotite monzogranite and associated muscovite-biotite granite intrude sillimanite-bearing Cambro-Ordovician sedimentary rocks in the Axial Belt (1, 2, 9). These granites occur in several composite plutons ranging in area from 100 to 1,000 km^2. A whole rock Rb-Sr age of 378 $\pm$ 7 Ma was obtained on one of the megacrystic plutons (10) and a whole rock-mineral age on muscovite-biotite granite yielded an age of 424 $\pm$ 24 Ma (11).

Upper Devonian to Lower Carboniferous granite underlies a total area of about 900 km^2 in New Brunswick. Three high level oval stocks of the Axial Belt intrude Cambro-Ordovician rocks in the central part of the province. These stocks comprise coarse-grained, equigranular, leucocratic, biotite and muscovite-biotite granite grading to fine-grained quartz-feldspar porphyry. The largest stock with an area of 150 km^2 gives an Rb-Sr age of 351 $\pm$ 8 Ma (12). Similar granite forms the eastern portion of the batholith in the Southern Belt where it has been dated at 345 $\pm$ 8 Ma (5). A small stock of quartz-feldspar northwestern margin of the pluton. The mafic rock is a quartz monzodiorite containing phenocrysts of plagioclase, hornblende, and minor pyroxene. An Rb-Sr isochron age of 384 $\pm$ 10 Ma was obtained on the stock (23).

Several mafic plutons intrude Ordovician to Devonian volcanic and sedimentary rocks of the Central Volcanic Belt of Newfoundland. These typically range from 50 to 200 km^2 in size and have narrow contact aureoles indicative of high level emplacement. The largest mafic complex with an area of 1,300 km^2 is composed dominantly of gabbro intruded by minor diorite and granodiorite, and a relatively large amount of biotite granite (24). Ar-Ar dates of 420 $\pm$ 8 Ma and 390 $\pm$ 15 Ma have been obtained on the gabbro and granite, respectively (25). Ultramafic rocks are associated with some of the mafic complexes (26) whereas others are composed entirely of hornblende-biotite granodiorite and tonalite (27).

Numerous plutons of megacrystic biotite granite and equigranular muscovite-biotite granite, ranging in size from 100 to 5,000 km^2, occur within Cambro-Ordovician eugeoclinal sedimentary rocks exposed between the Precambrian rocks of the Avalon Peninsula and the Central Volcanic Belt (27).

The megacrystic granite contains microcline crystals from 2 to 8 cm long. A mafic dyke swarm separates the time of emplacement of younger, massive, megacrystic plutons from older cataclastic ones (28). Rb-Sr ages on the megacrystic granite range from 437 ± 20 Ma to 359 ± 5 Ma (25).

The two-mica granites range in composition from granodiorite to leucocratic garnetiferous monzogranite and are commonly surrounded by a zone of sillimanite-bearing migmatite (29). Some are older and others are younger than the deformed megacrystic plutons (28).

Two episodes of Upper Paleozoic peralkaline plutonism are evident in Newfoundland. The earlier episode is represented by a 6,000 km^2 batholith that intrudes Ordovician and Siluro-Devonian volcanics on the western margin of the Central Volcanic Belt. Roof pendants of extrusive equivalents indicate a high level of emplacement. Hypersolvus granite containing quartz, perthite, and alkali ferromagnesium minerals, is associated with a marginal zone of biotite and hastingsite biotite granite in the northeastern portion of the batholith. Peralkaline dykes locally cut the hastingsite-biotite granite. Diabase dykes and mafic globules within the granites are indicative of contemperaneous basaltic magmatism (30). The peralkaline and non-peralkaline phases respectively yielded Rb-Sr ages of 436 ± 5 Ma and 402 ± 9 Ma (25).

A peralkaline granite pluton with an area of about 200 km^2 intrudes porphyry just north of this batholith yielded an Rb-Sr age of 337 ± 15 Ma (5, 8).

Nova Scotia is divided into two tectonostratigraphic zones: northern mainland Nova Scotia and Cape Breton Island underlain largely by Precambrian rocks; and southern Nova Scotia underlain largely by a Cambro-Ordovician eugeoclinal sedimentary succession.

Southwestern Nova Scotia is underlain by a 7,000 km^2 arcuate batholith surrounded by Cambro-Ordovician sediments containing andalusite and cordierite within 1.5 km of the contact. Locally, the batholith intrudes stratified rocks as young as Early Devonian, and is overlain by Carboniferous sandstone.

The batholith is composed dominantly of massive biotite granodiorite varying to monzogranite, and containing megacrysts of perthite. The megacrystic phase is intruded by several stocks of equigranular, muscovite-biotite monzogranite constituting about 25 per cent of the batholith (13, 14). Magmatic andalusite and garnet are accessory minerals in the two-mica granites (15, 16). The megacrystic phase has been dated at 372 ± 2 Ma and the two-mica phase, at 364 ± 1 Ma (17).

A 400 km^2 composite pluton located along the coast to the south of the main batholith intrudes Cambro-Ordovician sedimentary rocks containing a regional metamorphic assemblage of andalusite, cordierite, and staurolite. Biotite tonalite forms the western part of the pluton, and muscovite-biotite tonalite veined by garnetiferous aplite and pegmatite forms the eastern portion (18). Small bodies of norite and hornblende-biotite diorite occur within the tonalite (19). Ar-Ar dates on the tonalite and diorite, and on a smaller pluton of muscovite-biotite granodiorite and granite found farther east, range from 300-350 Ma (20).

Upper Paleozoic plutonic rocks are not abundant on Cape Breton Island. A lobate pluton with an area of about 150 km^2 intrudes Precambrian gneissic and volcanic rocks in the northern part of the island. Pelitic sedimentary rocks interbedded with volcanics in the contact aureole contain staurolite and garnet (21). The pluton comprises muscovite-biotite monzogranite varying to granodiorite and is intruded by dykes of garnetiferous pegmatite and aplite. The granite possesses a weak fabric defined by lensoid quartz and alignment of biotite parallel to the northern trend of the pluton. A Rb-Sr isochron yielded an age of 403 $\pm$ 22 Ma for the pluton (22).

An oval composite stock with an area of about 3 km^2 intrudes platformal Cambrian sedimentary rocks in southern Cape Breton. The bulk of the stock comprises fine-grained porphyritic, hornblende-biotite monzogranite containing phenocrysts of plagioclase and alkali feldspar. The granite intrudes a fine-grained mafic body along the Precambrian volcanic and Cambrian sedimentary rocks in the southwestern part of the Avalon Peninsula (31). Its Rb-Sr age of 328 $\pm$ 5 Ma (25) shows that is represents a Carboniferous period of peralkaline igneous activity.

Summary and Conclusions

Upper Paleozoic plutons of the Maritime and Newfoundland Appalachians are dominated by peraluminous megacrystic biotite, and equigranular muscovite-biotite granites emplaced within thick eugeoclinal sedimentary sequences of Cambro-Ordovician age. Many of these plutons are surrounded by migmatite zones indicating that the melt was generated in situ, whereas others, unsaturated with water, were generated at depth and rose to their present position. High-level batholiths of biotite granite emplaced across the northern boundary between these eugeoclinal rocks and Precambrian volcanics of the Avalon Peninsula and correlative Precambrian rocks in southern New Brunswick are only mildly peraluminous (32, 33) and may have been derived from even greater depth.

Mantle-derived rocks are common only in the Central Volcanic Belt of Newfoundland and on the northwestern flank of the Axial Belt in New Brunswick. The small volume of associated granite reflects the absence of a thick continental basement in these areas.

Plutonic activity was more-or-less continuous from the earliest Devonian to Carboniferous. However, periods of voluminous peraluminous activity were separated by intervals of extension, as evidenced by emplacement of mafic dyke swarms and plutons of peralkaline composition.

Granitic plutonism which occurred in the northern Appalachians some 50 Ma after ophiolite obduction and closure of the Lower Paleozoic ocean cannot be attributed to the same subduction processes that produced granodioritic and tonalitic magmas on the western margin of North America during the Mesozoic. Instead, generation of granitic melt in the northern Appalachians may be related to renewed continental convergence and attendant crustal shortening following detachment of continental crust from underlying lithosphere in the Late Paleozoic (34).

REFERENCES

1. Martin, F. R.: 1970, Petrogenetic and tectonic implications of two contrasting Devonian batholithic associations in New Brunswick, Canada. Am. J. Sci. 268, pp. 309-321.
2. Fyffe, L. R., Pajari, G. E., and Cherry, M. E.: 1981, The Acadian plutonic rocks of New Brunswick. Maritime Sediments and Atlantic Geology 17, pp. 23-36.
3. Dunham, K. C.: 1950, Petrography of the nickeliferous norite of St. Stephen, New Brunswick. Am. Mineral. 35, pp. 711-727.
4. Pajari, G. E., Trembath, L. T., Cormier, R. F., and Fyffe, L. R.: 1974, The age of the Acadian deformation in southwestern New Brunswick. Can. J. Earth Sci. 11, pp. 1309-1313.
5. Fyffe, L. R., Pajari, G. E., and Cormier, R. F.: 1981, Rb-Sr chronology of New Brunswick. Geol. Soc. Am. Abstracts, Programs 13, p. 133.
6. Fyffe, L. R., and Cormier, R. F.: 1979, The significance of radiometric ages from the Gulquac Lake area of New Brunswick. Can. J. Earth Sci. 16, pp. 2046-2052.
7. Stewart, R. D.: 1979, The geology of the Benjamin River intrusive complex. Unpublished M.Sc. thesis, Carleton University, Ottawa, Ontario, Canada.
8. Pajari, G. E.: 1976, Field guide to the geology and plutonic rocks of southwestern New Brunswick and the Penobscot Bay

area of Maine. Canadian Plutonics Study Group, Project Caledonide Orogen, I.G.C.P., Department of Geology, University of New Brunswick, Fredericton, New Brunswick, Canada, 69 p.

9. McCutcheon, S. R., Lutes, G. G., Gauthier, G., and Brooks, C.: 1981, The Pokiok batholith: a contaminated Acadian intrusion with an anomalous Rb-Sr age. Can. J. Earth Sci. 18, pp. 910-918.
10. Fyffe, L. R.: 1982, Geology in the vicinity of the 1982 Miramichi earthquake, Northumberland County, New Brunswick. New Brunswick Department of Natural Resources, Open File Report 82-27, 37 p.
11. Poole, W. H.: 1980, Rb-Sr ages of the "sugar" granite and Lost Lake Granite, Miramichi Anticlinorium, Hayesville map-area, New Brunswick. Geol. Surv. Can. Paper 80-1C, pp. 174-180.
12. Crouse, G. W.: 1981, Geology of parts of Burnthill, Clearwater and McKiel Brooks. New Brunswick Department of Natural Resources, Map Report 81-5, 46 p.
13. Smith, T. E.: 1974, The geochemistry of the granitic rocks of Halifax County, Nova Scotia. Can. J. Earth Sci. 11, pp. 650-657.
14. McKenzie, C. B., and Clarke, D. B.: 1975, Petrology of the South Mountain batholith, Nova Scotia. Can. J. Earth Sci. 12, pp. 1209-1218.
15. Clarke, D. B., McKenzie, C. B., and Muecke, G. K.: 1976, Magmatic andalusite from the South Mountain batholith, Nova Scotia. Contrib. Mineral. Petrol. 56, pp. 279-287.
16. Allan, B. D., and Clarke, D. B: 1981, Occurrence and origin of garnets in the South Mountain batholith, Nova Scotia. Can. Mineral. 19, pp. 19-24.
17. Clarke, D. B., and Halliday, A. N.: 1980, Strontium isotope geology of the South Mountain batholith, Nova Scotia. Geochimica et Cosmochimica Acta, 44, pp. 1045-1058.
18. de Albuquerque, C. A. R.: 1977, Geochemistry of the tonalitic and granitic rocks of the Nova Scotia southern plutons. Geochmica et Cosmochimica Acta, 41, pp. 1-13.
19. de Alburquerque, C. A. R.: 1979, Origin of the plutonic mafic rocks of southern Nova Scoita. Geol. Soc. Am. Bull. 90, pp. 719-731.
20. Reynolds, P. H., Zentilli, M., and Muecke, G. K.: 1981, K-Ar and ^{40}Ar-^{39}Ar geochronology of granitoid rocks from southern Nova Scotia: Its bearing on the geological evolution of the Meguma Zone of the Appalachians. Can. J. Earth Sci. 18, pp. 386-394.
21. Wiebe, R. A.: 1975, Origin and emplacement of Acadian granitic rocks, northern Cape Breton Island. Can. J. Earth Sci. 12, pp. 252-262.

22. Barr, S. M., O'Reilly, G. A., and O'Beirne, A. M.: 1979, Geochemistry of granitoid plutons of Cape Breton Island. Nova Scotia Department of Mines and Energy, Report 79-1, pp. 109-141.
23. Barr, S. M., and O'Beirne, A. M.: 1981, Petrology of the Gillis Mountain pluton, Cape Breton Island, Nova Scotia. Can. J. Earth Sci. 18, pp. 395-404.
24. Strong, D. F.: 1977, The Mount Peyton batholith, central Newfoundland: a bimodal calc-alkaline suite. J. Petrol. 20, pp. 119-138.
25. Strong, D. F.: 1980, Granitoid rocks and associated mineral deposits of eastern Canada and western Europe: In D. Strangway, ed., The Continental Crust and its Mineral Deposits. Geol. Assoc. Can. Special Paper 22, pp. 741-769.
26. Cawthorn, R. G.: 1978, The petrology of the Tilting Harbour igneous complex, Fogo Island, Newfoundland. Can. J. Earth Sci. 15, pp. 526-539.
27. Strong, D. F., and Dickson, W. L.: 1978, Geochemistry of Paleozoic granitoid plutons from contrasting tectonic zones of northeast Newfoundland. Can. J. Earth Sci. 15, pp. 145-156.
28. Jayasinghe, N. R., and Berger, A. R.: 1976, On the plutonic evolution of the Wesleyville area, Bonavista Bay, Newfoundland. Can. J. Earth Sci. 13, pp. 1560-1570.
29. Currie, K. L., and Pajari, G. E.: 1981, Anatectic peraluminous granites from the Carmanville area, northeastern Newfoundland. Can. Mineral. 19, pp. 147-161.
30. Taylor, R. P., Strong, D. F., and Kean, B. F.: 1980, The Topsails igneous complex: Silurian-Devonian peralkaline magmatism in western Newfoundland. Can. J. Earth Sci. 17, pp. 425-439.
31. Teng, H. C., and Strong, D. F.: 1976, Geology and geochemistry of the St. Lawrence peralkaline granite and associated fluorite deposits, southeast Newfoundland. Can. J. Earth. Sci. 13, pp. 1374-1385.
32. Whalen, J. B.: 1980, Geology and geochemistry of the molybdenite showings of the Ackley City batholith, southeast Newfoundland. Can. J. Earth Sci. 17, pp. 1246-1258.
33. Cherry, M. E.: 1976, The petrogenesis of granites in the St. George batholith, southwestern New Brunswick, Canada. Unpub. Ph.D. thesis, University of New Brunswick, Fredericton, New Brunswick, 260 p.
34. Coman-Sadd, S. P.: 1982, Two stage continental collision and plate driving forces. Tectonophysics 90, pp. 263-282.

MAIN FEATURES OF VOLCANISM AND PLUTONISM IN LATE PROTEROZOIC AND DINANTIAN TIMES IN FRANCE

B. Cabanis

Universite Paris
Laboratoire de Geochemie Comperee et Systematique
75230 Paris, France

INTRODUCTION

This general synthesis is made with the collaboration of A. Autran, P. L. Guillot, and D. Santallier. It is based on a summary of volcanism in France prepared for the International Geologic Congress in Paris (1980) as noted in the bibliography (1, 2, 3, 4, 5, 6). The present synthesis includes some new and in part unpublished data on trace elements.

VOLCANISM

Introduction

The tectonic setting of recent volcanic provinces is characterized by distinctive patterns of major and trace chemical elements. For ancient volcanic rocks, major chemical elements have been used to reconstruct their geotectonic environment. However, these rocks generally are of both metamorphic and spilitic character. Some chemical elements especially the major ones, become mobile so that chemical data must be carefully interpreted. Recent studies based on more immobile trace elements (Cr, Co, Ni, Y, Zr) and rare earths have resulted in important volcanic environments (7, 8, 9, 10). At the present time, trace elements of ancient volcanics in the Armorican Massif and Central Massif are being analyzed by neutron activation (11, 12). Hygromagmaphile elements (U, Th, Ta, La) appear to be the most characteristic of tectonic settings of recent volcanics (13, 14) and to be quite stable in spilitized or metamorphosed volcanic successions (11, 12, 15). This method offers improvement over previous work and is recommended for study of metamorphosed and spilitized ancient

P. E. Schenk (ed.),
Regional Trends in the Geology of the Appalachian-Caledonian-Hercynian-Mauritanide Orogen, 187–192.

volcanic rocks.

Late Proterozoic Volcanics

In France, Precambrian volcanic rocks are found only in the Armorican Massif. Volcanism occurred before intrusion of aluminous granitic batholiths belonging to Normandy's Cadomian block. The time of extrusion was between 600 and 525 MA (usually the different Precambrian phases of volcanism are poorly dated). The volcanics lie along a northeast to southwest trend, that is from the extremity of Brittany (Baie de Douarnenez) to Cotentin including Tregor (Baie de Lannion-Paimpol, Baie de Saint Brieuc and Channel Islands). Only the Douarnenez and Paimpols series were dated around 640 MA. All of these rocks include predominantly basic units with pillow-lavas associated with minor acidic units (tuffs and keratophyres). At present, the chemical characters of these series, based upon major and trace elements, show a well-defined calc-alkaline character (Douarnenez, Paimpol, Lannion). The Bay of Saint Brieuc volcanics are interpreted as arc tholeites (South of Massif Armorican). Lastly, in Massif Central, some so-called "leptyno-amphibolite" group (acid and basic, metamorphic volcanites) which are poorly dated, may be of Precambrian or early Paleozoic age. Vergonzac and Gartempe's series show a calc-alkaline trend and perhaps belong to the Precambrian unit.

North of the Armorican Massif, calc-alkaline volcanics may be related to a southeast dipping, subduction zone located in the Channel Sea at the northern edge of the Icartian basement (7, 16).

Cambrian to Early Devonian Volcanics

Mainly acid volcanism took place during the Cambrian between 530 and 500 MA (rhyolites, ignimbrites sometimes associated with andesite. This volcanism is most probably situated in a platform environment corresponding to a Cadomian basement. These extrusives occur in the northern part of North Brittany, Vendee du Sud, and Normandy. They may be due to a later orogenic event during which crustal melting occurred along a zone of intraplate distension situated around the Cadomian range and lasting for 100 MA until the Early Ordovician. At the same time, tholeitic volcanism characteristic of a rift zone occurred in Maures, Limousin (leptyno-amphibolite groups of Haut Allier, Lyonnais, Rouergue, Marjevols) and Pyrennees-Montagne Noire where the Cadomian basement is not well-known.

During the Ordovician and Early Devonian (500-400 MA) extrusion of basic (pillow lavas, sills, pyroclastics) and minor acidic units suggest crustal rifting of a continental area. These rocks include the Crozon and Erquy Series in the northern part of Armorican Massif (Ordovician age); (2) the St. Georges s/Loire series (Siluro-Devonian age) in the southern part of Armorican Massif; (3) the Albigeois Series in the southern part of the Central Massif; and (4) Limousin Series in western part of Central Massif. All of these series show both tholeitic and alkaline characters. Some volcanic rocks with tholeitic and alkaline suites are known in the Central Armorican Domain (Bolazec) and are considered as due to back-arc rift or intraplate volcanism. For the same period many arguments (9) are in favour of a convergence zone dipping northeast and situated in Southern Brittany (high pressure metamorphism of Groix Island) and Central Massif (eclogite of Bas Limousin and Marjevols). However, where are the calc-alkaline volcanics which relate to this Devonian subduction zone? Until now, none have been observed in the Armorican Massif except, possibly, for a small part of the St. Georges/Loire and Chataigneraie's series. In the western Central Massif, tholeitic extrusives of Genis (Limousin) located on the Cordillera suggest rifting perhaps related to convergence; however, they may be allochthonous.

Devonian to Dinantian Volcanics

There are two volcanic provinces (4). The first includes the southern part of the Armorican Massif, the eastern part of the Central Massif, the Vosges, and the Alpes areas. Spilites and keratophyres characterize this time period. The major chemical elements indicate an environment similar to the present-day circum-Pacific. This activity could be due to thermal and chemical readjustments of the upper mantle after Siluro-Devonian subduction.

The second province includes the northern Armorican Massif, the southwestern part of Great Britain, the Schisto-Rhenan Massif, and the Harz and Thuringe provinces. Here the spilites and keratophyres have mixed alkaline and tholeitic characteristics. This volcanism may be due to intraplate rifting similar to the first magmatic event of Early Paleozoic age in the same areas.

PLUTONISM

Late Proterozoic and Cambrian Plutons

These were intruded continuously between 640 and 520 MA. Thus they probably mark the Cadomian Orogeny. The plutons consist of gabbro and lesser amounts of tonalite and granodiorite. They too lie along a southwest to northeast trend

(Tregor, British Islands, Cotentin). The Mancellian batholith is composed of aluminous granodiorites and monzogranites dated between 540-570 My; the small initial ratio of 87 Sr/86 Sr suggests a deep origin at the bottom of the continental crust. In this region, called the Dommonean-Mancellian, the plutonic rocks did not undergo Acadian or Hercynian metamorphism.

An intense episode of plutonism occurred at the beginning of Cambrian time in the area of Vendee, in the Central Massif, and in the southern part of the Armorican Massif. It is represented by various orthogneisses characterised by calc-alkaline and alkaline affinities (Moelan-Lanvaux pro parte, nappe de Champtoceaux, Arc du Thaurion pro parte, Tulle antiform, Montagne Noire). In Arc du Thaurion, dioritic gneiss is associated with eclogitised gabbro of the same age.

In the southern part of France, in Pyrenees and Maures, some orthogneiss are dated between (540-580 My) and show a calc-alkaline character.

Cambrian to Early Devonian Plutons

Calc-alkaline plutonism with tholeiitic and alkaline admixtures continue into the Ordovician. This is interpreted as due to progressive rifting, behind the Caledonide chain, of a continental crust.

In Brittany, lens-shaped alkaline-potash granites are oriented along an east-west trend. They are dated at 470-430 MA (Moelan-Lanvaux, Cholet-Thouars, Berry). In addition, granodioritic rocks with trondjemitic affinities and dated at 470 MA (Douarnenez, Brest) are interpreted as due to melting of continental crust in a thermal zone.

In the Central Massif, granitic rocks with similar calc-alkaline tendencies were intruded during the Ordovician (Arc du Thaurion, St. Yriex, Tulle, Saut du Saumon). During the Silurian, only alkaline plutons were emplaced (Aubazine). Particularly the Albussac gneisses, which shows a trondjemitic affinity similar to the Ordovician orthogneiss, may be interpreted in the same manner.

Devonian to Early Visean Plutons

At the end of the Devonian orogeny (late anatexis is dated at approximately 375-380 MA) numerous granitic bodies were intruded in the orogenic terrain. They include tonalites, biotite granites (Gueret), leucogranites, and potash granites.

These undeformed plutons are dated between 360-340 MA, and belong to early Hercynean phases. In northern Brittany, granitic plutons with a potash character and dated at 385 MA were converted to orthogneisses during Late Devonian to Early Visean time (Brittani phase).

REFERENCES

1. A. AUTRAN (1980) Evolution structurale du Proterozoique aux distensions post-hercyniennes. In colloque C7, Geologie de la France (pp. 10-18) 26e International Geological Congress, PARIS.
2. A. AUTRAN et J. COGNE (1980) La zone interne de l'orogene varisque dans l'ouest de la France et sa place dans le developpement de la Chaine hercynienne. In Colloque C6, Geologie de l'Europe (pp 90 à III) 26e IGC PARIS.
3. BEBIEN J. et GAGNY CL. Volcanites du Precambrien au Cretace et leur signification geostructurale. In Colloque C7, Geologie de la France (pp 99-135) 26e IGC PARIS.
4. BEBIEN J., GAGNY Cl. et ROCCI G. (1980) La place du volcanisme devono-dinantien dans l'evolution magmatique et structurale de l'Europe moyenne varisque au Paleozoique. In Colloque C6, Geologie de l'Europe (pp 213-225) 26e IGC PARIS.
5. BOYER Cl. (1979) Bilan et synthese des paleovolcanismes ante Devonien en France. Bull. Soc. Geol. France (7), t. XXI, no. 6, (pp 695-708).
6. J. LAMEYRE and A. AUTRAN (1980) Les granitoides en France. In Colloque C7, Geologie de la France (pp 51-97) 26e IGC PARIS.
7. AUVRAY B. (1979) Genese et evolution de la croute continentale dans le Nord du Massif Armorican. These, Rennes (680 p).
8. Piboule M (1979) L'origine des amphibolites approche geochmique et mathematique. Application aux amphibolites du Massif Central Frangais. These, Universite de Lyon.
9. CARPENTER M. S. N., PEUCAT J. J., PIVETTE B. (1982) Geochemical and geochronological characteristics of paleozoic volcanism in the St Georges-sur-Loire synclinorium, S. Armorican Massif, France. / Evidence for pre-hercynian tectonic evolution. Bull. BRGM (2e serie) Section I, no. 1-2, 1982.
10. MAILLET P. (1977) Etude geochimique de quelques series spilitiques du Massif Armorican. These, Universite de Rennes.
11. B. CABANIS, CHANTRAINE J., HERROUIN Y., TREUIL M (1982) Etude geochimique (majeure et traces) des spilites et dolesites de Bolazec. / Mise en evidence d'un domaine en distension crustale au devonien inferieur en Bretagne Centre-Ouest. Bull. BRGM (2e serie) Section I, no. 1-2, 1982.

12. CABANIS B., GUILLOT P. L., SANTALLIER D., TREUIL, M., JAFFREZIC H., MEYER G. (1982) Apport des elements traces a l'etude geochimique des metabasites du Bas-Limousin. A paraitre in Bull. Soc. Geol. France.
13. JORON J. L., TREUIL M. (1977) Utilisation des proprietes des elements fortement hygromagmatophiles pour l'etude de la composition chimique et de l'heterogeneite du manteau. Bull. Soc. Geol. France (7), 19, no. 6, (pp 1197-1205).
14. TREUIL M., VARET J. (1973) Criteres volcanologiques, petrologiques et geochmiques de la genese et de la differenciation des magmas basaltiques : exemple de l'Afar. Bull. Soc. Geol. France (7), 15, no. 5-6, (pp. 506-540).
15. JORON, J. L., CABANIS B., TREUIL M (1982) in press Bull. Cent. Recherche Explor. Prod. Elf Aquitaine.
16. LEFORT J. P. (1974) Le socle peri-armoricain. Etude geologique et geophysique de socle submerge a l'ouest de la France.

REGIONAL OVERVIEW OF METAMORPHISM IN THE SCANDINAVIAN CALEDONIDES

I. Bryhni

Mineralogisk-Geologisk Museum, Sarsgt. 1, Oslo, Norway

ABSTRACT

Metamorphic variations in the Scandinavian Caledonides are related to the general constitution of the orogen as a sequence of nappes emplaced upon the Baltic foreland. In the central and eastern belts, the present pattern of regional metamorphism largely coincides with the nappe pattern, and multiply metamorphosed, high-grade units of Precambrian or Caledonian age are locally juxtaposed with low-grade units. The basement below the nappes has been increasingly involved in the Caledonian orogeny north-westwards, where inclusions of eclogite in the gneisses indicate pressures related to a doubling of the crust.

The metamorphism probably took place during various climactic events from the Late Precambrian to Upper Devonian time. The most important were related to subduction/ocean closure during the Ordovician to Cambrian and continental collision in the Late Silurian.

INTRODUCTION

The Scandinavian Caledonides is made up of a sequence of nappes which together with the underlying parautochthonous sediments and Precambrian basement became altered during the Caledonian orogeny. The age of the mineral parageneses is in many cases still unknown. Dating is incomplete, but original Precambrian elements are probably as common as the true Lower Paleozoic rocks within the orogenic zone. The extent of superposed Caledonian metamorphism on these older constituents of the orogene is, however, still controversial.

P. E. Schenk (ed.),
Regional Trends in the Geology of the Appalachian-Caledonian-Hercynian-Mauritanide Orogen, 193–204.

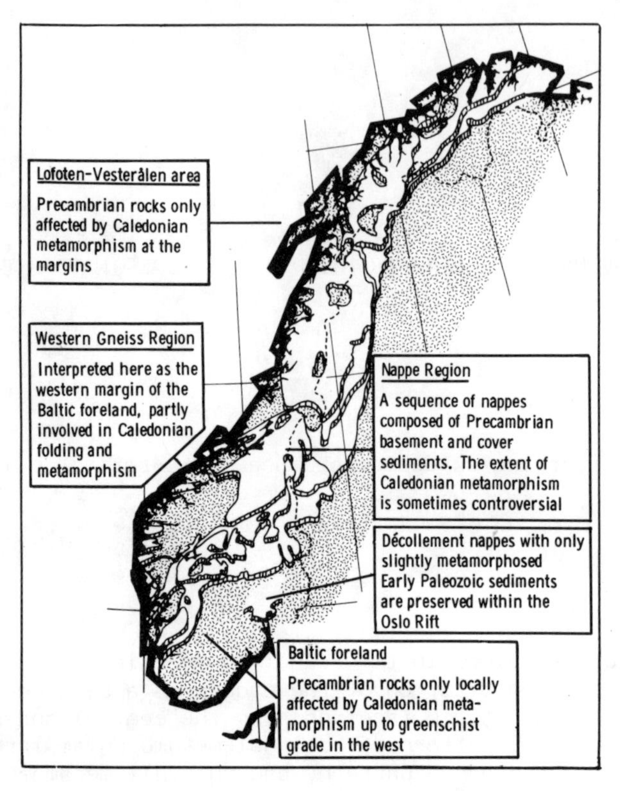

Fig. 1. Nappe model for the Scandinavian Caledonides.

The general features of the Scandinavian Caledonides (Fig. 1) is that of a Baltic foreland in the east with only weak, surficial metamorphism, a Nappe Region with rocks of Precambrian, Eocambrian and Lower Paleozoic age variously affected by Caledonian metamorphism, and the Western Gneiss Region with Precambrian rocks increasingly overprinted by Caledonian metamorphism northwestwards. Another large area of Precambrian rocks, the Lofoten-Vesterålen area, appear to be affected by superposed Caledonian metamorphism up to amphibolite facies only at the eastern margin below the nappes.

Much of the Caledonian metamorphism is transported in the sense that it took place closer to a suture now concealed off the present coast or during the emplacement of the nappes to their present position.

FACIES DISTRIBUTION

The grade of metamorphism in the Scandinavian Caledonides have been described in several recent reviews [1,2,3]. Facies boundaries (Fig. 2) are often, but not everywhere, parallel to thrusts and the tectono-stratigraphic sequences frequently show an upwards increasing grade. Dynamic metamorphism is widespread both at tectonic contacts and along other dislocations, where thick zones of cataclastic rocks can be found. Strain-induced metamorphism and heat from overriding crystalline nappes have in some cases lead to progressive metamorphism with the lower range of Barrowian type isogradeswhile other nappes are remarkably unaffected by Caledonian metamorphism.

Sub-greenschist facies

Sediments of the parautochthon or the decollement nappes along the eastern margin of the orogen have locally developed schistosity, partly recrystallized illite, high-reflecting carbonaceous matter or completely dehydrogenated kerogene and local recrystallization of clastic grains. The basement east of the Caledonian front is locally veined by prehnite or pumpellyite and may even have suffered brown to green biotite transformations [5,6]. Burial of higher nappes is the most likely cause of the very low grade metamorphism in the marginal zone.

Greenschist facies

The very low-grade sediments of the marginal decollement nappes of southeastern Norway appear to merge gradationally into stilpnomelane-, chlorite- and muscovite-bearing rocks with phyllitic appearance, but the contact to greenschist facies is elsewhere a thrust. Other zones of greenschist facies rocks occur in-between nappes of higher grade, probably Precambrian nappe units. Of particular interest are true greenschists and associated sediments in the highest tectono-stratigraphic levels of the Scandinavian Caledonides. Their low-grade metamorphism has probably been established in part already before obduction when the rocks still were in an oceanic environment.

Epidote amphibolite facies

Zones of progressive metamorphism have belts of epidote amphibolite facies between areas of greenschist and high pressure (kyanite type) amphibolite facies, and is elsewhere widely distributed as garnet bearing pelitic rocks. This facies is also represented in the schistose retrograde products of originally higher grade rocks. Plagioclase in equilibrium with epidote in one of the nappe sequences has been shown to undergo stepwise changes from An5-15 to An20-30 in the transitional stages between the greenschist- and the amphibolite facies [5].

Amphibolite facies

Amphibolite facies is represented over large areas underlain by overthrusted complexes and basement culminations. Kyanite is usually present in pelitic rocks although sillimanite is found at places, probably due to incomplete polymetamorphism. Many of the occurrences of amphibolite-facies rocks are of Precambrian age but also Cambro-Ordovician and late Silurian metamorphism may locally have lead to this grade.

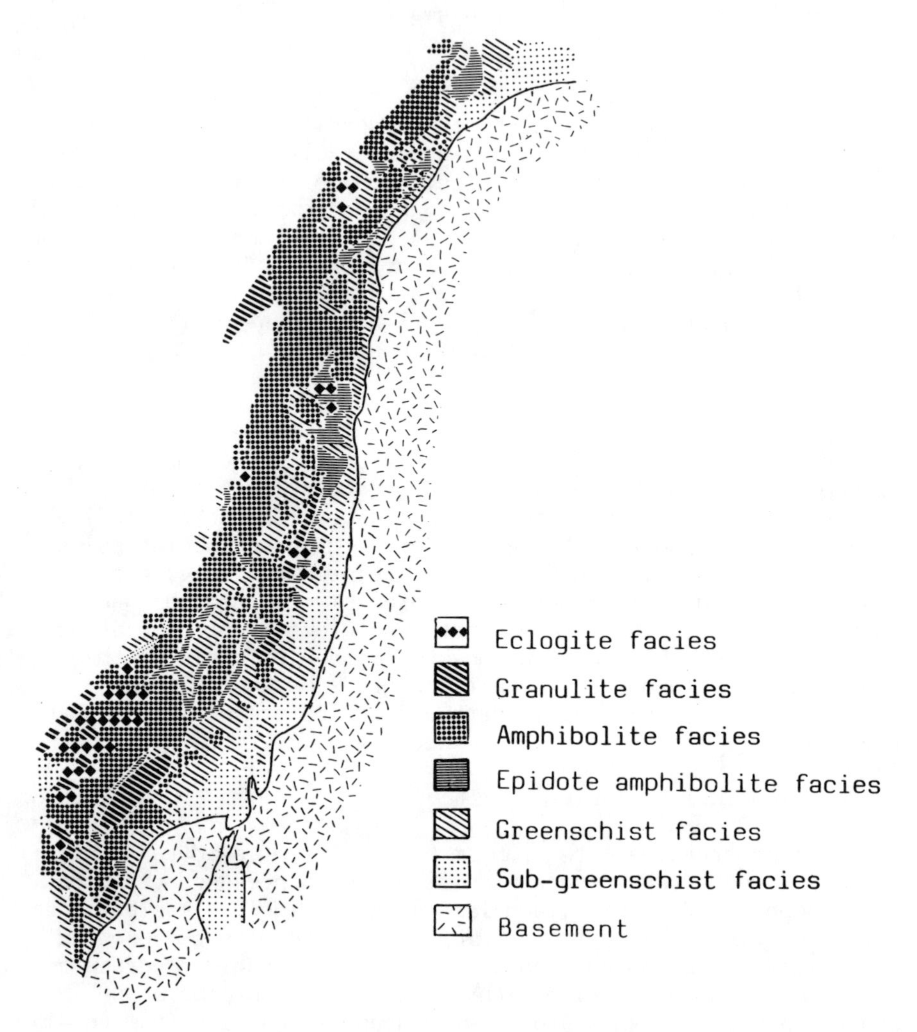

Fig. 2. Metamorphic map of the Scandinavian Caledonides. Facies conventions are those agreed upon by the IGCP-CO working group for metamorphism [4].

Granulite facies

Granulites of high-pressure (Ga-Cpx) and intermediate pressure (Opx-Cpx) are present in definite tectonostratigraphic horizons of the overthrust rocks of the Nappe Region and as constituents of the basement complexes like the Western Gneiss Region and the Lofoten-Vesterålen area. The age of granulite facies metamorphism is definitively Precambrian for some of the occurrences (e.g. in the Nappe region of South Norway) or disputed Precambrian or Late Silurian for others (e.g. in the Nappe Region of Sweden and in the Western Gneiss Region).

Eclogite facies

Eclogites and related high-pressure rocks are distributed over a large part of the Western Gneiss Region and occur also in overthrusted rocks along the whole length of the orogen. In the Western Gneiss Region, the eclogites appear to have formed at increasing pressure and temperature northwestwards, reaching maximum values near 800°C and 18-20 kb on the coast near Ålesund [6]. Compositional zoning and relict minerals have indicated a development of some of these eclogites from original amphibolites along an extremely steep P/T gradient (7-9 kb over 100-200°C) followed by a near adiabatic decompressition on the order of 10 kb [7].

Contact metamorphism, high-T regional metamorphism?

Andalusite-cordierite bearing rocks are present in areas close to gabbroid and trondhjemitic intrusives in some of the higher nappes containing eugeosynclinal Caledonian rocks. In the Trondheim Nappe Complex, the contact metamorphic minerals have largely vanished or been pseudomorphosed into kyanite, paragonite etc. during successive metamorphic events [8]. Relics of andalusite and staurolite + garnet also occur apparently unrelated to any intrusives and may conceivably be interpreted as indicative of an early regional thermal metamorphism [8,9].

AGE - POLYMETAMORPHISM

The Caledonian tectono-metamorphic development took place as a number of climactic events, the most important of which were in the Late Cambrian - Middle Ordovician and in the Late Silurian. The Trondheim orogeny of Vogt [10] was particularly important since it is marked over large parts of southern Norway and probably also in northern Norway by Ordovician conglomerates where the pebble material has been deformed, metamorphosed and uplifted before deposition [10,11]. The sequence above the conglomerate has been folded and metamorphosed during the late Silurian event which also involved large translations of nappes southeastwards onto the foreland and significant denudation in the central parts of the orogen. This event which was correlated with the Ardennian [10] is now known as the main Scandinavian or Scandian orogeny [12,13].

Two major stages of orogenic development is also known from the extreme north of Norway, where the Finnmarkian orogeny of Upper Cambrian - Lower Ordovician age preceeded the main, Late Silurian orogeny [14,15]. The two sets of tectono-metamorphic events are locally superposed and both have affected the Precambrian basement. An upper bracket for the Caledonian metamorphism is given by the Lower - Middle Devonian sediments which rest unconformably on older Caledonian rocks and which have suffered only relatively mild and locally distributed metamorphism during the Svalbardian orogeny in the Middle to Upper Devonian.

Late Precambrian metamorphism

It is not the intention here to discuss the various events of pre-Caledonian metamorphism which are represented within the belt of the Scandinavian Caledonides. Some of the late-Precambrian sediments and volcanites may, however, have retained vestiges of metamorphism obtained during the earliest stages of the Caledonian orogeny. Thus, the up to 6 km thick stratigraphic sequence of psammitic sediments of the Särv nappe in Sweden appear to have been metamorphosed up to epidote amphibolite facies during or prior to intrusion of the 665±10 Ma old dolerites which initiated the late Precambrian continental break-up [16]. Ophiolites formed at a later stage may have vestiges of early metamorphic stages represented by folded fabrics cut by sheeted dykes and by clastic grains of amphibole in the oceanic sediments. The early, metamorphism must have taken place after the ~665 Ma continental rifting but prior to obduction, which in this area was completed before about 465 Ma ago [17]. The possible late Precambrian metamorphism is also indicated by a few Rb-Sr dates around 700 Ma made on granitic rocks (Råheim pers.comm. 1978).

Middle Ordovician to Cambrian metamorphism

The Middle Ordovician or older metamorphism in southwestern Norway increases northwards from greenschist facies with stilpnomelane near Stavanger to low amphibolite facies with staurolite north of Bergen. The age of this metamorphic event is pre-Ashgillian from fossils in the sequence overlying the intra-Ordovician unconformity, but further dating rests on yet scanty radiometric determinations. Thus, Rb-Sr dates of about 465 and 535 Ma probably represent the upper and lower brackets for the pre-Ashgillian metamorphism in the southwestern Caledonides [17]. In the Trondheim region the early Caledonian metamorphic events affect Cambrian rocks but predate fossiliferous Arenig beds and trondhjemitic intrusives dated to about 450-480 Ma [18].

The Finnmarkian orogeny in North Norway produced metamorphic rocks up to middle amphibolite facies and locally also migmatites in Upper Cambrian - Lower Ordovician time, probably with peak of regional metamorphism at 535±17 Ma and cooling/uplift at 486±27 Ma [19].

Late Silurian metamorphism

The Late Silurian phase involved metamorphism, folding and south-eastwards translation of the major nappes onto the foreland. Vogt [10] showed that this most important orogenic phase in the Scandinavian Caledonides occurred at the Ludlow - Downton boundary and was comparable with the Ardennian phase in other countries. For the more marginal parts of the orogen, he suggested that the tectonic movements took place slightly later, in the Erian phase.

The Late Silurian stage of tectono-orogenic development in Scandinavia (main Scandinavian or Scandian phase) affected sedimentary sequences dated by fossils up to Upper Llandovery in Norway and Wenlock(?) in Sweden, and thrusting must have taken place subsequent to igneous events dated to 405-424 Ma [3,20]. The upper bracket for the Scandinavian phase is provided by the Old Red Sandstone sediments of Trøndelag which extend down into the Downton (Pridoli), possibly even into the Ludlow.

The grade of metamorphism varies much: fossiliferous strata have experienced mainly greenschist facies metamorphism, but are in the northernmost Norway locally transformed progressively through classical Barrovian isogrades into the kyanite and sillimanite zones [21]. In addition, large volumes of original Precambrian rocks may have experienced prograde reactions leading to granulitic and eclogitic rocks, followed by anatexis, migmatitization and finally a full range of retrograde effects. The basis for this interpretation, - which not everybody are prepared to accept yet -, is that the recently obtained U-Pb and Nb-Sm dates at 407-447 Ma give the age of formation of the eclogites [22]. A large body of mineral ages and K-Ar dates around 400 Ma give evidence of the vaning stages or the orogenic climax or the cooling following uplift.

Devonian metamorphism

The Old Red sediments are folded and locally thrusted during the Svalbardian orogeny [10] which probably took place at the Middle to Upper Devonian boundary. The metamorphism was mainly dynamic and related to the thrusts and faults which cut the sequences, but can reach lower greenschist facies at places. Phyllonites have been produced in the Old Red sediments and in the basement rocks adjacent to the movement zones.

Most lithologies of the adjacent area are represented among the clasts in the Old Red sediments which rest with profound angular unconformity upon Lower Paleozoic rocks. This is a manifestation of the large extent of denudation of the Caledonian mountain chain which took place in the interval between the Late Silurian orogeny and the Middle Devonian.

INTERPRETATIONS

The Scandinavian Caledonides can be related to the continental rifting, sea-floor spreading, ocean closure and final continental collision of an ideal Wilson cycle [15]. The metamorphism will be tentatively interpreted on this background:
The Late Precambrian metamorphism may have been related to the continental rifting (Fig. 3a) which produced thick accumulations of psammitic sediments, later to be intruded by dolerite swarms. Metamorphism was of low grade and probably connected to the intrusion of the dolerites about 665 Ma ago [16]. Subsequent sea-floor spreading and beginning ocean closure was accompanied by low-grade oceanic metamorphism which can now be seen in some of the ophiolite fragments. The Middle Ordovician to Cambrian metamorphism was related to ocean closure (Fig. 3b). This took place along subduction zone(s) off the present coast, probably situated several hundred kilometres northwest of the present external nappes. In the southwestern part of the Scandinavian Caledonides this stage may be recorded by immature island arc activity around 535 Ma, obduction of ophiolite upon continental crust and subsequent formation of Andean type volcanites at about 465 Ma ago [17]. The partly overlapping Finnmarkian orogeny in the northernmost part of Norway had probably reached an advanced stage at 535 Ma ago [19], indicative of a scissors-like closing of the Iapetus [15]. It would be surprising if the closing stage metamorphism took place as a single event along the whole length of Scandinavia although the Finnmarkian orogeny is broadly coeval with the Grampian phase on the British Isles. Vogt's [10] many and possibly diachronous disturbances or orogenies defined by conglomerates were probably more realistic than the recent attempts to fit the few available radiometric dates into a simple scheme of one single Middle Ordovician to Cambrian (e.g. Finnmarkian) orogenic stage valid for the whole Scandinavian Caledonides.

The Late Silurian metamorphism was related to continental collision whereby the leading edge of the Baltic plate became overridden by a thick succession of imbricated basement slices and more far-travelled thrust nappes (Fig. 3c). This was the main stage of regional metamorphism, dynamic metamorphism, strain in-

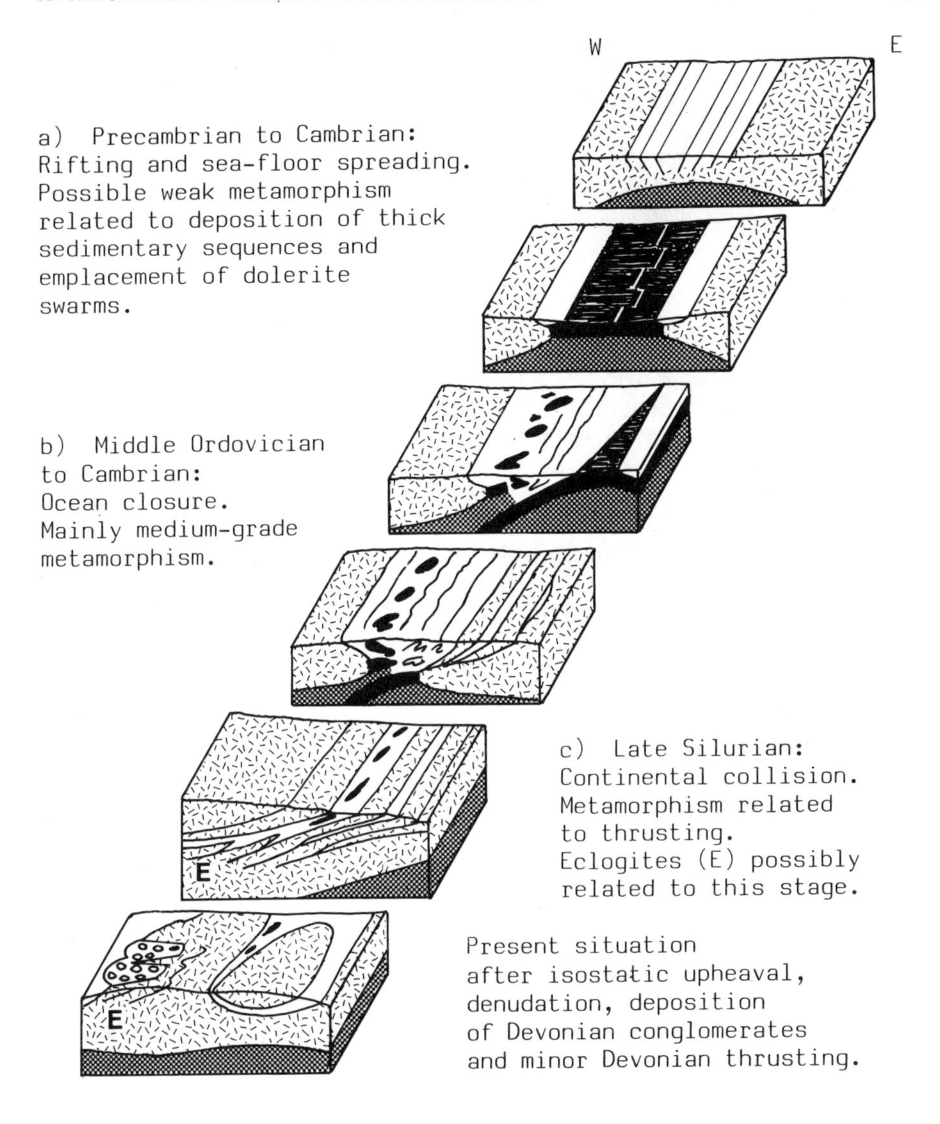

Fig. 3. Tectono-metamorphic stages in the development of the Scandinavian Caledonides. The lower block features the present situation across the Western Gneiss Region and the Nappe Region.

duced metamorphism and inversions of the metamorphic gradient along the whole belt of the Scandinavian Caledonides. Many of the nappes derived from the basement or its cover were transported as nappes with virtually no Caledonian metamorphism in their internal parts (e.g. theJotun Nappe Complex, Särv Nappe) while others experienced pre-thrusting medium-grade Caledonian metamorphism and possibly even granulite facies and eclogite facies at places (e.g. in the Seve Nappe).

The extensive formation of eclogites and granulites in the Western Gneiss Region can, - if they really are of Late Silurian age-, be related to the pressure increase and thermal blanketing effects of the overriding nappes. The northwestwards increasing PT-conditions of formation of eclogites indicates that the nappe overburden increased in that direction, possibly as a result of northwestwards subduction of the Baltic plate below the North American plate [6,7].
The crustal doubling caused by the imbrication and the overriding of nappes facilated partial melting at depth and made it easier for Caledonian structures to develop with complete obliteration of the pre-existing fabric. Rapid uplift and denudation must have followed the crustal doubling and the high-pressure assemblages became decompressed and subject to retrograde metamorphism.

The simple picture presented above leans heavily on the interpretation of the radiometric dates at 407-447 Ma as the real ages for eclogite formation [22]. It must be admitted, however, that it is hard to understand the close juxtaposition of eclogite-bearing gneisses and relatively low-grade Lower Paleozoic rocks along the coast if the eclogites really are of Late Silurian age.

The Devonian metamorphism can be related to extensional and compressional stress regimes which produced faults of various kinds, folds and thrusts. Deep sedimentary basins were formed and acted as sinks for the coarse continental detritus from the uplifted areas. The fact that most rock types from both the Lower Paleozoic sequences and the adjacent gneisses are represented among the clasts in the Middle Devonian conglomerates of western parts of South Norway testifies about the magnitude of the uplift and denudation in the time following the Late Silurian orogeny.

CONCLUSIONS

It is possible to map the regional distribution of metamorphic facies throughout the Scandinavian Caledonides, but it is not always certain which age the metamorphic assemblages are. The main stage of Caledonian development at the end of the Silurian is in some areas superposed upon rocks formed during the Middle Ordovician to Cambrian events, which in their turn may be superposed upon the results of Precambrian events. Much of the Caledonian belt in Scandinavia contains in fact Precambrian rocks which were overprinted to various extents by Caledonian metamorphism. The various climactic events of Caledonian tectono-metamorphic development can be put tentatively into a simple rifting, drifting, closure and collision model, but there are as yet quite unsatisfactory information about dates to bracket the various stages in this evolution.

REFERENCES

1. Bryhni, I., and Brastad, K.: 1980, J.Geol.Soc.London 137, pp. 251-259.
2. Andréasson, P.G., and Gorbatschev, R.: 1980, Geol.Fören. Stockholm Förhandl. 102, pp. 335-357.
3. Andréasson, P.G.: 1982, A convering article for dissertation for the Doctors degree in Natural Sciences, University of Lund, pp. 44.
4. Fischer, G.W.: 1980, U.S. Working Group and Specialist Study Groups Newsletter 2, pp. 6-9.
5. Kvale, A.: 1946, Bergen Mus. Årb. 1946, R. 1, pp. 1-201.
6. Griffin, W.L., Austrheim, H., Brastad, K., Bryhni, I., Krill, A.G., Krogh, E.J., Mørk, M.B.E., Qvale, H., and Tørudbakken, B.: (in press), In The Caledonide Orogen - Scandinavia and related areas (eds. Gee, D.G., and Sturt, B.A.), Wiley, New York.
7. Krogh, E.: 1982, Lithos, 15, pp. 305-321.
8. Guezou, J.-C.: 1979, Norges Geol. Unders., 340, pp. 1-34.
9. Dudek, H., Fediuk, F., Suk, M., and Wolff, F.C.: 1973, Norges Geol. Unders., 289, pp. 1-14.
10. Vogt, Th.: 1929, Norsk Geol. Tidsskr. 10, pp. 97-115.
11. Kvale, A.: 1960, Guide to excursions no. A7 and no. C4. Norges Geol.Unders., 212, pp. 1-43.
12. Gee, D.G., and Wilson, M.R.: 1974, Am. J. Sci., 275, pp. 1-9.
13. Gee, D.G.: 1975, Am. J. Sci., 275, pp. 468-515.
14. Ramsay, D.M., and Sturt, B.A.: 1976, Norsk Geol. Tidsskr., 56, pp.271-307.
15. Roberts, D., and Gale, G.H.: 1978, In Evolution of the Earth's crust (ed. Tarling, D.H.), Academic Press, London, New York, pp. 255-312.
16. Claesson, S., and Roddick, J.C.: 1983, Lithos, 16, pp. 61-73.
17. Furnes, H., Austrheim, H., Amaliksen, K.G., and Nordås, J.: Pers. comm. 1983.
18. Klingspor, I., and Gee, D.G.: 1981, Terra Cognita, 1, pp. 55.
19. Sturt, B.H., Pringle, I.R., and Ramsay, D.M.: 1978, J.Geol. Soc. London, 135, pp. 597-610.
20. Andersen, T.B., Austrheim, H., Sturt, B.A., Pedersen, S., and Kjærsrud, K.: 1982, Norsk Geol. Tidsskr., 62, pp. 79-85.

21. Andersen, T.B.: 1979, Unpublished Cand.Real. Thesis. University of Bergen, pp. 1-338.
22. Griffin, W.L., and Brueckner, H.K.: 1980, Nature, 285,pp.319-321.

METAMORPHISM IN THE BRITISH CALEDONIDES

D. J. Fettes

IGS, Edinburgh, UK

ABSTRACT

The British Caledonides fall naturally into two belts. The northern belt comprises middle Proterozoic to Middle Cambrian metasediments which have been subjected to the tectonothermal events of the Grampian orogeny (c. 490 Ma). The metamorphic grade of the rocks rises rapidly from the margins of the belt through the greenschist facies to the amphibolite facies. The southern belt comprises metasediments of Ordovician to late-Silurian age. They have been weakly metamorphosed during the late-Silurian.

INTRODUCTION

The British Caledonides falls into two belts whose metamorphic history and style are markedly different. For descriptive purposes these two belts are treated separately below. They are referred to here as the northern and southern belts, equivalent to the orthotectonic and paratectonic belts or the metamorphic and non-metamorphic belts of Read (37).

NORTHERN BELT

The northern belt is limited to the south by the Highland Boundary Fault and to the north-west by the Moine Thrust Zone (Fig. 1). It is composed of two great metasedimentary assemblages and their associated intrusives. These are the Moine, occupying the northern two-thirds of the belt, and the Dalradian in

P. E. Schenk (ed.),
Regional Trends in the Geology of the Appalachian-Caledonian-Hercynian-Mauritanide Orogen, 205–219.

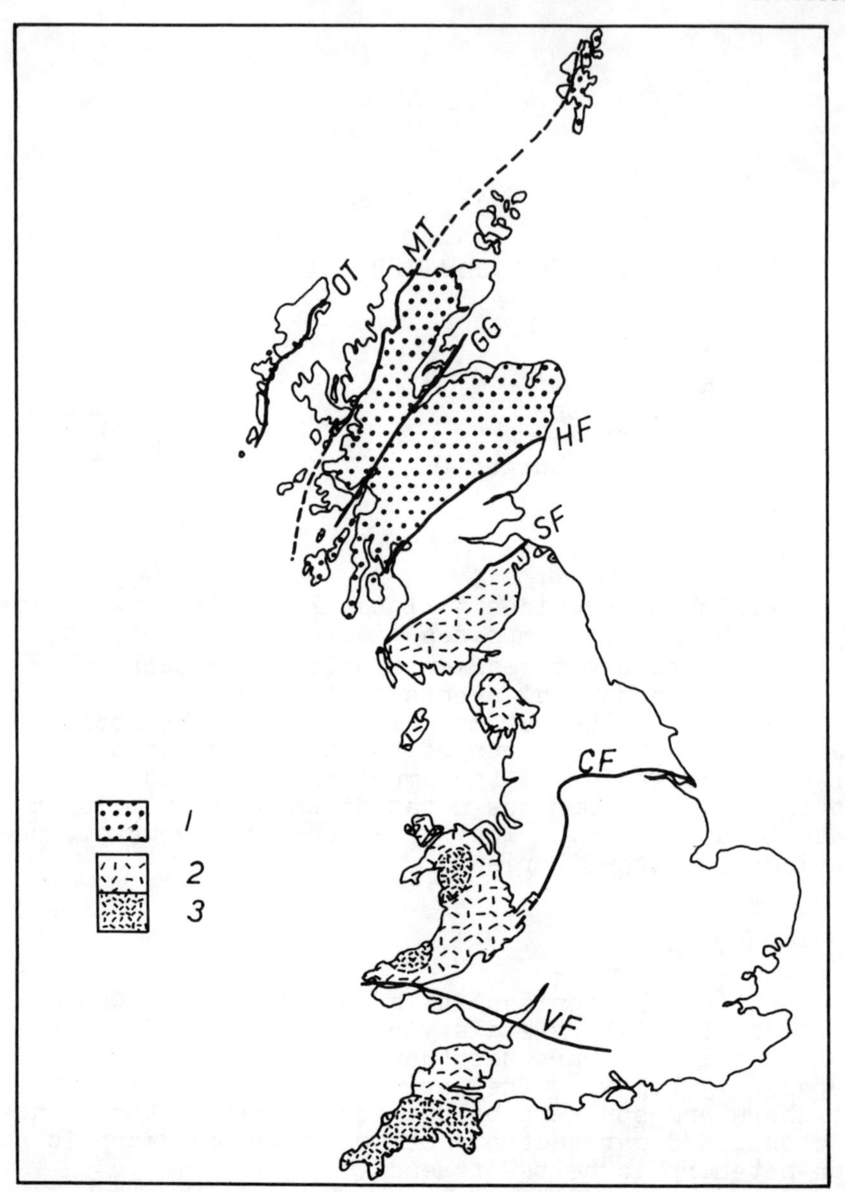

FIG. 1 Metamorphic rocks of the British Caledonides. 1, northern belt (see Fig. 2 for details). 2, southern belt, weakly metamorphosed. 3, southern belt, greenschist facies. OT, Outer Hebrides Thrust. MT, Moine Thrust. GG, Great Glen Fault. HF, Highland Boundary Fault. SF, Southern Uplands Fault. CF, southern limit of Caledonian (pre Devonian) effects, Variscan Front.

the south. The Moine is a largely monotonous sequence of psammite and pelite (26) of probable middle to late Proterozoic age although recent work (9, 33) has shown a complex polymetamorphic history. The Dalradian is a mixed sequence ranging from quartzites, pelites, black-schists and marbles in its lower part through to grits, psammites and pelites at the top (20). The sediments probably range from late Proterozoic to lowermost Middle Cambrian. Along the southern margin of the Dalradian and largely in faulted contact are a series of exotic rocks collectively termed the Highland Border Series. These include black phyllites, cherts, grits, amphibolites, spilitic lavas and serpentinites which may represent a tectonically emplaced ophiolitic slice (24). Limestones within this sequence have yielded fauna giving a lower Arenig age (10).

In the Shetland Isles ?foreland rocks in the extreme west give way eastwards through a complex tectonic zone (18) to a metasedimentary sequence interpreted as an attenuated Moine/Dalradian sequence (16).

Distribution of Facies

The metamorphic grade of the northern belt ranges from lower greenschist to upper amphibolite and from intermediate-pressure facies series to low-pressure facies series. The grade quickly rises inwards from the margins such that the greater part of the belt lies in the amphibolite facies (Fig. 2). In the Dalradian rocks of the Southern Highlands the grade is generally recorded by reference to the classical Barrovian (biotite → garnet → staurolite → kyanite → sillimanite) and Buchan (biotite → corderite → andalusite → sillimanite) assemblages. The metamorphic facies shown in Fig. 2 are based on the assemblages as defined elsewhere in this volume. In the Northern Highlands the composition of the Moine rocks has inhibited the widespread development of index minerals in pelites and the metamorphic grade is normally determined by reference to assemblages in calc-silicates. The earlier zonal scheme of Kennedy (27) was modified and detailed by Tanner (40) to give albite-zoisite-calcite-biotite → oligoclase-zoisite-calcite-biotite → andesine-zoisite-biotite → andesine-zoisite-hornblende → bytownite/anorthite-hornblende → bytownite/anorthite-pyroxene. Further work by Winchester (43) related the zonal mineral changes to rock composition based on the CaO/Al_2O_3 ratio of the rock. Correlation between the two zonal schemes based on pelitic and calc-silicate assemblages is still largely empirical. The construction of the metamorphic map (Fig. 2) necessitated certain arbitary decisions, namely, defining the base of the greenschist facies at the biotite isograd (in pelite), the base of the epidote-amphibolite facies at the oligoclase isograd, the base of the lower amphibolite facies at the andesine isograd, the base of the middle amphibolite facies at the bytownite isograd and the base of the upper amphibolite facies at the pyroxene isograd. It is hoped that subsequent data on the distribution of alumino-silicate polymorphs and other pelitic index minerals in the Moine and measurements on the absolute pressure and temperatures of crystallisation will refine the presently known facies pattern.

The distribution of the facies groups along with the nature and orientation of the associated isograd within the Dalradian rocks of the Southern Highlands is well documented (4, 14). Traditionally this area was divided into an intermediate-pressure facies series (Barrovian) in the central and south-west part and a low-pressure facies series (Buchan) in the north-east. More recently Fettes and others (15) showed a transition across the Highlands with decreasing pressure to the north-east. This was partly formalised by Harte and Hudson (22) who proposed a four-fold division within the eastern Dalradian, namely; 1) Barrovian, biotite – garnet – staurolite – kyanite; 2) Stonehavian, biotite – garnet – chloritoid + biotite – staurolite – sillimanite; 3) W Buchan, biotite – cordierite – andalusite – staurolite – kyanite; 4) E Buchan, biotite – cordierite – andalusite – sillimanite. Harte and Hudson (22) erected their Stonehavian sequence partly on the argument that chloritoid + biotite assemblages, abundant on the Stonehaven section (east coast Dalradian) wedged out to the south-west because of increasing pressure. Atherton and Smith (5) in reporting chloritoid + biotite pairs from the Central Highland Dalradian suggest that the wedging out of the Stonehavian sequence was controlled by changes in the bulk rock chemistry of the rocks and questioned the concept of a variation in the facies series. Harte (in reply to Atherton and Smith) noted that the facies series variation was independant of the bulk rock composition changes proposed. Harte (pers. comm.) has however, also suggested the possibility that the chloritoid + biotite pairs in the Central Highlands represent an **early**, relatively low pressure, phase in the progressive

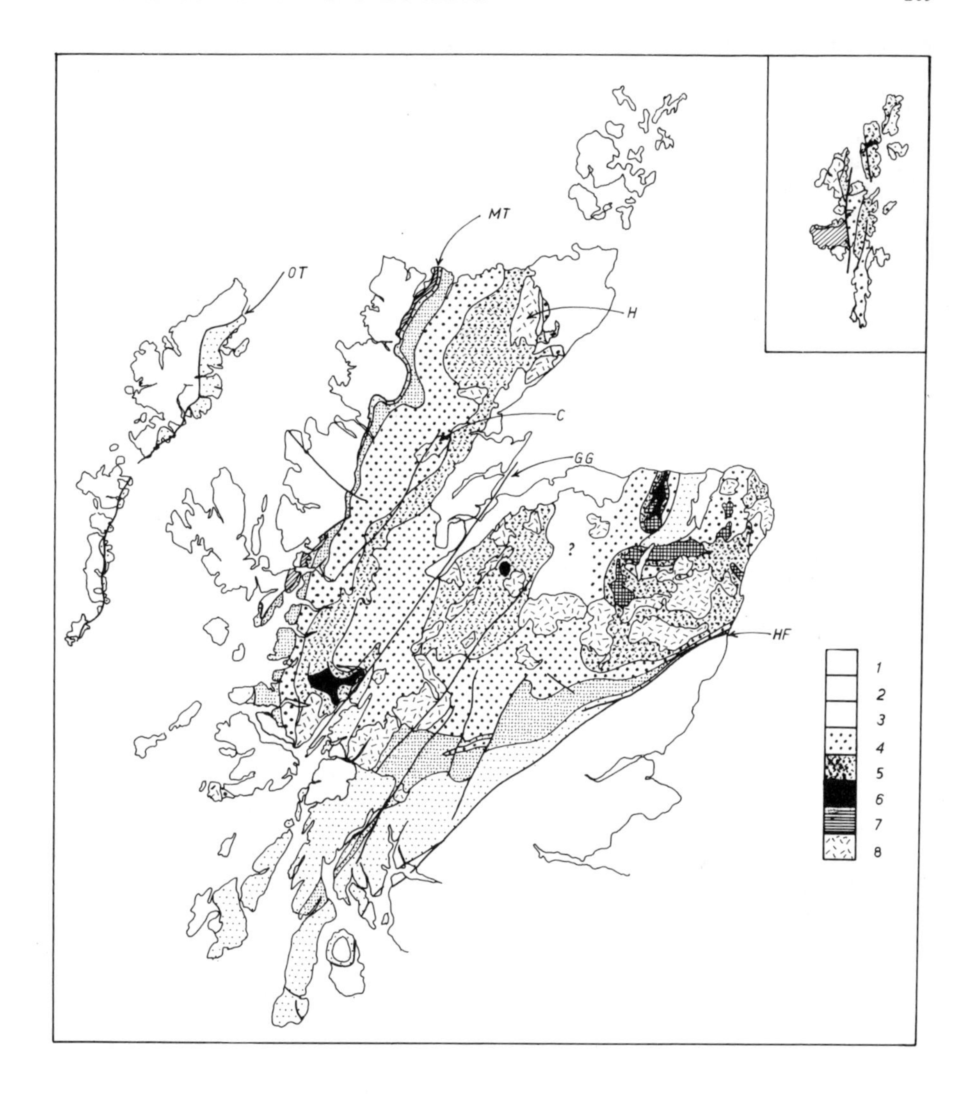

FIG. 2 Metamorphic facies in the northern belt. 1, weakly metamorphosed. 2, greenschist facies. 3, epidote-amphibolite facies. 4, lower amphibolite facies. 5, middle amphibolite facies. 6, higher amphibolite facies. 7, basic and dioritic intrusives of north-east Scotland. 8, granites. OT, Outer Hebrides Thrust. MT, Moine Thrust. GG, Great Glen Fault. HF, Highland Boundary Fault. H, Strath Halladale granite. C. Carn Chuinneag granite.

metamorphic sequence, the later phase of which was at a higher pressure than the equivalent phase at Stonehaven. That is, Harte is postulating a variation in P/T conditions through time as well as through space. This implies an increase in pressure during the later stages of the prograde metamorphism in the Central Highland Dalradian.

On the Banffshire coast near the junction of the Barrovian and Buchan zones, chiastolite crystals are found pseudomorphed by kyanite, indicative of an increase in pressure with time.

Within and adjacent to the Highland Border Series, rocks of anomalous metamorphic grade occur. Hornblende-schists (locally garnetiferous) are found within the Series adjacent to lower greenschist Dalradian rocks. And in the south-west, Dalradian rocks apparently show contact aureole effects against the Highland Border Series (24).

In Shetland the Caledonian rocks lie largely in the amphibolite facies (14).

In the Moine the distribution of facies groups is locally well documented (14) but there are large areas especially in the extreme north and east where information is sparse and current zonation is largely based on the work of Winchester (43). Possible variations in the facies series are poorly documented, partly because of the lack of work on the calc-silicate assemblages. It is probable, however, that there is at least as much variation within the Moine area as has been suggested within the Dalradian.

West of the Moine Thrust low grade Caledonian recrystallisation can be identified in the foreland rocks, particularly in the Torridonian rocks of southern Skye and in a broad belt associated with the Outer Hebrides Thrust.

Migmatites

Throughout the higher grade areas of the northern belt there is a localised but common development of so-called 'migmatites' and 'gneisses'; in reality these terms describe a complex series of phenomena including metamorphic segregation, partial melting, intrusive lit-par-lit and veined granites etc. These events are commonly associated with late fibrolite development.

Migmatites (quartzo-feldspathic lenses) in the eastern Dalradian are notably developed at one stratigraphic horizon (Ben Lui Schists) suggesting a strong lithological control over their formation (7). Mixed in with these migmatites are a complex series of granitic rocks which together form the 'older granite' of Barrow (6) (8). In the north-east Dalradian there is a series of basic and dioritic bodies whose intrusion is closely related in time to the main metamorphic crystallisation (13). The aureole effects of these masses has given rise to a variety of high grade metamorphic phenomena including, partial melting, hybridisation of magmas, sillimanite crystallisation etc (1, 2, 19). How extensive these aureole effects are, and to what extent the development of so-called gneisses (oligoclase-biotite) and other regional high grade effects can be attributed to these intrusives is uncertain. In the extreme north-east Dalradian there are zones of xenolithic gneisses and massive inhomogenous gneisses where the rock has apparently lost coherence due to extensive partial melting largely of semi-pelitic horizons. These zones appear, in part, to be associated with small basic intrusions the thermal effects of which may well have catalysed or promoted the metamorphic effects.

In the Central Highland Moine migmatities have been described by Ashworth (3). Elsewhere in the Moine migmatites may well represent pre-Caledonian basement (see below).

Polymetamorphism

One of the great problems in examining the metamorphic pattern in the Northern Highlands is the presence of polymetamorphic rocks. Although the inliers of Lewisian basement (c 2600 and 1800 Ma) are fairly well defined, the amount of ?Grenville (c 1000 Ma) and ?Morarian (c 720 Ma) deformation and metamorphism is highly uncertain. Current theories suggest that most of the Moine rocks north of the Great Glen as well as some areas immediately to the south-east of the Great Glen, have suffered pre-Caledonian metamorphism and have merely been recrystallised during the Caledonian. How the metamorphic zonation in these areas relates to the two events is still uncertain. In the south-western Moine Powell and others (36) suggest that the grade of the pre-Caledonian metamorphism changed rapidly from the lower greenschist facies (garnet zone) in the extreme west to the upper amphibolite facies (?sillimanite zone with extensive migmatite development) over most of the rest of the area. The Caledonian overprint also rose quickly from the lower greenschist (chlorite zone) facies in the west such that the greater part of the belt recrystallised in the amphibolite facies. In the northern Moine the polymetamorphic pattern is only locally known (14) and has recently been complicated by the suggestion (28) that the extreme north-east may well have escaped significant Caledonian thermal effects (see below). Moine rocks south-east of the Great Glen show a complex basement/cover relationship (32) with relatively low grade cover resting on gneissose (or migmatitic) basement. In none of these areas has widespread retrogression of the pre-Caledonian metamorphic pattern been identified.

Age of Metamorphism

The Caledonian metamorphism of rocks in the northern belt can be referred to the Grampian Orogeny. Stratigraphical constraints on the age of the metamorphism are poor. The youngest sediment affected by the main metamorphism is probably Lower to Middle Cambrian providing a maximum age of 530-540 Ma. A minimum age of 410 - 420 Ma is fixed by the post-kinematic granites and by the non-metamorphic cover of Lower Devonian (?Upper Silurian) Old Red Sandstone. Although in Shetland Old Red Sandstone sediments show low-grade recrystallisation (30).

Within the Dalradian area deformation was polyphasal. Although there are some significant regional differences the sequence can be broadly divided into two parts, namely, a primary or nappe forming and translation episode and a secondary deformation and uplift episode. A considerable amount of textural work has been done in relating metamorphic assemblages to the structural sequence (25, 7). In general terms the grade rose during the primary deformation peaked between the primary and secondary events and then fell off during the secondary event with variable degrees of retrogression. There has been considerable discussion whether the sillimanite/migmatite development is texturally separable from the development of other high grade assemblages (e.g. 21), although it is perhaps now accepted that they are all part of a continuous progressive metamorphic event (22, 7). A possible exception being the sillimanite development in the aureole of the basic intrusives of the north-east Dalradian (1, 14, see also 4).

In the central Highlands 'older granites' believed to be coeval with the metamorphic peak give ages of 514 + 6/-7 Ma and 491 ± 15 Ma (8). In the north-east the basic and dioritic rocks whose intrusion is thought to be shortly after the climax of metamorphism have given a date of intrusion of 489 ± 17 Ma (31). These dates seem to suggest an age of 490 - 495 Ma for the peak of metamorphism. Because significant recrystallisation probably began at c. 520 Ma, the progressive phase of metamorphism can therefore be given as 520-490 Ma.

The style of secondary deformation varies between areas but is generally associated with uplift and retrogression. In the south-eastern Dalradian, Dempster (pers. comm.) has suggested that the dominant uplift probably occurred around 460 - 440 Ma. The variable cooling ages from both isotopic and palaeomagnetic data (12, Dempster, pers. comm., Watts, pers. comm.) together with the differences in the degree of retrogression (4) across the Dalradian reflect differential and probably episodic rates of uplift.

The anomalously high metamorphic grade of parts of the Highland Border Series and the apparent contact metamorphism of adjacent low grade Dalradian rocks has been attributed by Henderson and Robertson (24) to dynamothermal metamorphism on the sole of an obducting ophiolitic slice. If the depositional age of the rocks is c 495 Ma (Lower Arenig) and because they apparently cut the major primary structures although affected by the secondary deformation, their associated metamorphism must fall in the period 490-460 Ma.

In Shetland migmatites roughly synchronous with the main metamorphism have given an age of 530 ± 25 Ma (17).

Over much of the Moine area the structural sequence is poorly known and correlation between local sequences is generally difficult, in consequence data on textural relationships between metamorphic assemblages and deformational phases is sparse. What information does exist is often complicated by the recognition of polymetamorphism and the separation of the various metamorphic phases into Caledonian and pre-Caledonian events. Uncertainty of correlations across major slide zones has also hindered the interpretation.

In the south-western Moine it seems probable that the earliest Caledonian (sensu stricto) movements were marked by movement on major slide zones. This was followed by two or three phases of ductile folding (35). Powell and others (36) have argued that the peak of Caledonian metamorphism was syn-sliding, that is, earliest Caledonian. Highton (pers. comm.) however working on the Moine rocks adjacent to the Great Glen has argued that the metamorphic peak post-dates the sliding and associated deformation in that area. Brewer and others (9) suggest an age of 467 ± 20 Ma as the minimum age for the peak of Caledonian metamorphism, they suggest that dates of c 450 ± 10 Ma may be cooling ages and they further suggest that cooling related to differential uplift may have continued until c 410 Ma in the Moine adjacent to the Great Glen. Powell and others (36) suggest a spread of 470-400 Ma for the Caledonian metamorphism. This is fixed on the assumption that the 467 ± 20 Ma date for the metamorphism was syn-sliding, that is, contemporaneous with the earliest Caledonian movements. Such a model (36) implies that the rocks were at substantial depths and temperatures at the initiation of movement, the sliding essentially acting as the precursor to uplift. There is however no deformational evidence of pre-metamorphic burial or tectonic thickening unless it is assumed that the rocks had remained buried since the Grenville (or ?Morarian) and that recrystallisation was perhaps initiated by early movements. In which case the 467 ± 20 Ma date also marks a cooling age. Interpretation of this age is particularly important in terms of implied diachronous metamorphism between the Moine and the Dalradian.

Smith (39) has shown remnant regional thermal gradients were still operating at the time of intrusion of a widespread suite of microdiorites between 450-420 Ma.

In the northern Moine Mendum (29) has argued that the main Caledonian metamorphism post-dates the early mylonites within the Moine Thrust Zone as well as associated slides within the Moine (although these may not correlate with the early Caledonian sliding described above). A date of 570 ± 10 Ma on the pre-metamorphic Carn Chuinneag intrusion (Fig. 2, 34) gives a maximum age to the metamorphic peak which is here marked by kyanite development. Farther north in the Strath Halladale (Fig. 2) region a series of thick granite sheets and associated intrusives which post-dates the local penetrative deformation and high grade metamorphic events has given a Rb-Sr age of 649 ± 32 Ma (28) suggesting minimal Caledonian tectonothermal effects. If both these dates are accepted then it follows that between the two areas (a distance of c 50-60 km through apparently uniform high grade psammites) the Caledonian thermal regime falls from kyanite grade (at Carn Chuinneag) to a maximum of lower greenschist facies, a difficult concept since no substantial zone of retrogression has been noted nor a zone of late stage tectonic disruption, which could juxtapose two high grade metamorphic terranes of different ages.

Discussion

Models for the metamorphism are still uncertain (14). Although a large amount of work is currently being carried out on the Dalradian rocks, apart from papers such as Harte and Hudson (22) and Wells and Richardson (42) little has been published. The latter authors propose temperatures of 550-620°C and pressures of 9-12 kb as conditions during peak metamorphism in the Central Highlands. They further suggest that the pressure was falling when the maximum temperature was reached. They (42) contrast these results with conditions in the Buchan region and suggest that there lower pressures were operative at peak metamorphism as a result of a thinner cover. Harte and Hudson (22) have argued however that in the Buchan region peak metamorphic conditions were influenced by heat introduction into the crust perhaps in the form of the 'older granites'. They also suggest (22) the rapid change in metamorphic conditions near the southern margin of the eastern Dalradian might reflect the long standing presence of a tectonic boundary in the area and the cooling effects of a subsiding sedimentary basin to the south.

It is interesting to note the extensive association of the north-east Dalradian with apparent high thermal conditions. There is a prolonged history of magmatitic activity in the area ranging from the older granites, through the younger basic and dioritic intrusives to the late granites. The latter falling into two groups; an earlier slightly foliated suite of late-kinematic granites and associated pegmatites dated at c 460 Ma which are restricted in occurrence to the north-east and a second suite of post-kinematic granites at c 420 Ma which although not restricted to north-east Scotland are notably abundant in that area (Fig. 2). In addition to the magmatitic activity the area is notable for the extensive development of the high grade metamorphic effects described above (migmatites, gneisses, etc) associated with the low-pressure facies series metamorphism. The underlying cause of this long-lived thermal activity is still, however, highly speculative.

SOUTHERN BELT

The 'Southern Belt' includes the Lower Palaeozoic rocks in the Southern Uplands of Scotland, the Lake District, Isle of Man, Wales and the Palaeozoic rocks of South-west England, together with a number of smaller inliers. The grade is in general very low only locally rising to the greenschist facies.

In the Southern Uplands a number of prehnite-pumpellyite localities have been recorded as well as a narrow zone of zeolite facies adjacent to the Southern Uplands fault (23). In Wales prehnite-pumpellyite localities have been recorded and two areas of greenschist facies identified partly on the basis of actinolite and chlorite in metabasic rocks and partly on illite crystallinity (Bevins, Gibbons, Robinson and Rowbotham, pers. comm.). In south-west England greenschist facies rocks have been identified on the basis of illite crystallinity (Robinson and Primmer, pers. comm.). Lower amphibolite grade (chloritoid, cordierite, garnet) rocks have been recorded on the Isle of Man by Simpson (38). High grade amphibolite rocks have also been described from Anglesey although their age is uncertain. In south-east Anglesey there is an extensive belt of glaucophane-schist. The presence of lawsonite at one locality in this belt indicates a fragment of blue-schist facies (Gibbons, pers. comm.). A small occurrence of glaucophane-schist and eclogite at Girvan in the Southern Uplands has also been referred to by Dewey (11) as a fragment of a 'blueschist metamorphic belt'.

The age of metamorphism in the southern belt is generally taken to be late-Silurian (c. 420 Ma), except in south-west England where it is Variscan. However if the evolving model of accretionary prisms is accepted for the Southern Uplands then the metamorphism would be progressive from the Ordovician through to the late-Silurian.

ACKNOWLEDGEMENTS
I should like to thank Drs. B. Harte and J. Mendum for

constructive comments. This paper by permission of the Director, IGS.

REFERENCES

(1) Ashworth, J.R. : 1976, Mineralog. Mag. 40, pp. 661-82.
(2) ____________ : 1979a, Geol. Mag. 116, pp. 445-56.
(3) ____________ : 1979b, in The Caledonides of the British Isles - reviewed. Geol. Soc. Lond., pp. 357-61.
(4) Atherton, M.P. : 1977, Scott. J. Geol. 13, pp. 331-70.
(5) ____________, and Smith, R.A. : 1979, Geol. Mag. 116, pp. 469-76.
(6) Barrow, G. : 1893, Q.J.geol.Soc.Lond. 49, pp. 330-54.
(7) Bradbury, H.J. : 1979, in The Caledonides of the British Isles - reviewed. Geol. Soc.Lond., pp. 351-56.
(8) ____________, Smith, R.A., and Harris, A.L. : 1976, J.geol.Soc.Lond. 132, pp. 677-84.
(9) Brewer, M.S., Powell, D., and Brook, M. : 1979, in The Caledonides of the British Isles - reviewed. Geol. Soc. Lond., pp. 129-37.
(10) Curry, G.B., Ingham, J.K., Bluck, B.J., and Williams, A.: 1982, J.geol.Soc. Lond. 139, pp. 453-4.
(11) Dewey, J.F. : 1971, Scott.J.Geol. 7, pp. 219-40.
(12) ____________ and Pankhurst, R. : 1970, Trans.R. Soc.Edinb. 68, pp. 361-89.
(13) Fettes, D.J. : 1970, Scott. J. Geol. 6, pp. 83-107.
(14) ____________ : 1979, in The Caledonides of the British Isles - reviewed. Geol. Soc. Lond., pp. 307-21.
(15) ____________, Graham, C.M., Sassi, F.P., and Scolari, A. : 1976, Scott. J. Geol. 12, pp. 227-36.
(16) Flinn, D., May, F., Roberts, J.L., and Treagus, J.E. : 1972, Scott. J. Geol. 8, pp. 335-43.
(17) ____________, and Pringle, I.R. : 1976, Nature, Lond. 259, pp. 299-300.

(18) ________, Frank, P.L., Brook, M., and Pringle, I.R. : 1979, in The Caledonides of the British Isles - reviewed. Geol. Soc. Lond., pp. 109-15.
(19) Gribble, C.D. : 1968, Contr. Mineral. Petrol. 17, pp. 315-30.
(20) Harris, A.L., and Pitcher, W.S. : 1975, Spec. Rep. Geol. Soc. Lond. 6, pp. 52-75.
(21) Harte, B., and Johnson, M.R.W. : 1969, Scott. J. Geol. 5, pp. 54-80.
(22) ________, and Hudson, N.F.C. : 1979, in The Caledonides of the British Isles - reviewed. Geol. Soc. Lond., pp. 323-37.
(23) Hepworth, B.C., Oliver, G.J.H., and McMurty, M.J. : 1981, in Trench-Forearc Geology. Geol.Soc. Lond., pp. 227-90.
(24) Henderson, W.G., and Robertson, A.H.F. : 1982, J. geol. Soc. Lond. 139, pp. 435 -50.
(25) Johnson, M.R.W. : 1963, Geologie Mijnb. 42, pp. 121-42.
(26) Johnstone, G.S., Smith, D.I., and Harris, A.L. : 1969, Am. Ass. Petrol. Geol. Mem. 12, pp. 159-80.
(27) Kennedy, W.Q. : 1948, Geol. Mag. 85, pp. 229-34.
(28) Lintern, B.C., McCourt, W.J., Storey, B.C., and Brook, M. : 1982, Newsletter geol. Soc. Lond. 11, pp. 12-3.
(29) Mendum, J.R. : 1979, in The Caledonides of the British Isles - reviewed. Geol. Soc. Lond., pp. 291-7.
(30) Mykura, W., and Phemister, J. : 1976, The Geology of Western Shetland. Mem. Geol. Surv. UK.
(31) Pankhurst, R.J. : 1970, Scott.J. Geol. 6, pp. 83-107.
(32) Piasecki, M.A.J., and van Breemen, O. : 1979, in The Caledonides of the British Isles - reviewed. Geol. Soc. Lond., pp. 139-44.
(33) ________, ________, and Wright, A.E. : 1981, Mem. Canad. Soc. Petrol. Geol. 7, pp. 57-94.
(34) Pidgeon, R.T., and Johnson, M.R.W. : 1974, Earth Planet. Sci. Lett. 24, pp. 105-112.
(35) Powell, D. : 1974, J. geol. Soc. Lond. 130, pp. 575-93.

(36) __________, Baird, A.W., Charnky, N.R., and Jordon, P.J. : 1981, J. geol. Soc. Lond. 138, pp. 661-74.
(37) Read, H.H. : 1961, Lpool Manchr geol. J. 2, pp. 653-83.
(38) Simpson, A. : 1964, Geol. Mag. 101, pp. 20-36.
(39) Smith, D.I. : 1979, in The Caledonides of the British Isles - reviewed. Geol. Soc. Lond., pp. 683-97.
(40) Tanner, P.W.G. : 1976, J. Petrology 17, pp. 100-34.
(41) van Breemen, O., Pidgeon, R.T., and Johnson, M. R.W. : 1974, J. geol. Soc. Lond. 130, pp. 493-507.
(42) Wells, P.R.A., and Richardson, S.W. : 1979, in The Caledonides of the British Isles - reviewed. Geol. Soc. Lond., pp. 339-43.
(43) Winchester, J.A. : 1974, J. geol. Soc. Lond. 130, pp. 509-24.

COMPILATION CALEDONIAN METAMORPHIC MAP OF IRELAND

C. Barry Long
Michael D. Max
Bruce W.D. Yardley*

Geological Survey of Ireland, 14 Hume Street, Dublin 2, Ireland.
* School of Environmental Sciences, University of East Anglia, Norwich NR4 7TJ, U.K.

INTRODUCTION

The Caledonian metamorphic rocks of Ireland occur in a series of inliers (Fig. 1), and for this reason a well-defined pattern of metamorphic zonation is not seen as clearly as elsewhere in the Caledonides. The metamorphic events occurred from mid Cambrian to Devonian times and locally overprint some earlier metamorphic events in inliers of pre-Caledonian basement. The Caledonian 'sensu lato' events can be grouped into three main tectono-thermal episodes: 1. GRAMPIAN (mid to late Cambrian through earliest Ordovician locally), 2. MID ORDOVICIAN (generally pre-Caradocian and broadly equivalent to Taconic), 3. CALEDONIAN 'sensu stricto' (mid Silurian through mid Devonian). These main episodes, and particularly the Grampian, are themselves composite and display sufficient internal variability to allow for the definition of sub-units.

The Grampian effects, with their pressure and temperature extremes and the complexity of the facies-series distribution, are greater than those of subsequent episodes, and are restricted to the orthotectonic Caledonides (Grampianides) exposed in inliers in NW Ireland. The various metamorphic zones in the orthotectonic Caledonides are regarded (1) as representing different depths as well as different crystallisation temperatures; the highest grade rock did not necessarily evolve along the P-T path suggested by the zonal sequence of metamorphic facies series. The narrowest zonal sequence (in Connemara) occurs in a region of

P. E. Schenk (ed.),
Regional Trends in the Geology of the Appalachian-Caledonian-Hercynian-Mauritanide Orogen, 221–233.

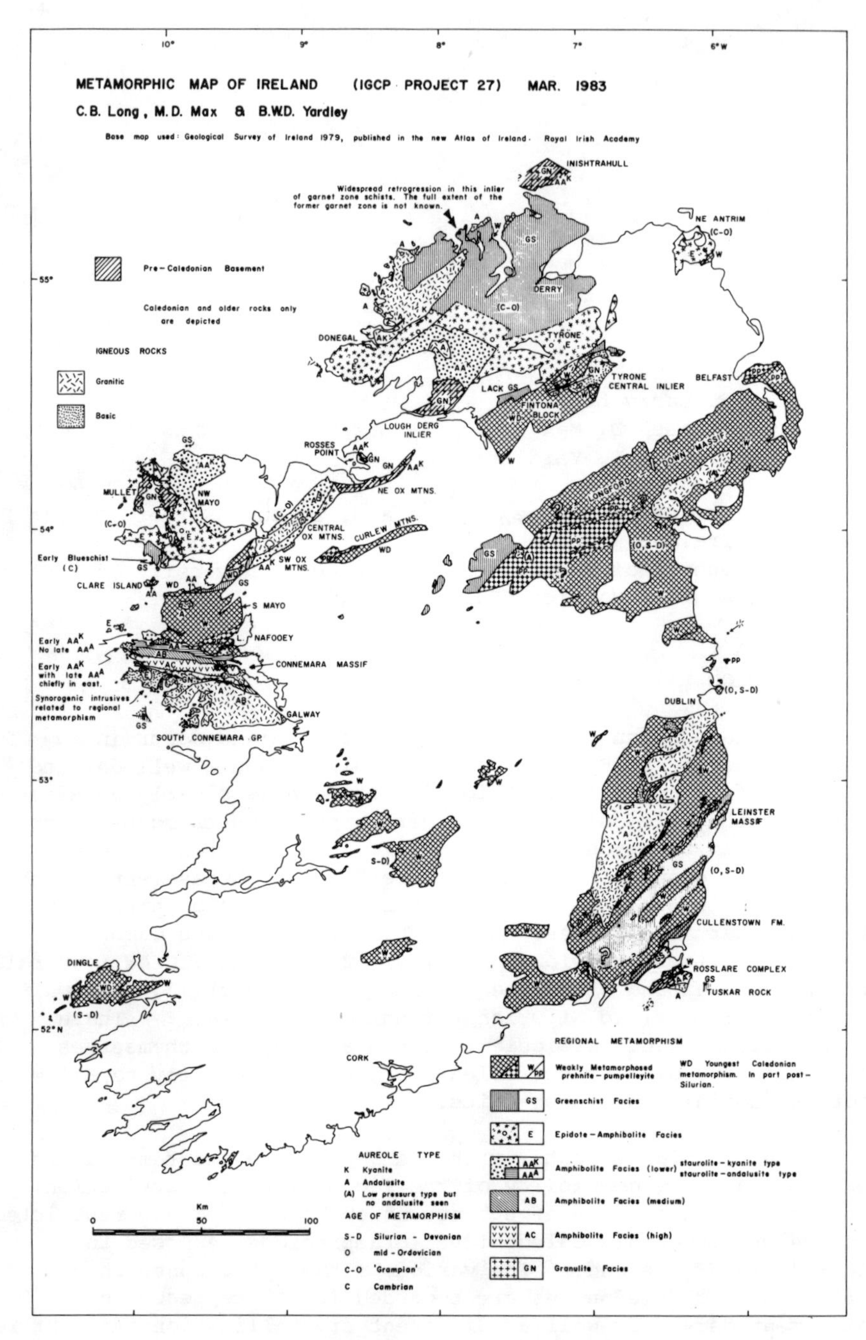

Figure 1. Metamorphic map of the Irish Caledonides. Pattern depicts the highest metamorphic grade reached.

post-metamorphic differential vertical movements. In the Grampian inliers peak metamorphism was generally of a very similar type. The oldest rocks affected by the Grampian orogeny are the shelf marine Dalradian supergroup metasediments, including the Younger Moine (= Grampian Group Dalradian) rocks of NW County Mayo, which span the late Proterozoic (upper Riphean and Vendian) and lower Cambrian (to possibly middle Cambrian on Clare Island). The Grampian metamorphic climax occurred under intermediate pressure (Barrovian) conditions of variable peak conditions, at about 500Ma, following the second phase of regional deformation (MP2). In Connemara a still later Grampian thermal peak was superimposed on this as pressures declined, giving rise to low pressure (Buchan) conditions during uplift and magmatic activity. On Achill Island assemblages formed under the Barrovian conditions were superimposed on 'blueschist' assemblages of an earlier high pressure greenschist facies event and very nearly totally obliterated them.

The paratectonic Caledonides were deposited in a very different environment from most of the sediments of the orthotectonic Caledonides. Whereas the Dalradian sequence was mainly deposited in shallow seas on a broad continental shelf where increasing tectonic instability is regarded as a precursor of the Grampian episode, sediments now seen over broad areas of the paratectonic Caledonides appear to have been deposited in a deeper water, oceanic environment where crustal thickening was being achieved through a complex process of wedge imbrication of an accretionary prism (2). Metamorphism both during sedimentation and imbrication of each tectono-sedimentary unit must therefore be regarded as a semi-continuous process, with local controls throughout the whole of the paratectonic zone subjected to these restricted tectono-thermal processes. The P-T ranges of these events were not marked, and almost everywhere their effects became coalesced with the early metamorphic history. A discrete mid Ordovician (Taconic) event is only well seen in the S Mayo trough in western Ireland, but may be present throughout the pre-Caradocian rocks elsewhere. A widespread low to moderate P-T metamorphic event, the Caledonian 'sensu stricto', has a fairly uniform prograde imprint on the paratectonic Caledonides, with a retrogressive imprint on the orthotectonic Caledonides, and is associated with the intrusion of nearly all the major granitoid bodies in the Irish Caledonides.

The metamorphic pattern within individual sub-zones and inliers often follows no clear zonal progression, though grades generally increase towards areas of pre-Caledonian basement. Boundaries between areas of differing metamorphic grade are commonly tectonic. There are many gaps in our present knowledge of the conditions of each major regional metamorphism and in the metamorphic zonal pattern of Ireland. Large areas have not been studied petrologically in detail (e.g. Donegal, Tyrone and Derry inlier).

REGIONAL FACIES DISTRIBUTION

Pre-Caledonian Basement

Granulite facies, pre-Caledonian basement occurs as tectonic inliers in the orthotectonic Caledonides; Caledonian metamorphic effects are largely retrogressive. Inishtrahull platform gneisses sustained amphibolite facies metamorphism at about 1700Ma. Relics of an even older (c. 2000Ma) granulite facies event may be preserved. Morarian? low amphibolite facies schist occupies the SE part of the Inishtrahull platform. The Erris Complex of NW County Mayo comprises: 1. the c. 2000Ma Annagh Division orthogneiss migmatites which were reworked at c. 1050 to 950Ma (Grenville) at granulite facies, and 2. the Inishkea Division mylonitic schists (c.750Ma) which were metamorphosed under amphibolite facies conditions, with later Caledonian amphibolite and epidote-amphibolite facies overprint. In the NE Ox Mountains, E. Rosses Point and Lough Derg inliers the Slishwood Division psammitic paragneiss, with typically quartz - K - feldspar (mesoperthite) - garnet - biotite (+ kyanite in the NE Ox Mountains), was metamorphosed under Grenville (c. 895Ma) granulite facies conditions with a complex history. Pressures greater than 10kbar and temperatures above $800^{\circ}C$ (with lower pressures to the NE in the Lough Derg inlier) were followed by Caledonian low amphibolite facies reworking. Ultrabasic pods within the NE Ox Mountains include an occurrence of 'regenerated periodite' with euhedral metamorphic olivine in serpentine, suggesting a complex metamorphic history with serpentinite being present prior to granulite facies metamorphism.

Undated granulite facies pelitic paragneiss xenoliths in Carboniferous volcanic agglomerates in the midlands W of Dublin may represent pre-Caledonian basement (3). These contain garnet (almandine 56%), sillimanite, perthitic K-feldspar, plagioclase (An 45-50), quartz, rutile and graphite. The K-feldspar is sanidine in some specimens. Traces of zircon and spinel are known. Basic lithologies contain hypersthene, garnet, plagioclase (An 60) and quartz. Retrogression affects most specimens. In SE Ireland metamorphic basement rocks comprise separate divisions: 1. the Rosslare Complex, dominantly comprising basic orthogneiss and micaceous paragneiss, has been metamorphosed at amphibolite facies conditions prior to intrusion of the 610Ma St. Helens gabbro. The complex may be as old as c. 2000Ma (4). Low amphibolite facies reworking at c. 500Ma was followed by greenschist facies retrogression. 2. the Cullenstown Formation metasediments, which are either entirely late Precambrian or in part Cambrian, have undergone greenschist facies metamorphism.

Orthotectonic Caledonides (Grampianides)

NE Antrim Inlier. Epidote-amphibolite facies was probably the highest Grampian grade attained (5), but there is little modern work.

Donegal, Tyrone and Derry Inlier. Shortage of modern work on the regional metamorphism also makes compilation difficult in certain areas here. The Dalradian schists close to their tectonic contact with the Lough Derg inlier, produced low amphibolite facies kyanite + staurolite + almandine + muscovite + biotite + plagioclase + quartz in pelites with P(total) = 6 to 7kbar and T c. 600°C (1, 6) during the Grampian event. Despite lithological limitations, it is considered that the kyanite - staurolite zone is more extensive than once believed, but the full extent to the NE is uncertain. A gradational decrease to epidote - amphibolite facies, which extends across the width of the inlier (1) is thought likely. NE trending faults influence the outcrop pattern. The N limits of the epidote - amphibolite facies zone are not well defined: at a few localities N and S of the Main Donegal granite several outcrops with regional garnet of unreported composition occur in pelitic lithologies in the map area shown as greenschist facies (7). There was possibly a more generally widespread epidote - amphibolite facies zone that has since been thoroughly retrogressed. A postulated garnet isograd on the Slieve League peninsula (1) is now believed to be a consequence of later retrogression which has been less effective in the W. Lower amphibolite facies probably obtained on the Rosguill peninsula in N Donegal where regional garnet is common, though staurolite is absent (7). In parts of the Inishowen peninsula the regional metamorphic grade possibly did not exceed chlorite subfacies. Chloritoid occurs in the Sperrin mountains area of County Tyrone, but details have not yet been published.

Central Tyrone Inlier. A suite of basic intrusives, most probably related to a similar suite in Connemara, occupies much of the inlier. Uncertainty exists as to the extent to which the sillimanite + cordierite + K-feldspar bearing schists should be regarded as a regional, rather than a thermal effect (8).

Lack Inlier. The highest metamorphic grade attained was possibly epidote - amphibolite facies. Garnet has been recorded at 3 localities. There is no modern work.

Ox Mountains and Rosses Point Inliers. This is a composite inlier with a Grenville basement inlier at its NE end (4, 6). At the SW end are undisputed Dalradian successions, and in the central portion, and extending to near the SW end, are probable Dalradian metasediments of the Ox Mountains succession. Two very small areas of Southern Highland Group (upper) Dalradian low greenschist facies (chlorite) turbidites occur near the SW end. Tectonic contacts separate the different structural and metamorphic units.

The epidote - amphibolite facies maximum prograde assemblages of the SW Ox mountains include pelites with garnet + biotite + chlorite, garnet + chloritoid + chlorite, or paragonite + chloritoid + chlorite, all with muscovite, quartz and plagioclase. They can be linked by comparison of metamorphic assemblages and history with the rocks of the adjacent NW County Mayo inlier. The low amphibolite facies prograde assemblage of the SW end of the central Ox Mountains, with staurolite + almandine in pelites, can be traced along strike to the NE and across the kyanite isograd. In addition to kyanite, earlier prograde armoured corroded inclusions of chloritoid in garnet are preserved (6). The P-T conditions of the kyanite - staurolite - almandine pelites, P(total) 6-7kbar and T c. 600°C, are indistinguishable from the conditions for formation of the kyanite - staurolite - almandine pelites adjacent to the Lough Derg inlier in S Donegal. The Manorhamilton pelite (6) is tectonically emplaced in the Grenville basement of the NE Ox Mountains and the pelites in the W part of the Rosses Point inlier (6) are in tectonic contact with the Grenville basement of the E part of the Rosses Point inlier. Both Manorhamilton and W Rosses Point also include similar kyanite - staurolite pelites. Retrograde chloritoid occurs in the central Ox Mountains and in W. Rosses Point.

In the extensively retrogressed (? Taconian) area of the NE part of the central Ox Mountains inlier rare patches of shimmer aggregate (Na-poor muscovite + paragonite ± magnetite ± chloritoid) probably represent earlier staurolite and kyanite, and garnet and biotite are thoroughly chloritised. Part of this area, however, may never have exceeded epidote - amphibolite facies.

NW Mayo and Achill Island Inlier. In this inlier a staurolite - kyanite zone in the N portion of the inlier, grades S into epidote - amphibolite facies (1, 9, 10). Lack of inland exposure obscures the exact location of the isograd. Chloritoid is present in N Achill. The S part of Achill Island, Achill Beg Island and a small area of adjacent mainland most probably did not exceed greenschist facies conditions during the Grampian orogeny.

In the S of Achill Island an early Grampian, pre-Barrovian 'blueschist' assemblage is preserved in a Dalradian volcanic metabasite (11). A boudinaged epidosite band with crossite occurs in predominantly albite - epidote - chlorite schists. The crossite is early and is directly replaced to a variable degree by barroisitic amphibole. Actinolite may mantle either the crossite or barroisite. These two alteration products are sometimes gradational. Stilpnomelane occurs with presumed crossite relicts which are now barroisitic and actinolitic amphibole. High pressure greenschist facies, rather than glaucophane schist facies, is considered most likely.

Clare Island. Two metamorphic inliers are known on Clare Island: 1. The Ballytoohy inlier of greenschist facies pelites, and spilites with chlorite and sometimes actinolite (12), and 2. the Kill Inlier which comprises two units: a. the Portruckagh Formation, metamorphosed in the greenschist facies, and b. the Deer Park Complex of uncertain though probably Dalradian age. In this, pelites are lacking, but semi-pelitic assemblages with garnet + biotite + andesine, and metabasite with diopside + green hornblende suggest low amphibolite facies conditions of regional metamorphism (12).

Clew Bay Inlier (Westport). The inlier is complex, with a greenschist facies area of Southern Highland Group Dalradian schists in tectonic contact with several smaller areas of probably low amphibolite facies Deer Park Complex (probably related to the Kill inlier of Clare Island). Sillimanite is reported, though not confirmed and, if present, could possibly be a local thermal contact mineral.

Connemara Inlier. Compilation work has drawn chiefly on much modern work (e.g. 13, 14, 15, 16). The Barrovian metamorphic grade generally rises from low amphibolite facies in the N, through amphibolite facies to sillimanite - cordierite - K-feldspar gneisses, and garnet - cordierite - sillimanite - plagioclase gneisses of the granulite facies in the S, close to and within the belt of intrusive migmatitic gneisses. P-T extremes are in excess of 700°C and 5-7kbar. Extensive metasomatism probably accompanied emplacement of the intrusive bodies of this belt. A small area of migmatites occurs in the NE near Kilbride Lough (13). The Barrovian metamorphism spanned D1 and the D2 nappe-forming deformational events, rising to low granulite facies in the S prior to the post-D1 intrusion of the syn-tectonic gabbroic suite. An MP2 pressure peak was attained during the later stages of emplacement of the migmatitic gneisses. Many sillimanite pelites lack K-feldspar and muscovite as a consequence of their compositions.

Metamorphically, Connemara can be contrasted with the other Dalradian inliers of Ireland. Following the Barrovian metamorphic peak, uplift and steepening of the thermal gradient resulted in a change to Buchan (low pressure) regional metamorphism, and pressures fell while temperatures continued to rise, leading to widespread development, chiefly in eastern Connemara, of late andalusite and cordierite (rarely coexisting) and sillimanite which replaces the andalusite (15). Even later metamorphic retrogression in N Connemara affected the low amphibolite facies zone giving rise to small areas of epidote - amphibolite facies with retrograde chloritoid in the Renvyle area. On Inishbofin and Inishshark (W of Renvyle Head) the epidote - amphibolite facies assemblages (also with chloritoid) may be either prograde, or retrograde representing thorough retrogression of low amphibolite facies assemblages.

Epidote - amphibolite facies rocks, chiefly of rhyolitic composition but of uncertain affinity, structurally underlie the Connemara migmatites and basic and ultrabasic bodies in the Delaney Dome tectonic window (in SW Connemara), which is ringed by epidote - amphibolite facies mylonites at their structural base.

Paratectonic Caledonides (Mid Ordovician and Caledonian 's.s.')

Weakly metamorphosed Ordovician and Silurian rocks occur in the Longford - Down (2) and Leinster (17, 18) massifs, S County Mayo, and in the more isolated exposures in the inliers of the midlands area, the SW part of the Curlew mountains, and the Dingle peninsula in the SW. Additionally, a few Devonian Old Red Sandstone inliers have been very weakly metamorphosed (e.g. Fintona block, SW Ox Mountains inlier, Curlew Mountains inlier and S County Mayo), late in their tectono-thermal histories.

Prehnite - pumpellyite facies assemblages have been recognised in the W end of the Longford - Down massif (19), in the Lough Nafooey Group basic volcanics (Tremadoc to Arenig) in S Mayo (20), at the SW end of the Curlew Mountains inlier and on Lambay Island NE of Dublin (19). Recognition of this metamorphic facies in Ireland is recent, but from work in the Southern Uplands of Scotland (2) and the known Irish occurrences it is to be expected that rocks of this grade will in future be found more widely.

Greenschist facies occurrences are known locally in the Ordovician and Silurian rocks of the W end of the Longford - Down massif, in S Mayo, in the Leinster massif (17), and also in the Ordovician South Connemara Group. The mainland and offshore Tuskar Group in SE Ireland (late Precambrian or Cambrian, with a minimum age of 550Ma) is now generally greenschist, but may earlier have attained lower amphibolite facies conditions. Supposed regional chloritoid has been recorded S of the Leinster granite.

CONTACT METAMORPHISM

Most of the Caledonian intrusives, the larger of which are

granitoid, were emplaced in both the ortho-and paratectonic Caledonides during the interval c. 420 - 390Ma. The granitoid plutons are mostly thermally disharmonious with their country rocks, and contact assemblages represent a fairly narrow and relatively low pressure range (in Donegal, 5kbar at 500°C decreasing with time to 3.5kbar at more than 700°C) indicative of a general though variable amount of uplift following the peak of regional metamorphism in the orthotectonic Caledonide areas prior to granite emplacement. Andalusite, sillimanite and cordierite are common aureole minerals, and rare kyanite is believed to have formed below the Al_2SiO_5 triple point. Complex aureole histories have frequently been described, hydrothermal events are generally late, and later retrogression is a common feature (7, 21, 22, 23). Chloritoid is a contact mineral in the aureoles of the Ardara granite (Donegal) and the Ballynamuddagh granite (SE Ireland).

The Ox Mountains granodiorite (c. 480Ma, Grampian) and the Oughterard granite (c. ?460Ma) of Connemara are apparently thermally harmonious with their envelope rocks, there being no obvious contact metamorphism, probably because the country rocks were still hot at the time of emplacement soon after the peak of regional Grampian metamorphism (6).

The Carnsore granite of SE Leinster, with a minimum age of c. 550Ma has andalusite and cordierite in its aureole and emplacement conditions have been estimated as P = 2 - 3kbar, T = 500 - 550°C. Also in the SE, the Saltees granite (c. 437Ma) has no proven aureole, though this could be thermally harmonious with regional amphibolite facies country rocks. The granitoid intrusive plutons of Tyrone are generally not well known. Certain bodies are probably mid Ordovician, though others are possibly pre-lower Ordovician. Sillimanite and cordierite are recorded in the hornfelses (8). Appinites associated with certain of the Donegal granites, the Newry granodiorite, the Leinster granite and the Ox Mountains granodiorite have narrow hornfelses with fibrolite and andalusite.

The now fragmented intrusive c. 500Ma syntectonic gabbroic suite and related gneisses of Connemara (and probably of the Tyrone central inlier), which intruded granulite facies schists, are thermally harmonious in a regional sense, though not locally, in that the pronounced aureole effects around the basic and ultrabasic bodies are clearly developed by contact metamorphism (24, 25, 26). The metamorphic and metasomatic effects of the Connemara migmatitic orthogneisses are regionally too widespread to be simple contact effects of this magmatism (27): both the intrusives and the high grade regional metamorphism are more likely to be consequences of a common cause. Desilication, dealkalisation and removal of alumina were caused by selective melting and subtractive metasomatism. Quartzo-feldspathisation and migmatisation of the gabbros and the country rock metasediments, leading to the development of large areas of migmatitic quartz-diorite gneiss (28), were accompanied by continuing amphibolite facies regional metamorphism and deformation (D2 and later). The intrusive and metasomatic phase, which broke up the gabbroic bodies, culminated in the pre-D3 emplacement of small granitoid bodies, which are now K-feldspar gneisses.

Pressures in the aureoles of the gabbros have been estimated as c. 5kbar and temperature estimates range up to 850°C for aureoles, and 1000°C for pelitic xenoliths. Aureole pelites contain various combinations of sillimanite, cordierite, andalusite, garnet and biotite. Strongly desilicated pelitic xenoliths contain combinations of corundum, magnetite, green spinel, cordierite, högbomite, and orthopyroxene.

No contact effects have been recorded around the c. 610Ma St. Helens Gabbro or the older Greenore diorite which intrude the Rosslare complex of SE Ireland.

INTERPRETATION

In the late Cambrian, after a period of relatively steady-state plate movements, changes in plate motions probably led to a NW (?) directed subduction zone being developed beneath the Dalradian basin. The early Grampian high pressure phase, represented now solely by the Achill island 'blueschist', occurred in a setting of reduced heat flow. Though possibly once more widespread, the 'blueschist' assemblages have been almost entirely destroyed throughout the western Dalradian by subsequent Grampian overprinting under Barrovian conditions.

The widespread post-D2 Barrovian conditions, which represent a fairly normal crustal thermal gradient, were followed by decreasing pressures, with continuing high temperature conditions in Connemara, associated with the synorogenic igneous activity of the early Ordovician volcanic arc root zone. The folded Dalradian nappe pile became basement to a Tremadoc and Arenig volcanic arc, with centres in Connemara and probably central Tyrone where the root zones are now exposed, and S Mayo, the Curlew mountains and Tyrone, where extrusives are exposed (29).

Further changes in relative plate motions brought about extinction of the volcanic arc by end Llanvirn times, cooling of the Dalradian rocks and initiation of the Longford - Down and Southern Uplands (Scotland) accretionary prism. Much of the Irish Dalradian underwent differential uplift and erosion during the lower Palaeozoic.

ACKNOWLEDGEMENTS

Thanks to Richard Smith (Geological Survey of Northern Ireland), Colin Ferguson, John Barber, Jonathan Gray and Ian Sanders. Published by permission of the Director of the Geological Survey of Ireland.

REFERENCES

1. Yardley, B.W.D.: 1980. Metamorphism and orogeny in the Irish Dalradian. J. geol. Soc. London 137, pp. 303-9.
2. Leggett, J.K., McKerrow, W.S., Morris, J.H., Oliver, G.J.H. & Phillips, W.E.A.: 1979. The north-western margin of the Iapetus Ocean. In Harris, A.L., Holland, C.H. & Leake, B.E. (Eds.) The Caledonides of the British Isles - reviewed. Spec. Publ. geol Soc. London 8, pp. 499-511.
3. Strogen, P.: 1974. The Sub-Palaeozoic basement in central Ireland. Nature, London 250, pp. 562-3.
4. Max, M.D. & Long, C.B., (in press): Pre-Caledonian Basement in Ireland and its structural significance. Geol. J.
5. Wilson, H.E.: 1972. Regional Geology of Northern Ireland. Geol. Surv. N. Ireland, H.M.S.O.
6. Yardley, B.W.D., Long, C.B. & Max, M.D.: 1979. Patterns of metamorphism in the Ox Mountains and adjacent parts of western Ireland. In Harris, A.L., Holland, C.H. and Leake, B.E. (Eds.) The Caledonides of the British Isles - reviewed, Spec. Publ. geol. Soc. London 8, pp. 369-74.
7. Pitcher, W.S. & Berger, A.R.: 1972. The Geology of Donegal: a study of granite emplacement and unroofing. Wiley and Sons, New York.
8. Hartley, J.J.: 1933. The geology of north-eastern Tyrone and the adjacent portions of County Londonderry. Proc. Roy. Irish Acad. 41B, pp. 218-85.
9. Max, M.D.: 1973. Caledonian metamorphism in part of north-west County Mayo, Ireland. Geol. J. 8, pp. 375-86.
10. Gray, J.R.: 1981. Regional Metamorphism in NW Mayo, Eire, and its Bearing on the Regional Geology. Ph.D. thesis, Univ. E. Anglia (unpubl.).
11. Gray, J.R. & Yardley, B.W.D.: 1979. A Caledonian blueschist from the Irish Dalradian. Nature, London 278, pp. 736-7.
12. Phillips, W.E.A.: 1973. The pre-Silurian rocks of Clare Island, Co. Mayo, Ireland, and the age of the metamorphism of the Dalradian in Ireland. J. geol. Soc. London 129, pp. 585-606.
13. Leake, B.E., Tanner, P.W.G. & Senior, A.: 1981. 1:63,360 scale Map of Fold Traces (F3 to F5) and Regional Metamorphism in the Dalradian Rocks of Connemara (based

on work of 31 authors and the G.S.I.) Univ. Glasgow.

14. Badley, M.E.: 1976. Stratigraphy, structure and metamorphism of Dalradian rocks of the Maamturk Mountains, Connemara, Ireland. J. geol. Soc. London 132, pp. 509-20.
15. Yardley, B.W.D.: 1976. Deformation and metamorphism of Dalradian rocks and the evolution of the Connemara cordillera. J. geol. Soc. London 132, pp. 521-42.
16. Yardley, B.W.D., Leake, B.E. & Farrow, C.M.: 1980. The metamorphism of Fe-rich pelites from Connemara, Ireland. J. Petrol. 21, pp. 365-99.
17. Wheatley, C.J.V.: 1971. Economic geology of the Avoca mineralised belt, SE Ireland and Parys Mountain, Anglesey. Ph.D. thesis, Univ. London (unpubl.).
18. Brück, P.M., Colthurst, J.R.J., Feely, M., Gardiner, P.R.R., Penny, S.R., Reeves, T.J., Shannon, P.M., Smith, D.G. & Vanguestaine, M.: 1979. South-east Ireland: Lower Palaeozoic stratigraphy and depositional history. In Harris, A.L., Holland, C.H. and Leake, B.E. (Eds.) The Caledonides of the British Isles - reviewed, Spec. Publ. geol. Soc. London 8, pp. 533-44.
19. Oliver, G.J.H.: 1978. Prehnite - pumpellyite facies metamorphism in Co. Cavan, Ireland. Nature, London 274, pp. 242-3.
20. Ryan, P.D., Floyd, P.A. & Archer, J.B.: 1980. The stratigraphy and petrochemistry of the Lough Nafooey Group (Tremadocian), western Ireland. J. geol. Soc. London 137, pp. 443-58.
21. Brindley, J.C.: 1969. Caledonian and Pre-Caledonian intrusive rocks of Ireland. In Kay, M. (Ed.) North Atlantic - Geology and Continental Drift Mem. Amer. Ass. Petrol. Geol. 12, pp. 336-53.
22. Leake, B.E.: 1978. Granite emplacement: the granites of Ireland and their origin. Geol. J. Spec. Issue 10, pp. 221-48.
23. Naggar, M.H. & Atherton, M.P.: 1970. The composition and metamorphic history of some aluminium silicate-bearing rocks from the aureoles of the Donegal granites. J. Petrol. 11, pp. 549-89.
24. Evans, B.W.: 1964. Fractionation of elements in the pelitic hornfelses of the Cashel - Lough Wheelaun intrusion, Connemara, Eire. Geochim. Cosmochim. Acta 28, pp. 127-56.
25. Evans, B.W. & Leake, B.E.: 1970. The geology of the Toombeola District, Connemara, Co. Galway. Proc. R. Irish Acad. 70B, pp. 105-39.
26. Leake, B.E. & Skirrow, G.: 1960. The pelitic hornfelses of the Cashel - Lough Wheelaun intrusion, Co. Galway, Eire. J. Geol. 68, pp. 23-40.
27. Senior, A. & Leake, B.E.: 1978. Regional metasomatism and the geochemistry of the Dalradian metasediments of Connemara, western Ireland. J. Petrol. 19, pp. 585-625.

28. Leake, B.E.: 1970 (for 1969). The origin of the Connemara migmatites of the Cashel district, Connemara, Ireland. Q. J. geol. Soc. London 125, pp. 219-76.

29. Yardley, B.W.D., Vine, F.J. & Baldwin, C.T.: 1982. The plate tectonic setting of NW Britain and Ireland in late Cambrian and early Ordovician times. J. geol. Soc. London 139, pp. 455-63.

A NORTHERN APPALACHIAN METAMORPHIC TRANSECT - EASTERN TOWNSHIPS, QUEBEC MAINE COAST TO THE CENTRAL MAINE COAST

C.V. Guidotti[1], W.E. Trzcienski[2] and M.J. Holdaway(3)

(1) Department of Geological Sciences, University of Maine, Orono, Maine 04473
(2) Département de géologie, Université de Montréal, Montréal, Québec H3C 3J7
(3) Department of Geological Sciences, Southern Methodist University, Dallas, Texas 75275

1. INTRODUCTION

A more or less continuous traverse starting near Montreal, Quebec and ending on the Atlantic coast near Rockland, Maine affords one the opportunity to look at a metamorphic section across an entire orogenic belt. In this section are exposed rocks that vary from non-metamorphosed platformal sediments to rocks that have undergone intense metamorphism associated with granitic batholith paragenesis. As well, in this section, one sees the effects of three metamorphic episodes complicated by tectonic discontinuities and juxtapositioning.

2. QUEBEC

The general geology of the Quebec section has been recently described (1) and is here only briefly summarized. The section can be divided into a number of distinct units. The northwesternmost unit of the transect is underlain by slightly deformed shelf-type sediments of Cambro-Ordovician age. Next come a sequence of Lower Cambrian to Lower Ordovician carbonates, sandstones, shales and conglomerates that constitute the external domain of St Julien and Hubert (1) followed by the internal domain composed of volcanics, ophiolite, shale and sandstone (Fig. 1). These units are overlain to the southeast by the Connecticut Valley - Gaspé Synclinorium rocks of Siluro-Devonian age. These latter rocks are composed of quartzites, impure carbonates and interbedded pelites. Into this sequence are intruded a number of Devonian granites. East of the synclinorium is another thin belt of Cambro-Ordovician

P. E. Schenk (ed.),
Regional Trends in the Geology of the Appalachian-Caledonian-Hercynian-Mauritanide Orogen, 235–247.

metavolcanics and metasediments that lie on the Chain Lakes Massif of Precambrian age.

Precambrian Metamorphism

Precambrian metamorphic rocks are limited in extent but do occur in the external domain (2) and probably in the internal domain as well (3) (Localities 1 & 2, Fig. 1). Both these localities are tectonic blocks that have been brought to the surface along fault planes. They consist of high grade amphibolites partially retrograded to lower greenschist facies. The only other high grade Precambrian rocks are those found in the Chain Lakes massif (Fig. 1) where K-feldspar-sillimanite gneisses have been described (4). The available radiometric data give ≈ 900 Ma (2) for one Quebec locality and (1500 Ma) for the Chain Lakes Massif (5).

Taconic Metamorphism

Much of the Quebec section north and west of the Connecticut Valley - Gaspé Synclinorium has received its most intense metamorphism during the Taconic Orogeny. Because of the major faulting and nappe structures associated with this orogenic event there are a number of metamorphic discontinuities, especially within the internal domain.

The least disturbed rocks are those of the westernmost sequence, the platformal rocks. Here, numerous gas wells along with crystallinity index and reflectance work suggest metamorphic temperatures below 175°C at less than 2 Kb pressure. In the external domain low grade prehnite-pumpellyite-bearing metavolcanics occur (Loc 3 & 4, Fig. 1). It has been suggested (6) that these rocks were metamorphosed at about 200°C and 2 Kb.

Within the internal domain, one finds the metamorphic history significantly more complex. The broad regional gradient is of increasing metamorphism from northeast toward the southwest as characterized by actinolite, albite, chlorite, quartz, epidote, opaque-bearing metavolcanics in the northeast (3) and hornblende-biotite amphibolite in the southeast (7). Metapelites are characterized by white-mica, quartz, chlorite, opaque, plagioclase ± biotite ± stilpnomelane ± chloritoid. Within this regional gradient, however, there is in the Richmond-Asbestos area (Loc 5 & 6, Fig. 1) blue, crossitic amphibole (8) and highly aluminous pumpellyite (6) suggesting a high pressure anomaly. This anomaly lies within the same belt of rocks that further to the south in Vermont contain glaucophane and omphacite (9).

Another prominent anomaly is found at Belmina Ridge (Loc 7, Fig. 1) where a high grade, banded, amphibolite occurs associated with the Thetford Mines ophiolite. Feininger (10) has

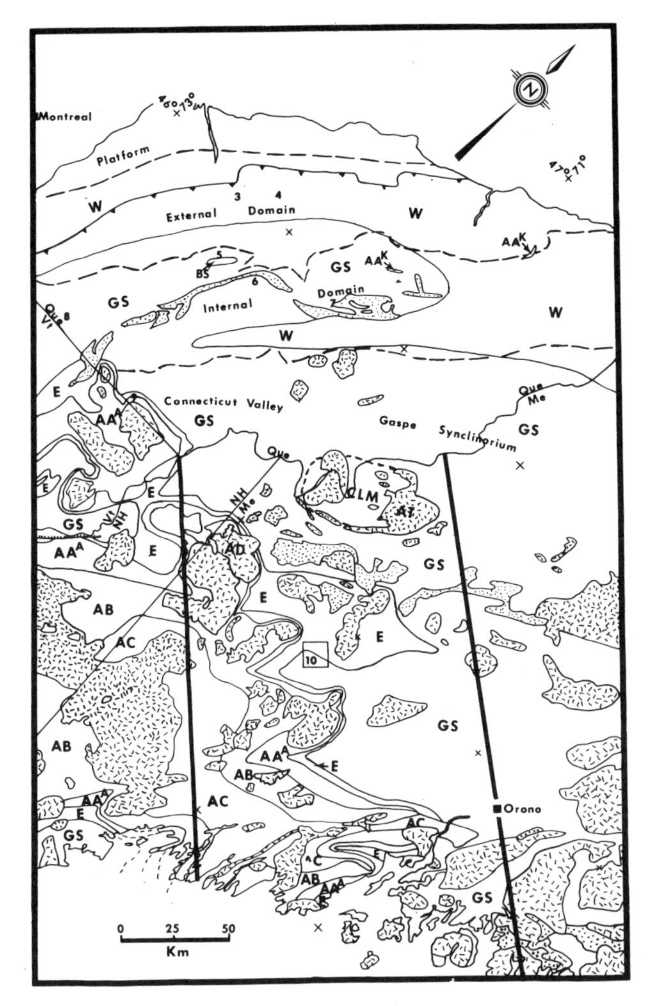

Fig. 1 Metamorphic compilation of the Eastern townships, Quebec, Central Maine and adjacent areas. Metamorphic facies designations: W - weakly metamorphosed (also prehnite and pumpellyite, GS - greenschist characterized by chlorite + actinolite or biotite, E - epidote amphibolite characterized by FE-garnet in pelites or Na-plagioclase + epidote + hornblende in basic rocks, AA - lower amphibolite characterized by staurolite and kyanite (AA^k) or andalusite (AA^A) in pelites and diopside in basic rocks, AB - medium amphibolite characterized by sillimanite in pelites, AC - upper amphibolite characterized by sillimanite + K-feldspar in pelites, GN - granulite characterized by ortho- + clinopyroxene in basic rocks, BS - blueschist, here characterized by blue amphibole. [granite symbol] - granites, [gabbro symbol] -gabbros, CLM-Chain Lakes Massif, AT- Attean Quartz Monzonite, AD-Adamstown Granite, [10] - Kingfield and Anson Quadrangles (see text).

interpreted this garnet-hornblende-pyroxene amphibolite as a thick slab (≈ 800 m) of oceanic crust welded to and forming the sole-plate of the Thetford ophiolite. He has estimated that the metamorphic temperature varied from about 500°C at the base of the plate to ≈780°C at the top of the plate while the pressure was between 5 and 7 Kb. The overlying ultramafic rocks are highly serpentinized and offer little information as to their metamorphism.

East of the ophiolite belt a metamorphic low extends southward along the eastern margin of the internal domain. St Daniel Formation rocks are found here and to date they have been found to contain no metamorphic index minerals or recrystallization.

Between the Connecticut Valley - Gaspé Synclinorium and the Chain Lakes Massif is found another thin belt of Cambro-Ordovician rocks that are as well low greenschist facies rocks. The metavolcanics contain albite, actinolite, chlorite, quartz and opaque while in the metapelites the common assemblage is white mica, chlorite, quartz, albite, opaque ± biotite ± spessartine garnet.

A recent review of available radiometric age dates (11) indicates that only along the western margin of the Cambro-Ordovician sequence does the radiometric data agree with the stratigraphic data. Elsewhere metamorphic ages were found to have been affected by the later Acadian event. Any attempt to date to determine a Taconian-Acadian boundary has been fruitless. In Vermont, however, Laird and Albee (9) have convincingly shown that a Taconian high pressure metamorphism was overprinted by an Acadian high temperature metamorphism.

Acadian Metamorphism

The Connecticut Valley - Gaspé Synclinorium is composed of thin-bedded quartzites and interbedded pelitic and carbonate-bearing metasediments characterized most commonly by assemblages containing various combinations of quartz, muscovite, biotite, calcite, plagioclase, chlorite, opaques and graphite. Only in the aureoles around the Devonian granites has contact metamorphism produced higher grade assemblages. In a recent study of the Stanhope Pluton (Averill in Vermont) on the Quebec-Vermont border (Loc 9, Fig. 1) Erdmer (12) has drawn isograds based on the first appearance of andalusite, staurolite and sillimanite in the metapelites. In the interbedded calc-schists similar isograds are drawn for the first appearance of tremolite, K-feldspar, clinozoisite and diopside. All these isograds run more or less parallel to the granite-country rock contact and occur within the first three kilometers of the contact.

3. MAINE

The discussion of the southern portion of the metamorphic transect being considered in this paper will be largely concentrated on the area lying between the two subparallel lines which on Fig. 1, extend from the southeastern border of Quebec, southeasterly to the central coast of Maine. For the purpose of providing a brief description of the geologic setting, we can consider the geology along a line lying roughly between the two lines described above.

Starting at the NW and going to the SE one passes through 30 to 40 miles of Cambro-Ordovician clastic metasediments and some meta-volcanic rocks. These strata have been deformed into tight NE trending folds. Continuing to the SE a very broad terrane of Siluro-Devonian strata is encountered. It extends for over 100 miles to within 30-40 miles of the coast.

The Siluro-Devonian strata are mainly metasandstones, graywackes, and shales with minor amounts of limestones and volcanics. Steeply dipping, northeast trending strata are the rule and reflect tight anticlines and synclines. However, there is evidence of earlier, broad-scale recumbent folding upon which the tight anticlines and synclines are now superimposed (13).

About 30 miles from the coast one again finds a terrane underlain by Cambro-Ordovician strata which consist of metamorphosed clastic sediments and volcanic rocks. As along other parts of the line being described, the sedimentary and volcanic strata are strongly folded.

In a broad structural sense, the above described pattern of Cambro-Ordovician to Siluro-Devonian-and back to Cambro-Ordovician strata as one crosses the State of Maine is a reflection of the Merrimack Synclinorium (14,15) trending through the center of the region.

Another large scale geologic feature included between the lines shown on Fig. (1) is the Chain Lakes Massif (CLM) (4). It is an enigmatic terrane of clastic rocks (probably Precambrian) which has experienced a metamorphic history differing from that of the surrounding Paleozoic rocks.

Fig. (1) shows a fairly complex pattern of metamorphic isograds which in a very general fashion trend ESE across the State. These isograds are believed to be essentially all part of one, "nearly synchronous" event (designated as M_3 below) which affected the whole state. In detail, there may be some time differences among the different regions affected by M_3. On a scale which would include the whole of the New England Appalachians, the isograds shown on Fig. (1) form the northern boundary of a very extensive high grade terrane extending several hundreds of miles southward.

It is apparent from Fig. (1) that plutonic bodies are common in the area being discussed and as discussed further below, there is some spatial interrelationship between plutonism and M_3 metamorphism. The plutons range from gabbros to granites, some of the latter being binary types (S-granites (16)). Of particular note is the strong tendency for the binary granites to be located in the areas of high metamorphic grade and for the non-binary granitic plutons and gabbros to be located in the low-grade areas. Although exceptions do occur, that pattern is nonetheless distinct. Moreover, as previously discussed (17) there is a tendency for the granitic plutons in the high-grade areas to give isotopic ages near 380 M.Y. whereas those intruding the low grade terrain tend to have ages near 360 M.Y. One ophiolite sequence has also been recognized (the Boil Mountain Complex (18)) and occurs along the southern border of the CLM. It is probably Cambro-Ordovician in age. With the exception of the Ordovician Attean Quartz Monzonite and Adamstown granite (see Fig. 1) all of the other large plutonic bodies have Devonian ages.

An Overview of the Metamorphism in the area of concern:

Numerous petrologic studies have been made in this area during the past 30 years. The vast majority of these studies have involved pelitic schists and it is the work on such rocks that forms the basis of the following treatment. Probably the first study using a modern approach was that of Green (19). If one considers the boundary between AA and AB on Fig. (1), it is apparent that four, NE trending lobes of metamorphism can be identified in Maine. Numerous workers have studied various parts of the three western lobes and this work has been summarized (17). Thus far, relatively little work has been done on the eastern lobe and so subsequent discussion of it should be considered as quite tentative.

In the paper of Holdaway et al. (17), at least four Acadian metamorphic events were recognized. These include:

(1) Broad scale, low grade regional metamorphism (M_1).
(2) Broad scale, low to moderately high grade regional metamorphism (M_2).
(3) Broad scale, low to high grade regional metamorphism (M_3). Isograds of this event are shown on Fig. (1).
(4) Local contact metamorphism where later Acadian plutons have intruded the lower grade portions of the M_3 terrane (M_4).

Although not yet demonstrated, it is presumed that M_1-M_3 have also affected the rocks in the easternmost metamorphic lobe.

For example, Wones (Pers. Comm.) has recognized two high-grade regional events near the north end of that lobe. (However, his interpretation of the ages of these events differs from that presented herein). Also, Bickel (20) in the central part of the lobe, and Guidotti (reconnaissance work) in the southern part of the lobe have observed metamorphic textural features which are identical with those developed in rocks affected by M_2 and M_3 in the three western lobes. Hence, until clearly demonstrated to be otherwise, it is presumed that M_1-M_3 have affected the easternmost lobe also.

Two other aspects of the metamorphism in the area considered which should be mentioned are (a) The events recorded by the Chain Lakes Massif, and (b) possible Taconic metamorphic effects. The CLM indicates that it was metamorphosed to very high grade in the Precambrian and then retrograded to greenschist facies later - possibly during the Taconic and/or Acadian Orogeny, such that its present grade is similar to that found in the surrounding Paleozoic rocks (21, 4).

Some definite Taconic metamorphic effects can be recognized. These include the contact effects adjacent to the Ordovician plutons mentioned above. In addition to these contact affects the possibility exists that the Cambro-Ordovician rocks near the Quebec border were regionally metamorphosed to greenschist facies during the Ordovician Taconic event as well as by similar grade Acadian events. Moreover, as developed earlier the Cambro-Ordovician strata only 50 miles to the NW in Quebec do seem to have been affected by Taconic metamorphism. Hence, it is not unrealistic to speculate that this event affected the similar age strata in Maine. Work is presently in progress on trying to discern how much areal overlap has occurred between Taconic and Acadian greenschist metamorphism.

Description of the Specific Acadian Metamorphic Events M_1, M_2, M_3, & M_4:

In discussing these events it is expedient to start with M_3 and then work back sequentially through the other regional events. M_4 may then be considered out of sequence in a separate fashion because it involves only local contact effects around plutons which intrude low to medium grade portions of the M_3 terrain.

M_3, The Youngest Regional Metamorphism:

The isograds of this event are shown on Fig. (1) and it can be seen that the grades involved range from greenschist facies (GS) in the NE to upper amphibolite facies (AC) to the SW. The (AB): (AC) boundary corresponds to an extension of the K-feldspar ± sillimanite isograd as recognized by Evans and Guidotti (1966).

The (AA): (AB) boundary corresponds to the first appearance of sillimanite which is followed at slightly higher grade by the disappearance of staurolite. This isogradic reaction has been mapped by numerous workers in Maine (19, 22, 23). Because of the bulk compositions present, in virtually all areas the incoming of an Al-silicate (sillimanite in M_3) is via the AFM projection tie line flip; staurolite + chlorite going to Al-silicate + biotite.

The three western lobes of M_3 in Maine show a close spatial relationship with large plutons of binary granite. However, it would be inappropriate to infer simple contact metamorphism because when considered in detail the isograds are in some cases found to be cut by the plutons. Moreover, in the higher grades of M_3, the isograds show relatively little spatial relationship to the plutons.

Along the NE noses of the three western lobes the isogradic surfaces dip fairly steeply (40-50°) to the NE. But, in some of the areas between the lobes the isogradic surfaces have quite low dips. Indeed, Guidotti (22) and Cheney and Guidotti (24) describe instances where the isogradic surfaces are essentially flat lying.

In the case of the easternmost lobe, there is no particular relationship between the distribution of metamorphic grades and plutons. In part this is reflected by greater convolutions in the isograd patterns. However, part of this increase in the complexity of the isograd patterns (e.g. the contact of (E) with (AC)) appears to be due to structural complications produced by post-metamorphic faulting.

As mentioned previously, the four lobes of M_3 are presumed to be essentially isochronous but it is possible that some time variation occurs from lobe to lobe. Regardless of whether M_3 is everywhere exactly synchronous, it always seems to have occurred as a completely post-tectonic, static event. This is shown clearly by the fact that minerals that can be recognized as having grown during M_3 cross-cut and overprint all of the foliations present in the rocks. And because in most places M_3 has been superimposed on rocks previously affected by the M_2 regional event, the result has been development of numerous types of pseudomorphic textures. Where M_3 attained a grade higher than M_2, the pseudomorphs formed by prograde processes (25). Where M_3 was lower grade than M_2 the pseudomorphs formed by retrograde processes. The papers of Guidotti (26), Novak and Holdaway (23) and Holdaway et al. (17) discuss in some detail the nature of these pseudomorphs and the mineralogic reactions which produced them.

Finally it should be noted that Fig. (1) shows an enormous area of (AC) rocks -- i.e. K-feldspar + sillimanite zone. This peak of M_3 metamorphism probably occurred at a temperature near 630°C and a pressure of about 4 Kb. Many of the rocks in the AC zone and some in the AB zone are best described as migmatites. Pegmatites are common in both zones and in the AC zone there are also innumerable small granitic plutons present. A notable feature of the enormous AC zone is that none of the rocks have apparently attained true granulite facies. It seems unlikely that this observation can be explained by all of the exposed rocks being at the same tectonic level. A more reasonable explanation is that extensive partial melting and the abundant small granitic plutons and pegmatites effectively provided a thermal buffer which prevented temperatures from rising to those of the granulite facies.

M_2-Regional Metamorphism:

M_2 is recognized as an earlier Acadian regional metamorphic event. Because its recognition is based mainly on the development of pseudomorphic textures it can be identified clearly only in areas where the appropriate prograde or retrograde pseudomorphs are present. In practice this amounts to those areas where M_2 attained at least staurolite grade. For example, M_2 and M_3 greenschist grade rocks can not be distinguished. On the other hand, rocks in much of the (AB) zone and all of the (AC) zone of M_3 are so recrystallized that the evidence for an earlier M_2 event has been destroyed.

Nonetheless as shown in the report of Holdaway et. al. (17) the areal extent of M_2 can be determined fairly accurately. Based upon various published and unpublished petrographic data (the latter from work in progress by CVG and MJH and their students) it appears that much of the area affected by the higher grades of M_3 has previously been affected also by M_2 at least at staurolite grade. In some cases M_2 staurolite grade was attained in areas subsequently affected by sub-staurolite grade M_3. Nonetheless, as a generalization, we can consider the staurolite zone of M_2 and M_3 as being very roughly coincident in terms of the areas involved. This suggestion probably applies to much of northern New Hamphsire also. The highest grade attained in M_2 was probably somewhat lower than the peak attained in M_3. The peak probably involved a T somewhat less than 600° and a P less than the 4 Kb suggested for M_3. It involved the establishment of the andalusite + staurolite + biotite three phase field on an AFM projection and in some rocks involved the disappearance of staurolite but not the appearance of sillimanite. There is little or no evidence for development of migmatites during M_2, nor has there been found any direct evidence which can demonstrate that M_2 was associated with plutonic activity. However, in most of the Kingfield and Anson Quadrangles (see Fig. 1), the grade increases rapidly over a few kilometers thereby suggesting the possibility of a shallow heat source such as batholith.

M_2 appears to have been similar to M_3 in that it was also a post-tectonic event. This suggestion is based mainly on evidence from the area covered in the report of Holdaway et. al. (17) but similar evidence is found in the eastern and southern parts of the high-grade regions shown in Fig. 1. In particular, it is common to find staurolite, andalusite and garnet which can be designated as M_2 in origin and which has a marked cross-cutting or overprinting relationship with all of the fabric features present. In some cases the cross-cutting and/or overprint relationships are expressed by pseudomorphs rather than the original phase but this does not change any of the implications of M_2 as a static event.

M_1 Regional Metamorphism:

The evidence for M_1 is indirect. It rests on the fact that in much of central and western Maine a slip cleavage which is overprinted by the M_2 megacrysts, cuts a well developed foliation that is axial planar to the major folds. It seems likely that this well developed axial plane cleavage required at least some metamorphism for its formation. However, because no megacrysts seem to have formed in association with this foliation, it is presumed that M_1 did not exceed greenschist facies.

M_4 Contact Metamorphism:

In the absence of isotopic dates on a given pluton, its associated contact metamorphism can be identified as M_4 only when it affects those low to medium grade rocks which clearly belong to the M_3 terrane (as opposed to low grade rocks which might be attributed to M_1 or M_2). In such cases, late Acadian plutonic rocks (~360 M.Y.; (17)) are surrounded by distinct contact aureoles. Cordierite, andalusite, and sillimanite are common in these aureoles but staurolite and garnet are rare or absent. In the high-grade portions of the M_3 terrain, the plutons (~380 M.Y. ages; (17)) appear to be nearly synchronous with the high-grade metamorphism and so no contact effects resulted.

Some general considerations

The previous discussion has provided some details of the four separate Acadian metamorphic events which have thus far been recognized in Maine. Brief mention has also been made of the possibility of a greenschist grade Taconic regional metamorphism of the Cambro-Ordovician strata occurring near the mutual border of Maine, New Hampshire, and Quebec. Similarly, the as yet poorly known metamorphic history of the Chain Lakes Massif was briefly mentioned.

It is emphasized that both M_2 and M_3 were static, post-tectonic events and that M_3 followed M_2 in terms of timing. In some cases (26) it is possible to demonstrate a definite time interval between M_2 and M_3 during which the rocks cooled down to temperatures approximating those ambient for the prevailing depth of burial at ~ 4 Kb. However, as discussed in Holdaway et. al. (17), in other cases M_2 and M_3 may have been sufficiently close in time that no definite cooling between the M_2 peak and M_3 peak can be recognized. In those cases involving cooling between M_2 and M_3, the evidence for such a happening rests on direct, detailed observations. As described by Guidotti (26) one can trace rocks from areas in which minerals produced by M_2 have been downgraded by M_3, to areas in which minerals produced by M_2 have been prograded by M_3. Moreover, in such cases one can map the "hinge line" along which M_2 and the superimposed M_3 were essentially at the same grade of metamorphism. In those areas where M_3 appears to have followed closely on M_2 (Augusta area (23), the M_2 and M_3 staurolite isograds appear to coincide.

It should be noted that in western Maine both M_2 and M_3 bring in an Al-silicate (andalusite and sillimanite respectively) by the same basic AFM reaction. This reaction involves breaking the staurolite-chlorite join and forming in its place a join connecting Al-silicate-biotite. Inasmuch as this reaction occurs in the andalusite stability field for M_2 and in the sillimanite field for M_3 it is necessary that the reaction during M_2 occurred at a pressure somewhat lower (e.g. $\frac{1}{2}$ to 1 Kb) than the M_3 equivalent reaction.

Because M_2 and M_3 were both static events with no evidence of tectonic activity during the time interval separating them, one must appeal to an atectonic means for increasing the pressure. Thus, the pressure increase might best be achieved by surficial accumulation of a volcanic pile. Such a volcanic pile may have been associated with rising plutons which eventually become intimately associated with M_3. Moreover, it should be noted that further east in Maine, the stratigraphic record does include thick sections of Devonian age volcanic rocks (27).

Although Fig. (1) does show some of the metamorphic terrain in northeastern Vermont, only little mention has been made of this area. Only recently has new work been initiated on that part of New England (9). Based on the descriptions of the igenous and pelitic metamorphic rocks given by Woodland (28), it seems likely that the plutonic and metamorphic relations in the part of Vermont shown on Fig. (1) are similar to those some 50 miles to the east in Maine. Moreover, based on study of mafic metamorphic rocks, Laird and Albee (9) arrived at a comparable conclusion.

Finally, it is useful to mention some major questions for which there are no meaningful answers as yet. The first concerns the age (or ages) of the vast area of greenschist facies rocks that are present in Maine (see Fig. 1). In the case of the Cambro-Ordovician rocks near the Quebec border, the metamorphism could be Taconic and/or any of the Acadian regional events, M_1 to M_3. In the case of the areas of Siluro-Devonian strata, the greenschist facies metamorphism could be any or all of M_1 - M_3.

A second and related question is the ages of the numerous contact aureoles that have formed where plutons have intruded the greenschist terrain described above. Some of these aureoles are very likely contemporaneous (~380 M.Y.) with the high-grade M_3 event which occurred to the south. Others have formed at a later time (~360 M.Y.) and can be designated as M_4. Distinguishing an M_3 aureole from an M_4 aureole is commonly very difficult and only now is work on this question receiving much attention. To date M_4 aureoles have been clearly identified only when isotopic ages are available for the pluton or when one can demonstrate that rocks intruded have previously been affected by <u>both</u> M_2 and M_3. To demonstrate the latter usually requires a situation where high grade (staurolite and/or andalusite) M_2 mineralogy has been downgraded on a <u>regional</u> basis to M_3 greenschist facies and then intruded. An example of such a case is described in the north central part of the area covered in the study of Holdaway et. al. (17).

Acknowledgements

Research support for this project was provided by National Science Foundation grants EAR7704521 and EAR7902597 to C.V. Guidotti and EAR7612463 and EAR8025891 to M.J. Holdaway and Natural Sciences and Engineering Research Council of Canada grant A8716 to W.E. Trzcienski.

REFERENCES

1. P. St. Julien and C. Hubert, Amer. Jour. Sci., 275-A, 337, 1975.
2. A. Vallières, C. Hubert and C. Brooks, Can. Jour. Earth Sci., 15, 1242, 1978.
3. T.C. Birkett, D.Sc.A. Dissertation, Ecole Polytechnique de Montréal, 1981.
4. E.L. Boudette and G.M. Boone, Geol. Soc. Amer. Memoir 148, 79, 1976.
5. R.S. Naylor, in: Annual Review of Earth and Planetary Sciences, ed. by F.A. Donath, Annual Reviews Inc., Palo Alto, 1975.

6. W.E. Trzcienski, J. and T.C. Birkett, Canadian Mineral., 20, 203, 1982.
7. R. Poulin, M.Sc. thesis, Ecole Polytechnique de Montréal, 1974.
8. W.E. Trzcienski, Jr., Can. Jour. Earth Sci., 13, 711, 1976.
9. J. Laird and A.L. Albee, Amer. Jour. Sci., 281, 127, 1981.
10. T. Feininger, Can. Jour. Earth Sci., 18, 1978, 1981.
11. C. Gariépy and C. Hubert, Geol. Soc. Amer., Abstracts with Programs, 13, 134, 1981.
12. P. Erdmer, Contrib. Mineral. Petrology, 76, 109, 1981.
13. P.H. Osberg, in: The Caledonides in the U.S.A.: Geological Excursions in the Northeast Appalachians, ed. by J.W. Skehan and P.H. Osberg, Proj. 27, IGCP, 37, 1979.
14. M.P. Billings, The Geology of New Hampshire; Part II - Bedrock Geology, New Hampshire State Planning and Development Commission, Concord, N.H., 1956.
15. P.H. Osberg, R.H. Moench and J.L. Warner, in: Studies in Appalachian Geology: Northern and Maritime, ed. by E-an Zen, W. White and J.B. Thompson, Jr., John Wiley and Sens, N.Y., 1968.
16. B.W. Chappell and A.J.R. White, Pacific Geol., 8, 173, 1974.
17. M.J. Holdaway, C.V. Guidotti, J.M. Novak and W.E. Henry, Geol. Soc. Amer., 93, 572, 1982.
18. E.L. Boudette, Ph.D. Dissertation, Dartmouth College, 1978.
19. J.C. Green, Amer. Mineral., 48, 991, 1963.
20. C.E. Bickel, Ph.D. Dissertation, Harvard University, 1971.
21. E.L. Boudette, in: New England Intercoll. Geol. Conf. guidebook for field trips in the Rangeley Lakes - Dead River Basin region, western Maine, ed. by G.M. Boone, Kennebec Jour. Press, Augusta, Me., 1970.
22. C.V. Guidotti, Jour. Petrology, 11, 227, 1970.
23. J.M. Novak and M.J. Holdaway, Amer. Mineral., 66, 51, 1981.
24. J.T. Cheney and C.V. Guidotti, Amer. Jour. Sci., 279, 411, 1979.
25. C.V. Guidotti, Amer. Mineral, 53, 1368, 1968.
26. C.V. Guidotti, New England Intercoll. Geol. Conf. guidebook for fieltrips in the Rangeley Lakes - Dead River Basin region, western Maine, ed. by G.M. Boone, Kennebec Jour. Press, Augusta, Me., 1970.
27. D.W. Rankin, in: Studies in Appalachian Geology: Northern and Maritime, ed. by E-an Zen, W. White and J.B. Thompson, J., John Wiley and Sons, N.Y., 1968.
28. B.G. Woodland, Vermont Development Dept. Bull., 28, 1965.

REALMS OF REGIONAL METAMORPHISM IN SOUTHERN NEW ENGLAND, WITH EMPHASIS ON THE EASTERN ACADIAN METAMORPHIC HIGH

Peter Robinson

Department of Geology and Geography, University of Massachusetts, Amherst, Massachusetts, U.S.A. 01003

ABSTRACT

Metamorphism in southern New England may be broadly divided into seven time-space realms as follows: 1. Grenvillian high amphibolite to granulite facies metamorphism in North American basement of the Berkshire, Housatonic, and Manhattan regions; 2. Late Precambrian low grade and contact metamorphism in the Dedham terrane, with local regional development to high ampnibolite facies in the Fall River, Milford-Selden Neck, and Massabesic regions, and to granulite facies in the Pelham dome; 3. Ordovician high amphibolite to granulite facies metamorphism in the Nashoba terrane; 4. complex Taconian high pressure to high temperature metamorphism localized along the Berkshire-Manhattan axis, associated with eastward subduction of the edge of North America beneath island arc and eastern basement terranes; 5. a western Acadian ("Vermont") high reaching high amphibolite facies and localized along a series of gneiss domes, separated by the Connecticut Valley metamorphic low (chlorite and biotite zones) from 6. an eastern Acadian ("New Hampshire") high, reaching granulite facies and centered east of the gneiss domes of the Bronson Hill anticlinorium; and 7. Pennsylvanian-Permian regional metamorphism up to sillimanite grade, best documented in southern Rhode Island, but apparently an overprint in much of central southern New England.

MAPS SHOWING DISTRIBUTION AND INTENSITY OF METAMORPHISM

Seven time-space realms of metamorphism in southern New England are illustrated in a series of four simplified metamorphic maps of the region, that show the distribution and intensity of Precam-

P. E. Schenk (ed.),
Regional Trends in the Geology of the Appalachian-Caledonian-Hercynian-Mauritanide Orogen, 249–258.

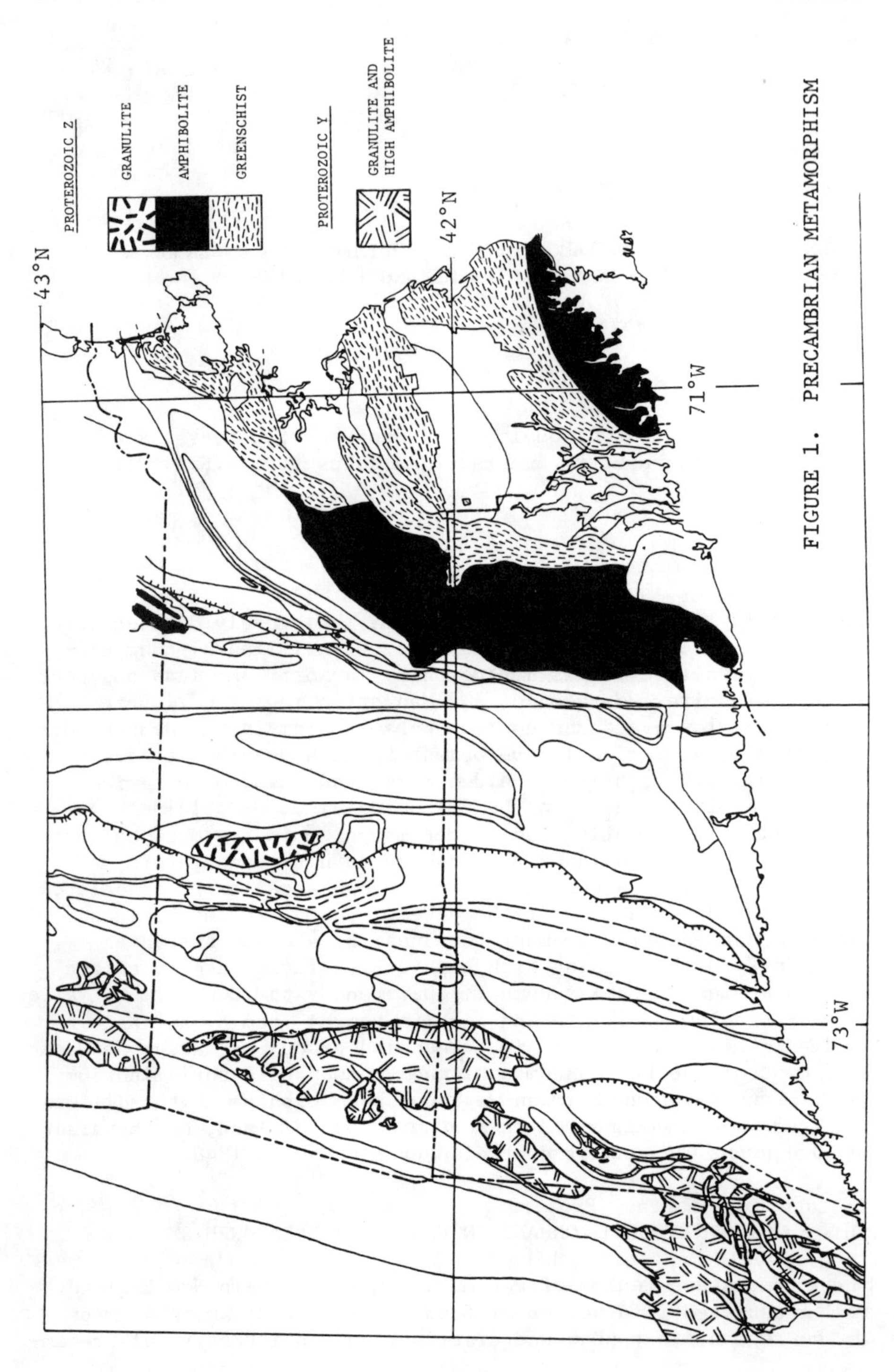

FIGURE 1. PRECAMBRIAN METAMORPHISM

brian metamorphisms (Figure 1), of Ordovician metamorphisms (Figure 2), of Devonian metamorphisms (Figure 3) and of Pennsylvanian-Permian metamorphism (Figure 4). In each map the metamorphism shown is only the most intense metamorphism experienced by the rocks, and weaker metamorphisms, either younger or older are not shown. Exceptions to this rule are in the western parts of Figures 2 and 3, where less intense Ordovician and Devonian metamorphisms are shown overprinting Proterozoic Y (Grenvillian) metamorphism; in central Massachusetts in Figure 3 where the less intense Acadian metamorphism is shown overprinting granulite facies metamorphism of probable Proterozoic Z age; and in northeastern Massachusetts where sillimanite grade Acadian metamorphism is shown overprinting Proterozoic Z amphibolite facies metamorphism in Figure 3, and where both of these are overprinted by Pennsylvanian-Permian sillimanite grade metamorphism in Figure 4.

For the Precambrian metamorphisms, the classifications of metamorphic grade are generalized into greenschist facies, amphibolite facies and granulite facies, due to the scarcity of bulk compositions that would permit more detailed subdivision. For the younger metamorphisms the intensities can be classified according to mineral assemblages in pelitic schists. The lowest grade zones of chlorite (CHL), chloritoid (CTD) and biotite (BIO), commonly not easily separated, are lumped together. These are followed by pelitic schists bearing garnet (GAR) and then kyanite-staurolite (KYA-STA) in regions of relatively high pressure, or andalusite-staurolite (AND-STA) or andalusite only (AND) in regions of relatively lower pressure (1). The sillimanite zones are divided in three parts: zones of sillimanite-staurolite (SIL-STA) or sillimanite-muscovite (SIL-MUS), zones of sillimanite-muscovite-Kfeldspar or sillimanite-Kfeldspar (SIL-KSP, SIL-MUS-KSP), or zones of sillimanite-Kfeldspar-garnet-cordierite (SIL-GAR-CRD).

SEVEN REALMS OF METAMORPHISM

The seven realms of metamorphism are as follows:

1. Grenvillian (Proterozoic Y) high amphibolite to granulite facies metamorphism in North American basement of the Green Mountain, Berkshire, Housatonic, Manhattan and Hudson Highland regions on the west side of Figure 1. Much of this is believed to have been synchronous with metamorphism in the nearby Adirondack Mountains.

2. Late Precambrian (Proterozoic Z) low grade and contact metamorphism in the Dedham terrane in the eastern part of Figure 1; with local development of amphibolite facies to the west and southeast; and with an isolated area of relict granulite facies rocks in the Pelham dome of the Bronson Hill anticlinorium. In some interpretations all these occurrences are considered to have been on an

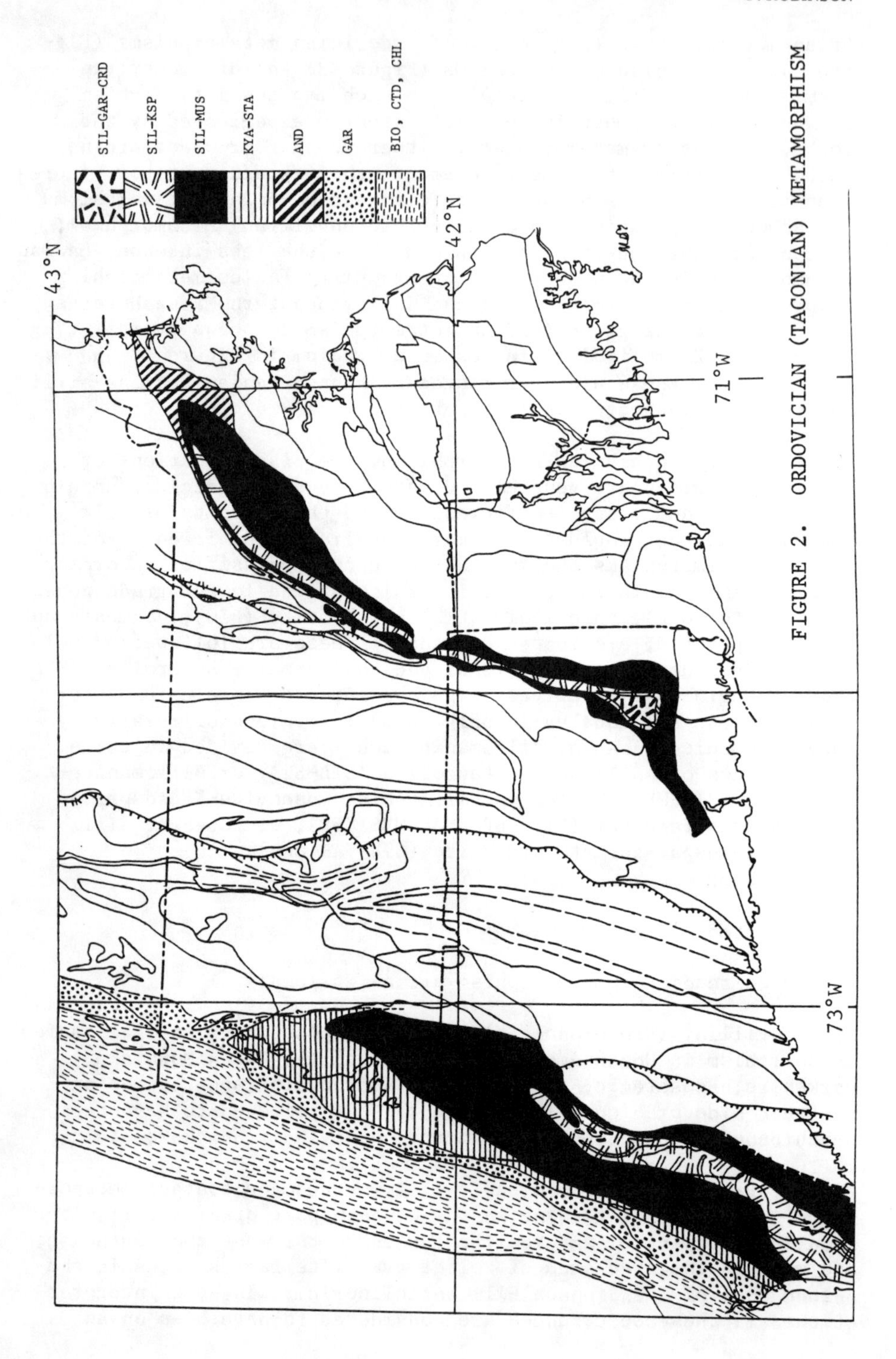

FIGURE 2. ORDOVICIAN (TACONIAN) METAMORPHISM

"Avalon plate". Recent discoveries of Devonian granites (2) cutting the amphibolite facies area in western Rhode Island are not shown.

3. Ordovician metamorphism to sillimanite and locally to sillimanite-garnet-cordierite grade in the Nashoba zone in the eastern part of Figure 2. New zircon ages (R. E. Zartman, pers. comm. 1980) of 490 m.y. on migmatitic granite associated with this metamorphism and of 430 m.y. on unmetamorphosed quartz diorite cutting the metamorphic rocks, seem to pin down this metamorphism as Ordovician, although the age of the rocks is uncertain (3). Although of Ordovician age, this metamorphism is probably unrelated to the classic Taconian orogeny to the west.

4. Complex Taconian high pressure to high temperature metamorphism localized along the Berkshire-Manhattan axis, and associated with eastward subduction of the edge of North America beneath the island arc and eastern basement of the Bronson Hill plate. Early high pressure phases that produced glaucophane and omphacite (4) are best preserved in northern Vermont, and in southern New England are heavily overprinted by a higher temperature probably collisional phase (1,5). The boundary between this Taconian metamorphism and the Acadian overprint to the east is very difficult to map, and is based on new detailed K-Ar studies (6). However, in the Connecticut Valley region, where Acadian metamorphism is low grade, Ordovician rocks below Devonian cover show similar assemblages (7).

5. A western Acadian ("Vermont")high (Figure 3) reaching kyanite and sillimanite zones and localized along a series of gneiss domes (8), that involved large scale vertical transport of deep-seated rocks including North American Grenvillian basement in some domes, and eastern basement of the Bronson Hill plate in others. This high is separated by the Connecticut Valley metamorphic low (chlorite and biotite zones) from the eastern Acadian high.

6. An eastern Acadian ("New Hampshire") high, reaching the sillimanite-garnet-cordierite zone (Figure 3) and centered east of the gneiss domes of the Bronson Hill anticlinorium (9, 10, 11).

7. Pennsylvanian-Permian regional metamorphism up to sillimanite grade (12) best documented in the Narragansett Pennsylvanian basin of Rhode Island (Figure 4). This metamorphism is thought to extend westward along the Connecticut coast, overprinting the effects of Ordovician and Devonian sillimanite-grade metamorphisms (13). Its influence is also seen in four isolated fault-bounded basins of Pennsylvanian rocks near Worcester, Massachusetts (14) metamorphosed to garnet or chlorite-chloritoid grade, and a contact aureole around the Pennsylvanian-Permian granite of Milford, New Hampshire (15). K-Ar and Rb-Sr ages on micas from a broad area of central and southern New England have been suggested as possible evidence of this metamorphism, though also used as evidence of erosional

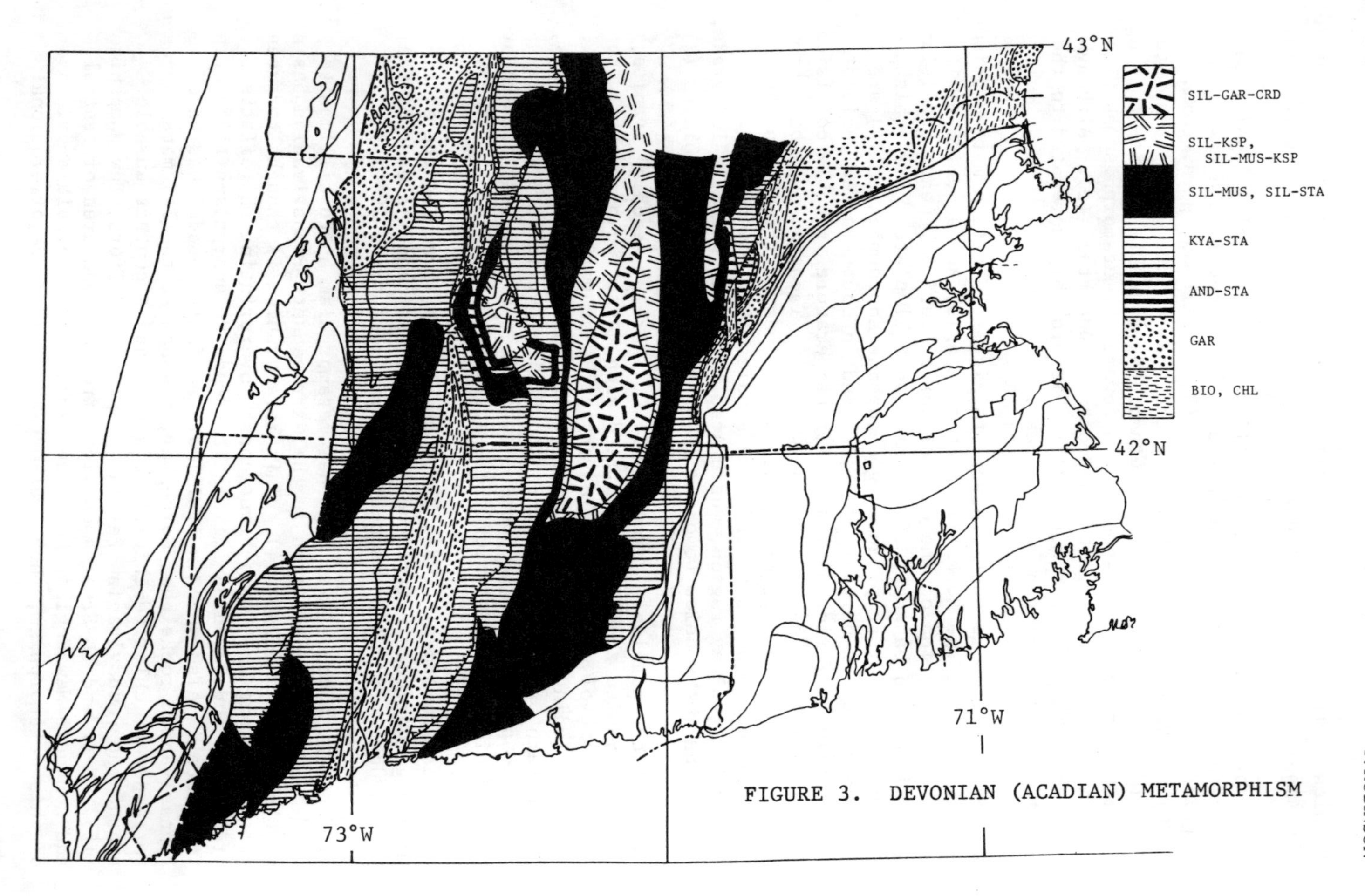

FIGURE 3. DEVONIAN (ACADIAN) METAMORPHISM

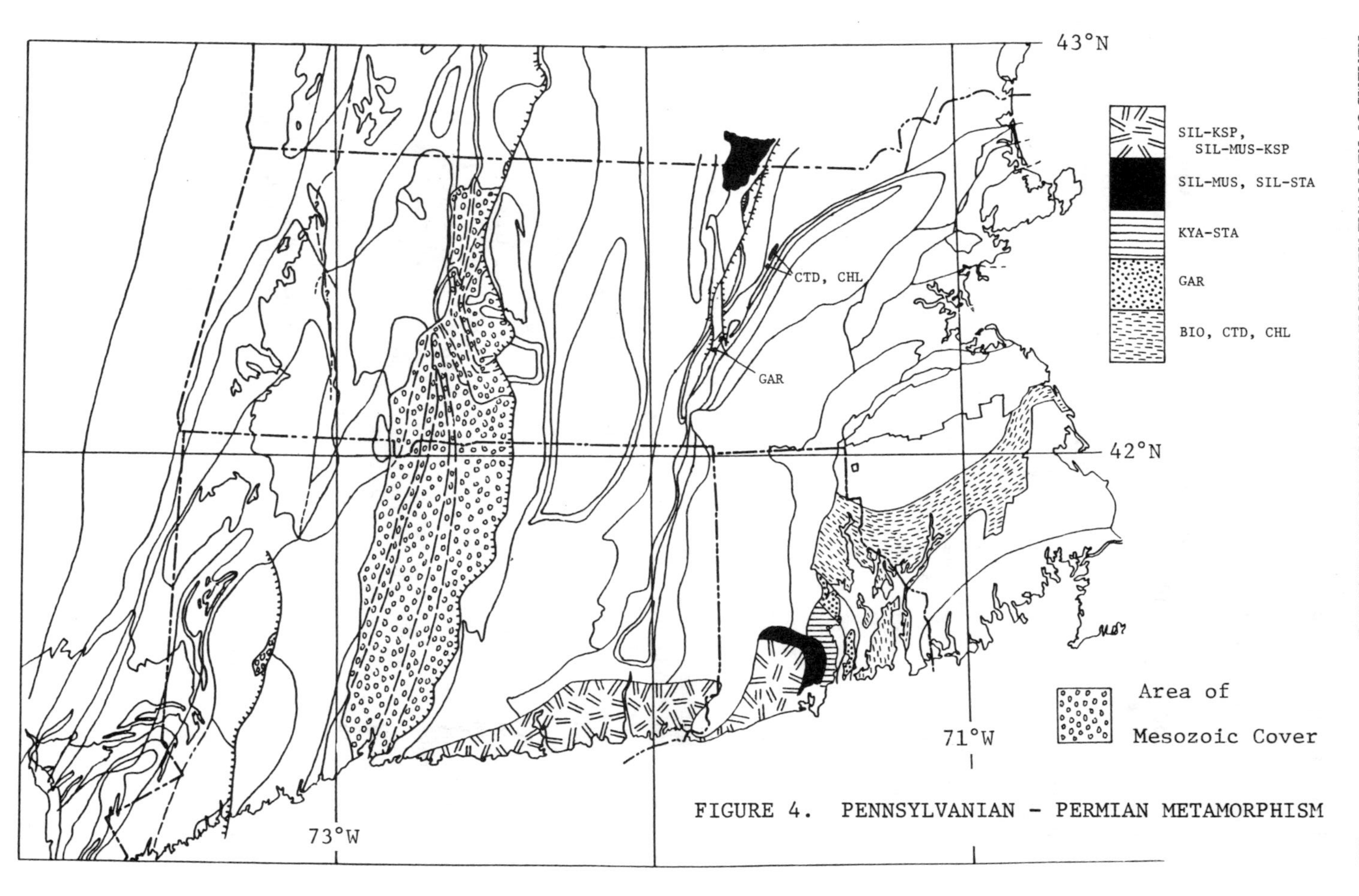

FIGURE 4. PENNSYLVANIAN - PERMIAN METAMORPHISM

unloading (16).

EASTERN ACADIAN METAMORPHIC HIGH

The eastern Acadian high occurred in a region dominated by three major episodes of Acadian deformation (3, 10, 11, 17): 1. regional nappes with east to west overfolding of tens of kilometers, 2. west to east backfolding of previous axial surfaces on a scale of tens of kilometers, with development of a powerful east-west trending linear fabric and synchronous ductile mylonite zones, and 3. a pattern of tight folds and linear fabrics associated with gravitationally induced rise of gneiss domes in the Bronson Hill anticlinorium. On the west side, the metamorphic high has an overhang of hotter rocks overfolded onto cooler rocks as a result of the early nappes. On the east side the high also overhangs, probably as a result of backfolding.

Before and during the early nappe stage the region was intruded by a variety of sheet-like calc-alkaline plutons ranging in composition from gabbros through voluminous biotite tonalites to granites (18), generally yielding intrusion ages around 400 m.y. (19). Evidence for early low pressure metamorphism is preserved in the widespread occurrence of sillimanite pseudomorphs after andalusite and at least one relict contact metamorphic aureole adjacent to augite-hornblende diorite (20). Peak metamorphic conditions were attained early in the backfold stage over a broad region east of the gneiss domes, resulting in widespread sillimanite-orthoclase-garnet-cordierite assemblages in pelitic schists and rare orthopyroxene-augite-orthoclase-garnet assemblages in felsic volcanics. Detailed geothermometry and geobarometry suggests peak metamorphic temperatures up to 730°C and pressures of about 6.3 kbar, with abundant evidence for local fluid-absent melting. At many locations the peak metamorphic fabric is cut by ductile mylonites containing the east-west linear fabric of the late backfold stage. The planar and linear fabric of the mylonites is deformed by north- and northeast-trending minor folds and axial plane foliation associated with the dome stage of deformation. Mineral assemblages produced during recrystallization of mylonites in this deformation suggest temperatures near 550°C at pressures possibly as high as 7-8 kbar. Complementary to this evidence of increasing pressure of metamorphism east of the gneiss domes, development of late cordierite and cordierite + corundum recation rims on sillimanite in gedrite gneisses in the domes is evidence of tectonic unloading related to doming.

The late Acadian Belchertown pluton (see Figure 3, small area of rectangular isograds west of eastern high) truncates isoclinal folds of early stages. A primary core of orthopyroxene-augite quartz monzodiorite yields a zircon age of 380 m.y. (21). The outer part of the pluton has been metamorphically hydrated to hornblende gneiss with development of planar and linear fabric identical to the dome-stage fabric of the country rocks. A K-Ar age of 361 m.y. on metamorphic hornblende from this gneiss suggests the end of dome-stage recrystallization. Thus, the Acadian orogeny in southern New England was complex and protracted over a period of at least 40 million years.

ACKNOWLEDGMENTS

The rationale for the metamorphic maps was worked out in sessions of the Caledonide Project Metamorphic Working Group, and details were discussed in particular with fellow North American compilers W. E. Trzcienski, G. W. Fisher, and C. V. Guidotti. The Massachusetts part of the map was produced with direct input of other members of the U. S. Geological Survey working group for the Bedrock Map of Massachusetts, E-an Zen, Editor; Richard Goldsmith, N. M. Ratcliffe, and R. S. Stanley, Compilers; and D. R. Wones and N. L. Hatch, Contributors. The base map was compiled for joint papers with Leo M. Hall who also contributed boundaries in western Connecticut and adjacent New York. Other significant information was provided in discussions or joint research with J. B. Thompson, Jr., John Rodgers, R. J. Tracy, M. T. Field, R. D. Tucker, Bruce O'Connor, G. W. Leo, L. D. Ashwal, J. O. Guthrie, D. J. Hall, K. T. Hollocher, C. K. Shearer, D. E. Maczuga, R. E. Zartman, John Aleinikoff, J.B. Lyons, J. C. Cheney, John Sutter, J. C. Hepburn, D. P. Murray, E. S. Grew, J. D. Peper, J. S. Pomeroy, and G. R. Robinson. Research was supported by grants from the U. S. Geological Survey and the U. S. National Science Foundation Programs for Geology and Geochemistry. To each of these persons and institutions I express my grateful acknowledgment.

REFERENCES

(1) Thompson, J. B., Jr. and Norton, S. A.: 1968, in Zen, E-an and White, W. S., eds., "Studies in Appalachian Geology, Northern and Maritime", Interscience, New York, pp. 319-341.
(2) Hermes, O. D., Gromet, L. P. and Zartman, R. E.: 1981, Guidebook, New England Intercollegiate Geological Conference, pp. 315-338.
(3) Hall, L. M. and Robinson, Peter: 1982, Geological Assoc. of Canada Special Paper 24, pp. 15-41.
(4) Laird, Jo and Albee, A. L.: 1981, Amer. Jour. Sci. 281, pp. 127-175.
(5) Ratcliffe, N. M.: 1975, Guidebook, New England Intercollegiate Geological Conference, pp. 186-216.
(6) Sutter, J. F., and Ratcliffe, N. M.: in press, Geol. Soc. Amer. Bull.

(7) Hatch, N. L.: 1982, Geological Assoc. of Canada Special Paper 24, pp. 67-85.
(8) Zen, E-an, Goldsmith, Richard, Robinson, Peter, Ratcliffe, N. M., and Stanley, R. S.: 1983, "Bedrock Geologic Map of Massachusetts", U. S. Geological Survey.
(9) Tracy, R. J., Robinson, Peter, and Thompson, A. B.: 1976, American Mineralogist 61, pp. 762-775.
(10) Robinson, Peter: 1979, in Skehan, J. H., and Osberg, P. H., eds., "Guidebook, Caledonides in the U. S. A., Excursions in the Northeast Appalachians", Boston College, pp. 126-174.
(11) Robinson, Peter, Tracy, R. J., Hollocher, K. T., and Dietsch, C. W.: 1982, Guidebook New England Intercollegiate Geological Conference, pp. 289-339.
(12) Murray, D. P., Hepburn, J. C., and Rehmer, J. A.: 1979, in Skehan et al., eds., "Evaluation of Coal Deposits in the Narragansett Basin, Mass. and R. I.", U. S. Bureau of Mines, Final Report for Contract #J0188022, pp. 39-47.
(13) Lundgren, L., Jr.: 1966, J. Petrology 7, pp. 421-453.
(14) Grew, E. S.: 1976, Guidebook, New England Intercollegiate Geological Conference, pp. 383-404.
(15) Aleinikoff, J. N.: 1978, "Structure, petrology, and U-Th-Pb geochronology in the Milford quadrangle, New Hampshire", Ph. D. thesis, Dartmouth College, 247 p.
(16) Zartman, R. E., Hurley, P. M., Krueger, H. W., and Giletti, B. J.: 1970, Geol. Soc. America Bull. 81, pp. 3359-3374.
(17) Robinson, Peter, Field, M. T., and Tucker, R. D.: 1982, Guidebook, New England Intercollegiate Geological Conference, pp. 341-373.
(18) Robinson, Peter, Shearer, C. K., and Maczuga, D. E.: 1981, EOS 62, p. 436.
(19) Lyons, J. B. and Livingston, P. E.: 1977, Geol. Soc. America Bull. 88, pp. 1808-1812.
(20) Shearer, C. K. and Robinson, Peter: 1980, EOS 61, p. 389.
(21) Ashwal, L. D., Leo, G. W., Robinson, Peter, Zartman, R. E., and Hall, D. J.: 1979, Amer. Jour. Sci. 279, pp. 936-969.

METAMORPHISM IN THE U.S. APPALACHIANS--OVERVIEW AND IMPLICATIONS

George W. Fisher

Department of Earth and Planetary Sciences
The Johns Hopkins University
Baltimore MD 21218 USA

The metamorphic map of the U.S. Appalachians compiled as a part of IGCP Project 27 reveals an intricate pattern of variation in metamorphic grade, reflecting regional differences in depth of erosion and heat input, modified by late faulting. The effect of faulting is particularly clear in the Southern Appalachians: a series of major faults carry the amphibolite facies rocks of the Blue Ridge 100 km or more to the west over greenschist facies rocks, now exposed in the Grandfather Mountain Window; the Brevard Zone forms a nearly linear belt of low-grade rocks marking the western boundary of the Piedmont; and the Goat Rock and Towaliga faults combine to carry a major sheet of Piedmont rocks over basement rocks exposed in the Pine Mountain Window. In the northern Appalachians the effects of faulting are subtler but still present: near the Vermont - New Hampshire boundary the Ammonoosuc listric normal fault is associated with a pronounced belt of low-grade rocks; in eastern Massachusetts the Clinton - Newberry fault zone marks a major metamorphic gradient; and near the Maine coast the Norumbega fault zone telescopes metamorphic zonation.

Mentally removing the effects of these faults, a generalized picture of Appalachian metamorphism emerges. The highest pressures (kyanite - K-feldspar and other assemblages suggest 8 to 10 kb) come from the central part of the belt, near Philadelphia (8) and in southwestern Connecticut (6). In the northern Appalachians pressure decreases steadily to the north and northeast, as shown by the alumino-silicate isobar (7) and related assemblages (1). In the Southern Appalachians pressure decreases rapidly to the south, with andalusite - sillimanite appearing just 200 km southwest of Philadelphia (3), then

P. E. Schenk (ed.),
Regional Trends in the Geology of the Appalachian-Caledonian-Hercynian-Mauritanide Orogen, 259–262.

increases very slightly and remains roughly constant through the middle amphibolite facies rocks which dominate the southern Piedmont. The variation in the depth of erosion implied by these pressure differences correlates closely with depth to basement in the Appalachian basin to the northwest (2, Fig. 3); it appears as if there is a crude mass balance in which the volume of cover rocks locally removed by erosion is roughly equal to the volume of sediment deposited in the basin to the northwest.

The age of peak metamorphism within the Appalachians spans a period of roughly 300 million years. Rocks metamorphosed during the late Precambrian are preserved in eastern Massachussets and southeasternmost Pennsylvania. During the Paleozoic, the peak of metamorphism in New England moved gradually eastward: an Ordovician pulse centered in western Connecticut, Massachussets and Vermont; a Devonian pulse culminated in central Connecticut and Massachussets and continued northeast through New Hampshire to central Maine; and a Pennsylvanian pulse centered in Rhode Island and southeastern Massachussets. A similar picture may hold within the southern Appalachians, but relations are far less clear than in the north owing to uncertainties about the ages of the rocks. All that we can say with confidence is that an Ordovician metamorphism affected much of the western Piedmont, and that there are two patches of Pennsylvanian metamorphism along the eastern edge of the Piedmont. A Devonian stage of metamorphism can be recognized in Alabama, but its extent to the northeast is poorly understood owing to the lack of recognizable Devonian sediments. Meckel (5) provided a clear upper limit to the time of metamorphism in the northern Piedmont by showing that detrital kyanite in Mississippian sedimentary rocks of the Valley and Ridge province was derived from the Piedmont, implying deep erosion by Mississippian time.

Because of the systematic northward decrease in pressure in the northern Appalachians, New England provides a crude cross section of a metamorphic belt, from rocks buried as deeply as 27 km (southern Connecticut) to rocks buried only 5 or 6 km (central Maine). The region is an ideal place to study relations between metamorphism, heat flow, emplacement of igneous rocks, and rock deformation.

During metamorphism, heat is supplied both by thermal conduction in response to the ambient geothermal field (commonly involving thermal gradients on the order of 25 C/km) and by heat transfer associated with the emplacement of magma. Heat can be transferred from a rising magma to its wall rocks both by cooling the magma and by allowing it to crystallize, releasing the latent heat of crystallization; for most magma compositions the latent heat constitutes a very large proportion of the total heat available for heating the wall rocks. The amount of heat transferred must reflect the quantity of magma moving through the system and its rate of flow. The rate of flow of a magma body is largely determined by its viscosity, which in turn depends upon the proportion of crystals present. Once the proportion of crystals exceeds roughly 50 percent, the viscosity increases dramatically (4), slowing or stopping the magma altogether. When this happens the magma releases all of its remaining heat, providing a concentrated influx of heat to the rocks nearby.

Consequently, we should expect to find an intricate relation between magma composition (especially volatile content) and the style of metamorphism. The possibilities are many, but three scenarios seem especially likely. Highly fluid magmas may be expected to move rapidly through an orogenic belt, adding relatively little heat to the rocks. Such a belt would be characterized by extensive volcanism and relatively high pressure, low temperature metamorphism. Magmas which are very viscous or which tend to crystallize rapidly may freeze deep within the crust, leading to a terrain characterized by high temperature, high pressure metamorphism and the presence of voluminous plutonic rocks. However, because the ambient temperature was high prior to intrusion the spatial relation between metamorphic isograds and plutons may be obscure. Finally, magmas which are fluid enough to rise high within the crust but not fluid enough to rise to the surface will have the most pronounced effect upon metamorphism; they will rise rapidly through the lower part of the crust, causing only minor heating there, and will release the bulk of their heat at a shallow level, quite possibly producing a thermal inversion in which shallow rocks are hotter than those more deeply buried.

Each of these three styles of metamorphism seems to be present within New England: the Ordovician metamorphism appears to have been of the high pressure, low temperature type, the Devonian metamorphism in southern New England of the high temperature, high pressure type, and the Devonian metamorphism in northern New England of the low pressure, high temperature type. From northeastern Vermont across to coastal Maine there is a strong association between Devonian plutons and the sillimanite isograd, clearly indicating that the heat for metamorphism was derived from the plutons. Viewed in cross-section , the sillimanite isograd in southwestern Maine forms a closed loop just above a large pluton near Sebago, strongly suggesting a thermal inversion. This area seems to provide an ideal opportunity to investigate the relation between plutonism and metamorphism at a regional scale.

References Cited

(1) Carmichael, Dugald M., 1978, Metamorphic bathozones and bathograds: A measure of the depth of post-metamorphic uplift and erosion on the regional scale. Amer. Jour. Sci., v. 278, pp. 769-797.

(2) Colton, G.W., 1970, The Appalachian basin - its depositional sequences and their geologic relationships; in Fisher, G.W. and others, eds., Studies of Appalachian Geology. Central and Southern, Interscience, pp. 5 - 47.

(3) Fisher, G. W., 1970, The metamorphosed sedimentary rocks along the Potomac River near Washington, D.C.; in Fisher, G.W. and others, eds., Studies of Appalachian Geology: Central and Southern. Interscience, pp. 299 - 316.

(4) Marsh, B.D., 1981, On the crystallinity, probability of occurrence, and rheology of lava and magma. Contrib. Mineral. Petrol., 78, pp. 85-98.

(5) Meckel, Lawrence D., 1967, Origin of Pottsville Conglomerates (Pennsylvanian) in the Central Appalachians. Geol. Soc. America Bull., v. 78, pp. 223-258.

(6) O'Connor, B.J., 1973, A petrologic and electron microprobe study of pelitic mica schists in the vicinity of the staurolite-disappearance isograd in Philadelphia, PA and Waterbury, CT; unpublished Ph.D. dissertation, Johns Hopkins University, Baltimore, MD, 446p.

(7) Thompson, J.B. and Norton, S.A., 1968, Paleozoic regional metamorphism in New England and adjacent areas. in Zen, E-an and others, eds., Studies in Appalachian Geology; Northern and Maritime, Interscience, N.Y., pp. 319-328.

(8) Wagner, M.E., 1982, Taconic metamorphism at two crustal levels and a tectonic model for the Pennsylvania-Delaware Piedmont (abs.). GSA Abstracts with Programs, v. 14, no. 7, p. 640.

MAIN METAMORPHIC FEATURES OF THE PALEOZOIC OROGEN IN FRANCE
(French contribution n° 24 to the I.G.C.P. Project n° 27)

SANTALLIER Danielle S. co-ordinator.

Laboratoire de Géologie. 123, rue A. Thomas
87060 Limoges Cedex, France.

Abstract : The plurifacial Paleozoic metamorphism of France comprises four main phases which succeeded to one another under regularly decreasing pressure conditions. The first High Pressure phase, mostly Silurian or Ordovician, followed a Lower Paleozoic distensional period. The second phase, Devonian, is correlated with the first tectonic culmination ; it is an Intermediate Pressure one with kyanite and staurolite as index minerals. Nearly so is the third Lower Carboniferous phase. The fourth phase, of Upper Carboniferous age, resulted of the final continental collision ; it is a Low Pressure one.

This synthetic and very general work was elaborated by the whole French "Metamorphism" Groupe and is based upon recent, sometimes unpublished, datas gathered by A. Beugnies (Belgium) for the Ardennes Massif ; C. Audren, B. Cabanis, J. Chantraine, J.R. Darboux, P. Jegouzo, J. Le Métour, J. Marchand, S. Paradis, J. Rolet, J.P. Sagon, M. Ters, C. Triboulet and R. Wyns for the Armorican Massif ; S. Bogdanoff, F. Carme, R.P. Ménot, R. Oliver and G. Vivier for the Alpine Cristalline Massifs ; A. Autran, S. Bogdanoff, P. Collomb, M. Demange, J.P. Floc'h, J. Grolier, P.L. Guillot, J. Marchand, J.M. Quenardel, P. Rolin and D. Santallier for the Central Massif ; A. Autran for the Pyrenees. A. Piqué helped to establish a general scheme including deformation and magmatism.

1. INTRODUCTION

The French Paleozoic Chain is a polyphased one. This is why, for such a long time, people have taken for granted that Precambrian

P. E. Schenk (ed.),
Regional Trends in the Geology of the Appalachian-Caledonian-Hercynian-Mauritanide Orogen, 263–274.

formations must play an important part in the "Hercynian" domain. In fact, there are few enough ascertained tectonic and metamorphic events of Ante-Paleozoic age in the French so-called Hercynian Chain.

Apart of some presumed Precambrian cores in the Pyrenees and Maures Massif (49), the only evidences of Precambrian Orogens relicts are all located in the Armorican Massif : these are small Icartian slices (Fig. 1) of 1 800 to 2 000 M.Y. old gneisses, a large Cadomian domain (intruded circa 600/540 M.Y. by post-tectonic granitoids) in Brittany, and a plurifacial cadomian metamorphism in Southern Armorican Massif (Wyns). The Cadomian and Paleozoic metamorphisms are easily distinguished in Brittany, for the Cadomian basement (Fig. 1) has not been overprinted by later Hercynian events except for a very thin outer rim in the western part. It exhibits its own tectonic and metamorphic features (grading from epizone up to anatectic catazone) and its own calc-alkaline magmatism. In Central Brittany, the Paleozoic events affected indifferently Brioverian (Upper Precambrian in Brittany) and Paleozoic deposits, but the only visible structuration seems to be a merely Hercynian s.l. one. In Southern Armorican Massif (Vendée) on the contrary, both Cadomian and Paleozoic polyphased orogenic cycles are superposed although the overprinting is not clearly visible everywhere in the concerned area.

2. THE MAIN PALEOZOIC METAMORPHIC FEATURES. Their relations with deformations and magmatism :

Because of the polyphased character, the Paleozoic orogenic features appear very complex nowadays. Roughly speaking, four main metamorphic events, themselves polyphased, can be identified (Fig. 1), which succeed one another in a regularly enough decreasing pressure regime. These main metamorphic events are generally closely correlated with main tectonic episods, except at least for the first one, and with various magmatic products.

At the beginning of the orogenic cycle, mostly during Cambrian and before the distension episod or just at the beginning of it, the first manifestations in the future orogenic domain were purely magmatic : emplacement of acid and intermediate plutonites (540/520 M.Y.) very probably correlated with the finishing Cadomian orogen (mainly in Southern Armorican Massif and Central Massif), and of middle Cambrian rhyolitic lavas in Brittany. After that, the first clearly distensive magmatic products (alcaline low-alumina volcanism and potassic alcaline granites) appeared in Vendée, Southern Central Massif and Maures. The distension process culminated during Ordovician. It generated acid and basic tholeiitic volcanics, acid to basic plutonites generated by deep-seated crustal anatexis, and probably most of the eclogites

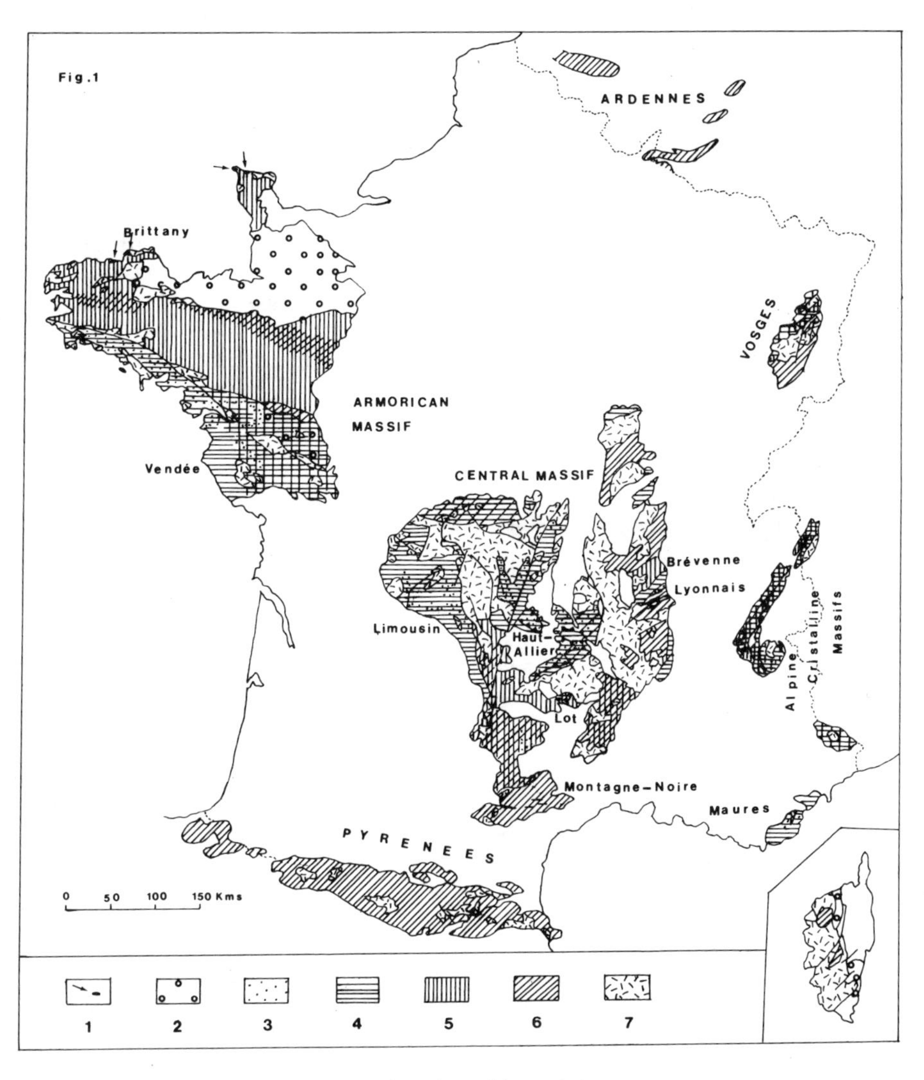

The Paleozoic Orogen in France. Main metamorphic periods.

1 - Icartian relicts. 2 - Precambrian basement. 3 - Silurian (and may be Ordovician) early H.P. phases. 4 - Devonian phase mostly Int. P. 5 - Very Upper Devonian and ante Upper Visean phase. 6 - Namuro-Westphalian phase mostly L.P. 7 - Post-metamorphism granitoids.

protoliths although this last point is still discussed by some geochronologists.

2.1. The first metamorphic phase.

$\varphi 1$ predated the main deformations. It is a High Pressure one. The corresponding rock types (mainly acid granulites with or without aluminium silicates, so-called khondalito-kinzigitic paragneisses and scarce marbles) are exposed on a large scale in Central Massif (mostly central (1, 2, 3) and north-eastern parts of it) as well as in Vosges (4) and north-western coast of Vendée (Fig. 2). In the mentionned areas, the great abundance of acid granulites, although more or less retrograded lately, implies that a largely developped continental crust has been involved in the early stages of the Paleozoic Orogen. In Western Central Massif (Limousin) as well as inland in Vendée, retrograded acid granulites are also found as rare and small tectonic slices in lately metamorphosed mesozonal country-rocks. In these regions too, the evidence for a continental crust during early Paleozoic Times is a remarkable fact.

Other relictual High Pressure rocks are well represented in the French Paleozoic Chain. These are more or less amphibolitized eclogites either associated with acid granulites as in the Central part of Central Massif (1,3) and Vosges, or emplaced as tectonic slices in originally mesozonal country-rocks as in Western (10, 12) and Southern (6,9) Central Massif, Southern Armorican Massif, Alpine Cristalline Massifs and Maures. According to their different geographic locations, those eclogites may have different compositions and geochemical characters corresponding to their originally different geodynamic settlements. For example, they are considered as parts of an old oceanic crust in Southern Armorican Massif (30,46) while they were originally emplaced in an opening rift site in Western Central Massif (12).

Independently of the protoliths ages, all these High Pressure facies were probably not contemporaneous. Some were Silurian (in Southern (28) and Western Central Massif) ; other were Ordovician (6) or may even be assigned to the Cambro-Ordovician boundary in some places (29). Concerning the relationships of High Pressure metamorphism with possibly contemporaneous deformations, an important point must be emphasized : although almost rubbed out by later recrystallization, some evidences of syn-crystallization deformations have been recently reported in Southern Armorican Massif eclogites (30) and in Central Massif acid granulites (7) and eclogites (Marchand, unpublished observations). Thus the earliest tectonic manifestations of the Paleozoic Orogen in France may in fact have predated the Devonian culminating phase.

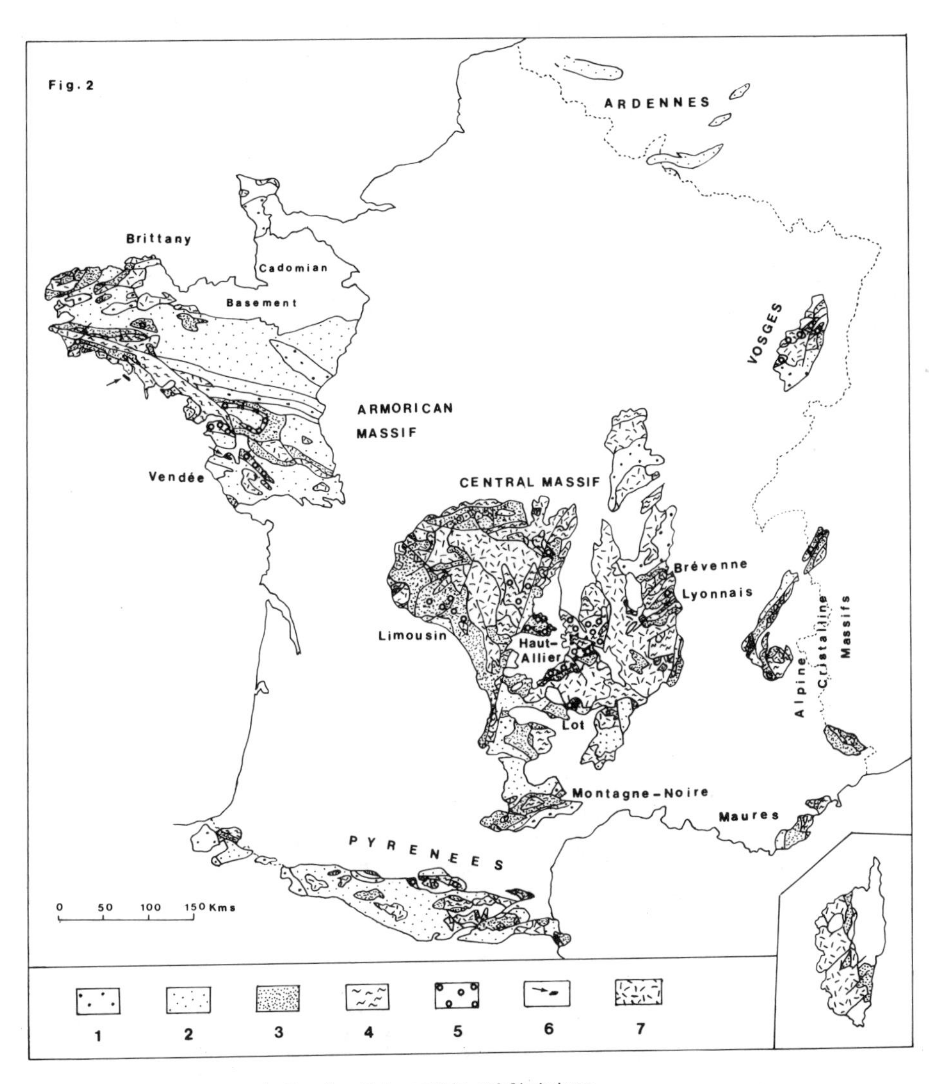

The Paleozoic Orogen in France. Metamorphic subdivisions.
1 - Anchizone. 2 - Greenschist facies. 3 - Mesozonal amphibolite facies. 4 - Catazonal amphibolite facies. 5 - Granulite facies and eclogites. 6 - Blueschist facies. 7 - Post metamorphism granitoids.

Some calc-alkaline and alkaline magmatic rocks have been emplaced during Ordovician and Silurian times. These include some now well-known orthogneisses in the Western Central Massif and South Brittany series. But suitable informations concerning possible equivalent magmatic activities in other areas are not still available. The three other metamorphic phases are always associated with main tectonic events.

2.2. The second phase.

φ2 took place during lower to middle Devonian times (Fig. 1). It was contemporaneous of the first tectonic culmination. The Devonian orogenic belt shows strong deformation and syn-metamorphic shear-zones and thrusts, associated with regional scale flat cleavage.

φ2 metamorphism is particularly well exposed in the southern part of Armorican Massif (Vendée (18) and South Brittany (23,24, 32), in Western Central Massif (10,11,12,20,47,48) and probably in Maures Massif (49). Most probably, it is responsible of the first retromorphism of the granulite series in North-Eastern and Central parts of Central Massif (1) and in Vosges (4). In Haut-Allier (central part of Central Massif), it is contemporaneous with a major crustal thrust emphasized by retrograded granulites overlapping a prograde meso- or catazonal series (this major φ 2 thrust can be compared with an almost identical one occuring during φ 3 as exposed in next paragraph).

Most probably, φ 2 must be correlated with the first metamorphic event recorded in the Alpine Cristalline Massifs (5,17,39,42).

Everywhere, φ 2 has been described as an Intermediate Pressure type metamorphism which grades from epizone up to catazone inclusively and shows staurolite and kyanite as index minerals (Fig.2). The zoneographic succession is complete in Western Central Massif (10) but is generally incomplete in other regions. Compared with tectonic contemporaneous events, this φ 2 metamorphism generally lasted a fairly long time. In Western Central Massif for example, it is associated with two successive deformation phases P1 and P2, which emphasized a decrease in Pressure conditions since the crystallization of syn- to post-P1 staurolite- and kyanite-bearing gneisses, until that of post-P2 sillimanite- and cordierite-bearing migmatites. In Southern Armorican Massif, φ 2 metamorphism is characterized by a remarkable dualism : blue-schists coexist with relatively low-Pressure anatexites and their associated granites. This fact has been pointed out and refered to as a paired belt of Siluro-Devonian age (32) which must be dabated because :
- the blue-schists are allochthoneous (obducted slab of oceanic crust) and may come either from a southern mafic inter-plate suture (43) or from a northern orogenic shear limit (44) ;

- the geological relations between the High Pressure and the Low-Pressure belts are poorly known in spite of recent submarine structural studies (43) ;
- the High-Pressure and High-Temperature facies are not strictly contemporaneous, the former being older (420/400 M.Y.) than the later (376 M.Y.) (45). In fact, this problem seems to be a very general one in the French Paleozoic Chain. Cordierite-bearing migmatites have been described in most areas. In the Devonian domain, they are either part of the prograde series (Southern Armorican Massif, Western Central Massif) or retrograded earlier granulites (central part of Central Massif and Vosges). In any case, they represent the later event of φ 2 metamorphism and testify of a generally more or less decreasing pressure compared to the mesozonal correlated facies. These cordierite-bearing migmatites are not restricted to the Devonian phase but are also present in φ 3 (Northern Brittany) and φ 4 (South and South-East Central Massif) series.

Immediatly after the φ 2 main tectono-metamorphic event, the building of the Devonian belt was followed by the emplacement of post-deformation granitoids (Upper Devonian tonalites and huge amounts of deep-seated granites). Then the first erosive products, mixed with calc-alkaline volcanics, were deposited in Devono-Dinantian basins established both in the Devonian belt domain and on the surrounding northwards and southwards platforms.

2.3. The third phase.

φ 3 is refered to as Ante-Upper Visean. It occured mainly during Lower Carboniferous, but started as early as very Upper Devonian, just before Strunian, in some parts of Central Brittany, and after Strunian in North-Western Brittany. φ 3 metamorphism was linked with folds, associated strike-slip faults and syn-metamorphic thrusts (see below) developping cleavage and blastomylonites. It is either an Intermediate Pressure metamorphism like φ 2, or a slightly Lower-Pressure one. It is well developped and characterized in Brittany and Eastern Central Massif. It may be identified with the second Paleozoic metamorphic event described in the Northern Alpine Cristalline Massifs (17). But it is not so clearly identified in most cases in Central Massif because it can be easily confused with φ 2 (they occured nearly in the same P/T conditions) every times there are no good tectonic or radiometric controls available. φ 3 main characteristics are :

In Eastern Central Massif, the epizonal so-called "Brevenne series" unconformably overlies upon the mesozonal φ 2 "Lyonnais series" (Fig. 2) (33,34,50). Which gives one of the best evidences of the polyphased character of the French Paleozoic Orogen. The same phenomenon occurs in the Northern Alpine Cristalline Massifs where an epizonal series of meta-conglomerates, black shales and spilites unconformably rests upon retromorphosed φ 2 mesozonal rocks (38).

In North-Western Brittany, metamorphism occured during the development of a large shear belt. The metamorphic gradient decreases eastwards from anatexis in the coastal area down to epizone in the Cadomian basement. The metamorphic type is an Intermediate Low-Pressure one with chloritoid, biotite, almandine, staurolite and sillimanite as index minerals (36). In the epizonal central area, like in Central Brittany (next paragraph), the previous minerals are overprinted by a new generation of staurolite, andalusite and sillimanite : the main metamorphic event is overprinted by a more purely thermic one. As in Central Brittany (next paragraph), this is due to the emplacement of syn-metamorphic granitic plutons.

In Central Brittany, the metamorphic series is mostly an anchi- or epizonal one (Fig. 2) in which chloritoid (37) and kyanite have been locally described along some tectonic accidents. Some penecontemporaneous granitoids developped kilometers-broad contact aureoles. So, in their vicinity, the metamorphic series appears as intermediate between regional- and contact-types, giving to the Central Brittany series the false aspect of a Low-Pressure one.

In Southern Central Massif, a major syn-metamorphic thrust occured during φ 3 in the so-called "Lot series" (Fig. 2) (8,19). This thrust is quite similar to the above-mentionned φ 2 one in the central part of Central Massif. It also deals with retrograded granulites overlying a prograde mesozonal series. These main thrusts are a very important feature of the Central Massif geologic history and have been emphasized by several people (7,8,19, 28,29,31,40) during the last ten years. Although they were operative at different periods, they have several common characters : all were synchronous of a main metamorphic phase and affected more or less retrograded granulitic series ; all occured in high P/T conditions (meso- or catazonal ones) ; for all of them it has been more or less clearly spoken of reverse metamorphic gradient and they all have been compared to the Main Central Thrust in the Himalayan Chain. Three of them are already known (Fig. 1 and 2), two in the central part of Central Massif and one in the Southern part of it.

φ 3 tectono-metamorphic phase was followed by an important calc-alkaline plutonism (for example in North and Central Brittany, Vendée and Limousin), especially by the emplacement of aluminous (S type) leucogranites.

2.4. The fourth phase.

φ 4 occured during Upper Carboniferous, mainly Namurian and

Westphalian, but could possibly persist until Stephanian included. Roughly speaking, the associated structures are facing North in Ardennes and on the contrary they are facing South in Montagne-Noire. φ 4 can either be the only metamorphic event (mainly in Pyrenees (8,13,27) and Ardennes) or be the last one and thus overprint one or more earlier phases (Fig. 1). φ 4 metamorphism is always a Low-Pressure one, with andalusite- and cordierite-bearing assemblages in the mesozonal and catazonal facies.

Locally, in Central Massif (Fig. 1), φ 4 facies show more or less the same peculiarities than the φ 3 Brittany series : the emplacement of Namuro-Westphalian granites induced a regional scale thermal metamorphism which overprints an Intermediate-Pressure φ 3 episod.

In Ardennes (which are nearly wholly located in Belgium), Cambrian rocks exhibiting true Caledonian deformation devoided of any associated metamorphism are overlaid by Lower Devonian stratas affected by a φ 4 greenschist metamorphism. In North-Eastern Central Massif and Southern Vosges a generally weak event (anchi- or epizonal) affected the Dinantian terranes. In Pyrenees (13,27), a complete zoneographic succession including anchizone up to Low-Pressure granulites (Fig. 2) has been described. Following Guitard (8), the corresponding isograds are disposed in the Paleozoic cover close to Precambrian cores. In North-Western Central Massif, a mesozonal φ 4 episod overprints and older (φ 2) phase of lower grade and in Eastern Central Massif it also overprints the φ 2 Intermediate-Pressure so called "Lyonnais series". In Montagne-Noire, φ 4 is the most widespread metamorphic event even if very scarce relictual kyanite and retrograded eclogites provide evidences of a longer and more complex history. The Low-Pressure succession is complete (9) including anchizonal up to cordierite-bearing migmatitic facies. In fact, two different interpretations of the Montagne-Noire metamorphic series are still facing each other. For Demange (9) the Low-Pressure evolution is the only important one and relictual kyanite and eclogites are too localized to advocate a large scale pre-φ4 event. For Quenardel, the Montagne-Noire series must be a polyphased one partly because of the relictual kyanite and eclogites, partly because the strong structuration in the catazonal orthogneisses may not be consistent with a Low-Pressure metamorphic evolution. In the second scheme, the older phase could be Devonian. In South-Eastern part of Central Massif, cordierite-bearing migmatites and their associated granite are well developped in an anatectic dome (Fig. 2) where cordierite- and sometimes andalusite-bearing assemblages crystallized at the expense of Intermediate-Pressure φ3 schists (22,26) southwards from the dome, or φ 1 granulites already retrograded during φ2 (westwards from the dome), or of φ2 "Lyonnais series" northwards from the dome (26,51).

Mainly, φ 4 is characterized by an important thermal activity and large scale anatexis phenomena, which induced the emplacement of leucogranites and calc-alkaline biotite-granites everywhere in the Paleozoic Orogen.

3. CONCLUSIONS

Finally, the main features of the French Paleozoic Orogen are the following :

1. From the tectonic point of view, and up to now,nothing important has been recognized during the Caledonian s.s. period, except in Ardennes. The lower Paleozoic is considered as a period of distensive activity in Western Europe in spite of the existence of High Pressure metamorphic activity during Silurian. But the recently discovered evidences of syn-tectonic crystallization in some eclogites and granulites may change this ascertainment.
2. Very diversified geodynamic sites at the beginning of the orogen : a typical continental crust in the Eastern half of Central Massif and in Vosges, a shelf in Central and Northern Brittany as well as in Pyrenees and Montagne-Noire, the opening of a rift into the above mentionned continental crust now well documented in Western Central Massif, and an oceanic crust in Southern Armorican Massif and Alpine Cristalline Massifs pro parte (41).
3. Widespread High Pressure facies during the Silurian phase or even earlier. According to their location in large and more or less retrograded units or in small tectonic slices, and to their initial geodynamic settlement, two genetic models have been proposed to explain their genesis : a subducted oceanic crust (30) or a large scale thrust into a continental crust (continental subduction of the thinned crust in the rift area) (8).
4. A decreasing pressure regime corresponding to the evolution of the Orogen since an initial peri-oceanic (Devonian) type, until an intra-continental-like type (Upper Carboniferous).

To summarize briefly the previous datas, a simple evolutive scheme can be established, with the main following steps :
During Ordovician, opening of a rift in a continental crust close to an already established oceanic zone. The rifting process evolving or not into an oceanic zone, depending upon the various locations.
During Silurian, again owing to different locations, oceanic or continental subduction. Generation of High-Pressure metamorphic facies devoided of large scale correlated deformations.
During Devonian, closure of the ocean (or of the rifted zone). First tectonic culmination connected to an Intermediate-Pressure type metamorphism. Obduction of the High-Pressure series by huge intra-crustal thrusts. Thickening of the crust and uplift.

During Carboniferous, continental collision s.s. inducing further intra-crustal thrusts. Important thermal activity in all the previous crustal thickening domain. The correlated metamorphism grades progressively from Intermediate-Pressure to Low-Pressure conditions.

REFERENCES

(1) Lasnier, B. : 1977, Thèse d'Etat, Nantes, 351 p.
(2) Marchand, J. : 1974, Thèse 3ème cycle Nantes, 207 p.
(3) Leyreloup, A. : 1973, Thèse 3ème cycle Nantes, 305 p.
(4) Fluck, P. : 1980, Sci. Géol. Mém. n° 22, 248 p.
(5) Bogdanoff, S. : 1980, Thèse d'Etat Orsay, 316 p.
(6) Nicollet, C. : 1978, Thèse 3ème cycle Montpellier.
(7) Burg, J.P. : 1977, Thèse 3ème cycle Montpellier, 106 p.
(8) Bard, J.P. and al. : 1980, I.G.C. (C7), Mém. B.R.G.M. n° 107, pp. 161-189.
(9) Demange, M. : 1982, Thèse d'Etat Paris, 2 vol.
(10) Santallier, D. et al. : 1978, Bull. Soc. Fr. Minéral. Cristallogr. 101, pp. 77-88.
(11) Guillot, P.L. : 1981, Thèse d'Etat Orléans, 3 vol.
(12) Santallier, D. : 1981, Thèse d'Etat Orléans, 2 vol.
(13) Autran, A. and al. : 1970, Bull. Soc. Géol. Fr. (7) XII, 4, pp. 673-731.
(14) Sagon, J.P. : 1976, Thèse d'Etat Paris, 671 p.
(15) Collomb, P. : 1970, Mém. Serv. Expl. Carte Géol. Dét. Fr. Paris, 419 p.
(16) Grolier, J. : 1971, Mémoire B.R.G.M. n° 64, 163 p.
(17) Carme, F. : 1973, C.R. Acad. Sci. Paris 277, pp. 2133-2136.
(18) Ters, M. : 1979, Bull. B.R.G.M. I, 4, pp. 293-301.
(19) Briand, B. and Gay, M. : 1978, Bull. B.R.G.M. I, 3, pp.167-186.
(20) Rolin, P. : 1981, Thèse 3ème cycle Orsay, 210 p.
(21) Triboulet, C. : 1977, Thèse d'Etat Paris, 138 p.
(22) Marignac, C. and al. : 1980, C.R. Acad. Sci. Paris 291, D, pp. 605-608.
(23) Audren, C. and Lefort, J.P. : 1977, Bull. Soc. Géol. Fr. (7) XIX, 2, pp. 395-404.
(24) Audren, C. and Le Métour, J. : 1976, Bull. Soc. Géol. Fr. (7) XVIII, 4, pp. 1041-1049.
(25) Cogné, J. : 1960, Mém. Serv. Expl. Carte Géol. Dét. Fr. Paris, 382 p.
(26) Chenevoy, M. and Ravier, J. : 1968, Bull. Soc. Géol. Fr. (7) X, 5, pp. 613-617.
(27) Guitard, G. : 1970, Mémoire B.R.G.M. n° 63, 316 p.
(28) Pin, C. : 1979, Thèse 3ème cycle Montpellier, 205 p.
(29) Suire, J. : 1982, Thèse 3ème cycle Clermont-Ferrand, 167 p.
(30) Godard, G. : 1981, Thèse 3ème cycle Nantes, 167 p.
(31) Burg, J.P. and Matte, P.J.: 1978, Z.dt.geol.Ges. 129,429-460.

(32) Peucat, J.J. and al. : 1978, Bull. Soc. Géol. Fr. (7) XX, 2, pp. 163-167.
(33) Peterlongo, J. : 1958, Annales Fac. Sci. Clermont n° 4, 187p.
(34) Autran, A. : 1978, Geol. Survey Canada Paper 78-13, pp. 159-175.
(35) Cabanis, B. and Fabriès, J. : 1972, C.R. Acad. Sci. Paris 275, D, pp. 2311-2314.
(36) Cabanis, B. and Fabriès, J. : 1975, C.R. Acad. Sci. Paris 280, D, pp. 1769-1772.
(37) Paradis, S. : 1981, Thèse 3ème cycle Brest, 167 p.
(38) Carme, F. : 1971, C.R. Acad. Sci. Paris 273, D, pp. 1671-1674.
(39) Le Fort, P. and Pêcher, A. : 1971, C.R. Acad. Sci Paris 273, D, pp. 3-5.
(40) Carme, F. : 1974, C.R. Acad. Sci. Paris 278, D, pp. 2501-2504.
(41) Bodinier, J.L. and al. : 1981, Contrib. Mineral. Petrol. 78, pp. 379-388.
(42) Le Fort, P. : 1973, Thèse Nancy.
(43) Lefort, J.P. and al. : 1982, Tectonophysics (in press).
(44) Bard, J.P. and al. : 1980, 26. I.G.C. Paris, C6, pp. 233-246.
(45) Peucat, J.J. : 1982, Thèse Rennes, 158 p.
(46) Peucat, J.J. and al. : 1982, Earth Planet. Sci. Lett. 60, pp. 70-78.
(47) Autran, A. : 1980, 26. I.G.C. Paris, C7, pp. 7-17.
(48) Autran, A. and Cogné, J. : 1980, 26. I.G.C. Paris, C6, pp. 90-111.
(49) Seyler, M. and Boucarut, M. : 1978, Bull. B.R.G.M., I, 1, pp. 3-18.
(50) Bourrier, M. and al. : 1979, 7ème Réun. Ann. Sci. Terre, Lyon.
(51) Chenevoy, M. : 1973, 98ème Congr. Soc. Sav. St Etienne, 1, pp. 425-434.

PROCEEDINGS OF THE DEFORMATION STUDY GROUP, CALEDONIDE OROGEN PROJECT

KEPPIE, J.D., Dept. of Mines and Energy, Halifax, Nova Scotia, Canada; GEE, D.G., Geological Survey of Sweden, Box 670, 75128 Uppsala, Sweden; ROBERTS, D., Norges Geologiske Undersokelse, Poste Box 3006, 7001 Trondheim, Norway; POWELL, D., Geology Dept., Bedford College, University of London, London, England; MAX, M.D., Geological Survey of Ireland, 14 Hume Street, Dublin 2, Eire; OSBERG, P., Geology Dept., University of Maine, Orono, Maine, U.S.A.; PIQUE, A., Institute de Geologie, Univ. Louis Pasteur, 67084 Strasbourg, Cedex, France; LECORCHE, J.P., CNRS Faculte de Sciences St-Jerome, 13397 Marseille, France

INTRODUCTION

The NATO Advanced Study Institute - IGCP Caledonide Orogen Project Meeting in Fredericton, New Brunswick, Canada, in August 1982, provided the opportunity for members of the Deformation Study Group to compare structural maps of various parts of the Caledonide Orogen. This allowed the legends to be standardized, for some general conclusions to be made about the spatial distribution, nature and trends of structures throughout the entire orogen and for some preliminary hypotheses to be formulated about their origin. These are presented here.

The structural maps of various parts of the Caledonide Orogen show the age, orientations and spatial distributions of major structures. To date, the only published map is the Structural Map of the Appalachian Orogen in Canada (1). Others are in various stages of preparation and generalized small scale black and white versions are presented elsewhere in this volume. In general, on the coloured maps, background colours indicate the time of the first deformation. The intensity of penetrative deformation is shown by different shades of the backgound colours, the darker shades representing higher strain. Ideally, the time of deformation is constrained by stratigraphic information. However, in many areas the

P. E. Schenk (ed.),
Regional Trends in the Geology of the Appalachian-Caledonian-Hercynian-Mauritanide Orogen, 275–278.

constraints are limited to isotopic age determination data. In this context, a preliminary standardized time scale was agreed upon as follows: Precambrian-Cambrian boundary at 570 Ma, Cambro-Ordovician boundary at 520 Ma with the Tremadocian Stage placed in the Ordovician Period, Ordovician-Silurian boundary at 435 Ma, Siluro-Devonian boundary at 415 Ma, Devono-Carboniferous boundary at 360 Ma, Permo-Carboniferous boundary at 290 Ma and Permo-Triassic boundary at 250 Ma. While this differs slightly from that used on the Structural Map of the Appalachian in Canada, it does not affect the colours assigned to the times of deformation.

CONCLUSIONS

Based upon the structural maps and tables showing the times of deformation in various parts of the orogen, five major periods of deformation may be recognized within the orogen as a whole Late Proterozoic-Early Cambrian (Cadomian-Avalonian-Panafrican), Late Cambrian-Early Ordovician (Grampian-Finnmarkian-Penobscotian), Mid-Late Ordovician (Taconian), Siluro-Devonian (Late Caledonian-Scandian-Acadian), and Permo-Carboniferous (Hercynian-Alleghanian).

The Late Proterozoic-Early Cambrian, Cadomian-Avalonian-Panafrican deformation (650-550 Ma) is limited to the south-eastern side of the orogen, excluding the Baltic Shield. It varies rapidly in intensity with narrow anastamosing and branching zones of intense deformation separated by blocks with weak deformation. The temporal coincidence of Cadomian deformation with rifting and the opening of Iapetus along the margins of the Laurentian and Baltic Shields suggests that these events occurred along the margins of different oceans, one tensional (Iapetus) and the other compressional.

The Late Cambrian-Early Ordovician, Grampian-Finnmarkian-Penobscotian deformation (530-490 Ma) appears to have occurred along or near the margins of Iapetus. The intensity of this deformation increases from mild in the northern Appalachians to intense in northwestern Ireland, northern Scotland, and western and northern parts of Scandinavia. In general, the rocks affected by this event were subsequently thrust over the cratonic margins.

The Mid-Late Ordovician (Llanvirn-Ashgill), Taconian deformation (470-435 Ma) is mainly restricted to the northwestern margin of the Appalachians, but it occurs locally in western Ireland and is inferred in the northern part of the Mauritanides. In the Appalachians, the Taconian deformation may be attributed to subduction in part oblique to the cratonic margin, terminating in volcanic arcs and trenches in Atlantic Canada and Eire.

The Mid-Silurian-Early Devonian, Late Caledonian-Ligerian-Scandian-Acadian deformation (425-400 Ma) is widespread throughout the Appalachians and Caledonides but is absent from the Mauritanides. The distinction between the Mid-Late Silurian Late Caledonian (Scandian) Orogeny and the Early-Mid Devonian Acadian Orogeny may be due to diachronism and/or differences in the dating method and different ages assigned to the Siluro-Devonian boundary. Thus, in the British Isles, the time of Late Caledonian deformation is bracketed by the youngest deformed strata of Mid and Late Silurian age and the oldest post-tectonic plutons yielding isotopic ages of 400-415 Ma. Here, post-tectonic strata are usually continental and poorly dated. However, in the northern Appalachians, post-Acadian plutons yielding similar ages cut deformed fossiliferous Early Devonian rocks. In the Scandinavian Caledonides, rocks as young as Downtonian are involved in the Late Caledonian deformation. Such comparisons lead to the conclusions that the Late Caledonian, Scandian and Acadian Orogenies are temporal equivalents and are correlatives along strike with little recourse to diachronism. The Late Caledonian-Acadian deformation varies considerably in intensity along the belt. It is most intense in Scandinavia, Newfoundland and New England and is of moderate intensity in Britain, Eire, France, Spain and Maritime Canada where there are local zones of intense deformation surrounding less deformed blocks. The deformation may be attributed to the closing of Iapetus in the northeast and the collision associated with the accretion of Avalonian terranes in the southwest.

The Permo-Carboniferous, Hercynian-Alleghanian deformation (350-250 Ma) was most intense in the southern and central Appalachians and in the Mauritanides. Northeast along the orogen, zones of Hercynian deformation splay out into zones of intense deformation surrounding mildly deformed blocks. This deformation is related to the collision of Africa and North America.

Several general observations regarding the nature of deformation in the Caledonide Orogen are as follows. The widespread deformational events appear to be relatively short-lived fabric-forming events followed by transpressive deformation, and may be attributed to a series of collisions as terranes are accreted with the cratons and with one another. Deformation during several of these events was clearly diachronous. Exceptions occur in (i) interpreted trench complexes such as the Southern Uplands of Scotland and part of the Merrimack Synclinorium where deformation is related to long-lived subduction and the formation of an accretionary wedge and (ii) areas where mechanisms involve gravity sliding and spreading.

Where large scale overthrusting occurs, such as in the southern and central Appalachians, the Mauritanides and the Scandinavian Caledonides, the deformational front does not coincide with the cratonic margin. In some places, such as the northern Appalachians, the deformational front appears to coincide with the edge of the craton.

Variations in the intensity of the various deformational events in the Caledonide Orogen provide some constraints upon palinspastic reconstructions for terranes involved in the accretionary collisions.

REFERENCE

(1) Keppie, J.D., Ruitenberg, A.A., Fyffe, L.R., McCutcheon, S.R., St. Julien, P., Skidmore, B., Beland, J., Hubert, C., Williams, H. and Bursnall, J.: 1982, Memorial University of Newfoundland Map No. 4.

TIMING OF DEFORMATION IN THE SCANDINAVIAN CALEDONIDES

Gee, D.G.[1] and Roberts, D.[2]

SGU, Box 670, S-751 28 Uppsala, Sweden.
NGU, Box 3006, N-7001 Trondheim, Norway.

ABSTRACT

Western Scandinavia is composed of a variety of Caledonian nappes transported eastwards on to the Baltoscandian Platform. Many of the higher tectonic units provide evidence of complex deformational histories that both preceded and accompanied nappe emplacement. Palaeontological evidence for the timing of emplacement of the nappes in the southern and central parts of the mountain belt requires that the main nappe translation occurred after Llandovery deposition. In the mountain front, nappe advance resulted in deposition of extramontane continental sandstones in the Wenlock, the sedimentation continuing into the latest Silurian in the Oslo area, prior to folding and thrusting. This deformational sequence is referred to as Scandian (or Scandinavian). In northern Norway, isotopic age-determination evidence on syntectonic plutons, intruded into the Baltoscandian miogeocline, indicates that deformation (530-490 Ma) commenced in the late Cambrian and continued into the early Ordovician. This early orogenesis, involving polyphase folding, metamorphism and nappe translation, is referred to as Finnmarkian. The youngest sedimentary rocks in the mountain front in Finnmark are of Tremadoc age requiring that the main nappe emplacement occurred during or after the Arenig.

In the higher nappes of the Scandian allochthon, a variety of isotope age-determination evidence is indicative of a complex pre-Silurian tectonic and metamorphic history. Major unconformaties locally separate late Ordovician from older successions. In several areas, the obduction of fragmented ophiolites can be shown to have occurred prior to the Ashgill

P. E. Schenk (ed.),
Regional Trends in the Geology of the Appalachian-Caledonian-Hercynian-Mauritanide Orogen, 279–292.

and in one case to be of pre-mid Arenig age. Thus, in the higher allochthonous units, containing the transported island-arc, back and fore-arc and Iapetus oceanic environments, present evidence favours a history of Caledonian deformation that was concentrated to an early orogenic episode (probably reaching a climax in the earliest Ordovician) and later (mid Silurian to early Devonian) depression, metamorphism and thrusting of these and the younger Ordovician and Silurian rocks on to the Baltoscandian Platform. Late-orogenic erosion of the nappe pile was followed by deposition of an Old Red Sandstone intermontane molasse of early Devonian and, perhaps locally, of Pridoli (latest Silurian) age; these sedimentary rocks were also subject to Caledonian (probably late Devonian) deformation.

INTRODUCTION

The mountains of western Scandinavia are composed of a variety of Caledonian nappes that have been thrust eastwards on to the Baltoscandian Platform. Many of the higher tectonic units have been transported distances of several hundreds of kilometres and provide evidence of complex deformational histories that both preceded and accompanied nappe emplacement. Northwest of the thrust front, the entire tectonic succession also shows evidence of deformation after emplacement, major antiforms and synforms parallel to the length of the orogen being particularly conspicuous features developed during this late-stage deformation. Late-orogenic erosion of the nappe pile in the interior of the orogen was followed by deposition of an Old Red Sandstone intermontane molasse of early-mid Devonian and, perhaps locally, of Pridoli (late Silurian) age; these sedimentary rocks were also subject to Caledonian (probably late Devonian) deformation.

Biostratigraphic and isotope age-determination evidence relevant to the timing of deformation are summarized in Fig. 1 and Table 1. Evidence from the southern and central parts of the mountain belt requires that the main nappe translation occurred after Llandovery deposition. In the mountain front, nappe advance resulted in deposition of extramontane continental sandstones in the mid-late Silurian, prior to folding and thrusting. This deformational sequence (1) starting perhaps in the late Llandovery, climaxing in the mid-late Silurian and continuing into the early Devonian, is referred to as Scandian (it has also been referred to as Scandinavian). In northern Norway, isotopic age-determination evidence on syntectonic plutons, intruded into successions correlated with those of the Baltoscandian miogeocline, indicates that deformation (530-490 Ma) probably commenced in the late Cambrian and continued into the early Ordovician. This early orogenesis, involving polyphase folding, metamorphism and nappe translation, is referred to as Finnmarkian (2).

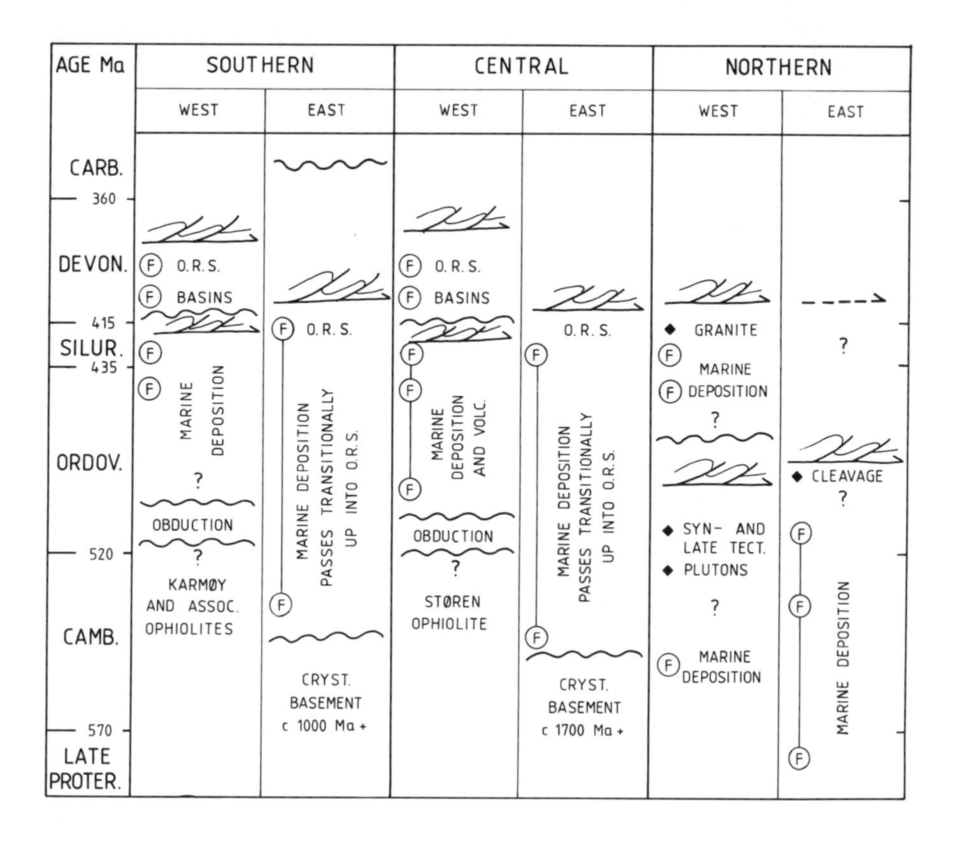

Table 1. Evidence for timing of Caledonian deformation in the southern, central and northern segments of the orogen in Scandinavia (F...fossil, age-det. symbol as on Fig. 1).

The tectonostratigraphy of the Scandinavian Caledonides can be conveniently treated in six categories: the Autochthon (incl. Parautochthon), Lower, Middle, Upper and Uppermost Allochthons and the intermontane Old Red Sandstone basins (Fig. 1). The autochthon comprises a basement complex composed of Precambrian crystalline rocks with a thin veneer of sediments largely of latest Precambrian and Cambrian age. The various allochthons are nappe complexes generally composed of related thrust sheets; most of them contain components of both Caledonian cover and sheets of Precambrian crystalline rocks showing various stages of Caledonian reworking. Some of the higher nappes contain intra--Caledonian basement elements (e.g. ophiolite complexes or

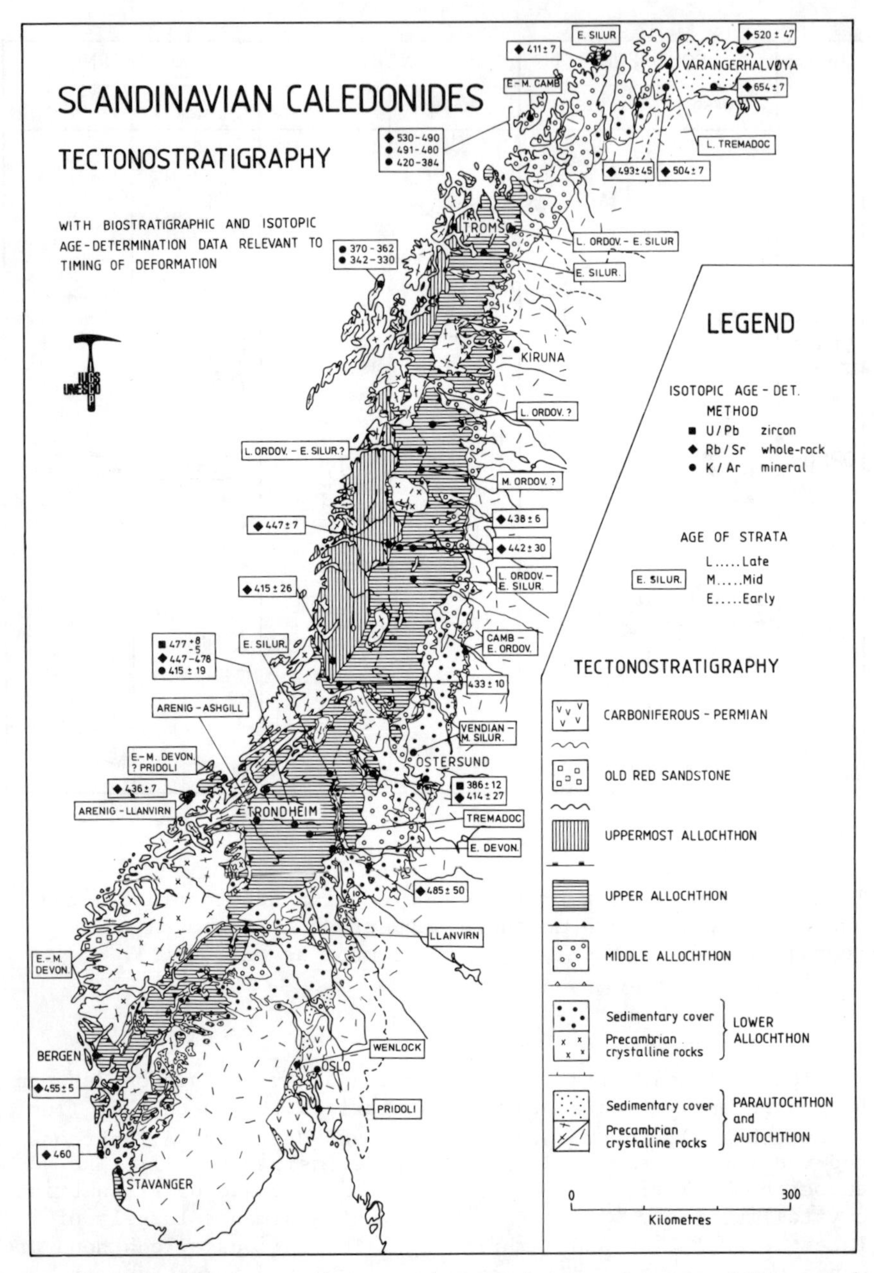

Fig. 1. This tectonostratigraphic map is based on an IGCP compilation, presented at the Uppsala Caledonide Symposium 1981. The UCS edition (scale 1:1 M) was drafted at the geological surveys of Norway and Sweden based on published sources and many unpublished contributions.

Finnmarkian-deformed metamorphic sequences) separated from overlying cover by major unconformities.

AUTOCHTHON, PARAUTOCHTHON AND LOWER ALLOCHTHON

The Caledonian successions of the Autochthon and Lower Allochthon along the thrust front are dominated by latest Precambrian and Cambro-Silurian sedimentary rocks. These units can be traced westwards, towards the interior of the orogen, across a variety of windows to the Norwegian west coast where the sedimentary successions are highly attenuated and probably mainly of Cambrian age. Within the Lower Allochthon the sedimentary cover dominates in the Caledonian front; further west it includes an increasing proportion of Precambrian crystalline rocks.

Along the mountain front the successions are folded above a basal décollement. In the Oslo area they range in age from mid Cambrian to latest Silurian. The strata are dominated by basinal shales and limestones passing upwards in the mid Wenlock from limestone through estuarine deposits into continental sandstones, derived from a landmass rising to the northwest (3, 4). Other evidence, referred to below, suggests that this elevated area marked the front of the advancing Caledonian nappes. Further north along the mountain front, late Precambrian and early Palaeozoic sedimentary successions in the lowermost tectonic units seldom contain strata younger than Ordovician. However, in the Jämtland region, strata below the Middle Allochthon do continue up into the latest Llandovery and perhaps into the Wenlock. Instability in the mid Ordovician is witnessed by the influx from the west of continentally derived greywackes into the Jämtland basin; this incursion was followed by a more stable interval in the late Ordovician and early Silurian prior to renewed instability and return in the late Llandovery to a greywacke facies. These Llandovery greywackes give way upwards through shales into sandstones of probable continental origin (5) which may be of Wenlock age; the facies changes would appear to reflect the onset of Scandian orogenesis further to the west.

North of Jämtland, in the Caledonian front, Silurian strata are not preserved in the successions of the Autochthon and Lower Allochthon. Early Ordovician limestones and greywackes of probable mid Ordovician age occur in Västerbotten and southernmost Norrbotten but further north the mid-late Cambrian black shales are the youngest strata preserved. In the far north on Digermulhalvøya (6), thick, very low grade, late Precambrian and lower Cambrian shallow-marine sandstones and shales pass up through a seemingly barren black shale interval into black shales and quartzites of Tremadoc and perhaps earliest Arenig age (B.-D. Erdtmann pers. comm. 1982).

Along the Caledonian front in Scandinavia various attempts have been made to date shales and phyllites in the lower tectonic units by the Rb/Sr whole rock method. All ages quoted here have been (re)calculated using a decay constant 1.42 x $10^{-11}a^{-1}$ for ^{87}Rb. In the Mjøsa area of southern Norway (7), red and green shales overlying the latest Precambrian tillites yielded an age of 612±10 Ma (E. Welin, pers. comm. 1982). Further north, in northern Jämtland, similar shales from above the tillites gave an age of 500 Ma (E. Welin, in 8). In northernmost Norway an age of 654±7 Ma has been reported from inter-tillite shales of the Nyborg formation (9), while 504±7 Ma has been reported from slates of the lowest Cambrian Stappogiedde Formation and 493±45 Ma for slates of the Friarfjord Formation of the Laksefjord Nappe (2). On Varangerhalvøya, north of the Trollfjord--Komagelv fault, cleaved mudstones of the Upper Riphean Kongsfjord Formation have recently yielded a Rb/Sr isochron age of 520±47 Ma (10). Whereas some of these dates (c. 650 Ma) are probably related closely to the time of deposition, others are believed to reflect the time of slaty cleavage formation. In the case of the 500 Ma date reported from northern Jämtland, the Cambro-Ordovician sedimentary history is known in detail from an extensive drilling programme (11, 12), and this particular date can thus hardly be related to a cleavage-forming event. However, in northern Norway the youngest fossiliferous strata reported in these tectonic units are of Tremadoc age and the 493-520 Ma ages are considered (2) to date the development of a pervasive slaty cleavage in the Arenig. This interpretation would require great diachroneity (c. 100 Ma) of componental movement along the thrusts at the base of the Lower Allochthon, the movement being of early Devonian (or younger, but pre-Carboniferous) age in the far south and of early Ordovician age in the far north.

MIDDLE ALLOCHTHON

The nappe complexes grouped into the Middle Allochthon are dominated by Precambrian crystalline rocks and late Precambrian sedimentary successions derived from the Baltoscandian miogeocline and transported eastwards from west of the Norwegian coast, distances of c. 300 km, on to the Baltoscandian platform. Metamorphism of these units is generally in greenschist facies. Locally, in the south, the late Precambrian strata (13) pass up through Cambrian sandstones and shales into mid Ordovician greywackes.

At various localities along the mountain belt, the sequences contain radioactive black shales (or phyllites) of probable late Cambrian age. In northernmost Norway (on Sørøy) limestones in the upper half of the succession contain archaeocyathids thought to be of mid Cambrian age (14). In this northernmost (Finnmark) area a variety of syn- and post-tectonic intrusions in the Seiland Igneous Province (15, 16, 17) have yielded Rb/Sr isochrons in the age range 530-490 Ma (18, 19, 2, 9), implying latest Cambrian to earliest Ordovician polyphase deformation, metamorphism and nappe translation. These phenomena are referred to the Finnmarkian phase of the Caledonian orogeny (20, 2). Regional aspects of this major orogenic event were synthesized by Roberts and Sturt (21).

Within the Middle Allochthon of Jämtland, age-determination work (22) on mylonites separating two of the nappes in this complex (the Särv Nappes and the underlying Tännäs Augen Gneiss Nappe) resulted in a Rb/Sr isochron of 485±50 Ma. This evidence, viewed along with the presence of the mid Ordovician greywackes in the underlying tectonic units, suggests the possibility that Finnmarkian deformation may have extended far south through the orogen influencing the rocks of the Baltoscandian miogeocline that were subsequently thrust on to the platform during Scandian orogeny.

UPPER ALLOCHTHON

The overlying Upper Allochthon contains amphibolite facies and higher grade units in the base, overlain by a wide variety of volcano-sedimentary rocks at least in part of Lower Palaeozoic age. Substantial components of pre-Caledonian crystalline rocks (e.g. Jotun Nappe) may occur in the lower part of these tectonic units. In the Swedish Caledonides the Seve Nappes contain both Precambrian crystalline rocks and cover sediments (23). The high grade metamorphism has been attributed to the Scandian orogenesis on the basis of a Rb/Sr isochron of 414±27 Ma on the granulite facies rocks of central Jämtland (24). Younger mineral ages (c. 385-395 Ma) indicate a cooling history extending into the early Devonian.

The overlying largely greenschist facies volcano-sedimentary rocks (24) in the Köli Nappes, derived from west of the Baltoscandian miogeocline, contain successions in the lower tectonic units ranging in age at least from the early Ordovician (Otta conglomerate, 24) to the early Silurian (late Llandovery). As in the underlying non-volcanic successions of the Lower Allochthon of Jämtland, the sedimentary record indicates a period of great instability (greywackes and fanglomerates) in the mid Ordovician, a more stable interval in the Ashgill and early Llandovery and a passage in the late Llandovery from black shales into greywackes, heralding the onset of Scandian orogenic movements further to the west. Prior to the deposition of the mid Ordovician greywackes of the Lower Köli Nappes there is evidence (27) of island-arc development and subsequent arc-splitting. In the higher Köli Nappes (the Storfjället Nappe of northern Västerbotten) successions that have yielded crinoids are metamorphosed locally up to upper--amphibolite facies, migmatized and intruded by major gabbros. The migmatites have yielded a Rb/Sr isochron of 442±30 Ma (28); a related syntectonic granite (the Vilasund granite) has been dated (29, 30) to 438±6 Ma by the same method. These lines of evidence suggest important mid Ordovician orogenic activity at least in some areas. Further south in central Trøndelag, within the Gula Complex, syn- and late-tectonic trondhjemites have yielded zircon (U/Pb) and Rb/Sr ages ranging from c. 480-450 Ma (31), the scattering is interpretated to be result of the influence of subsequent (Scandian) nappe movements, and the age of intrusion is most probably early Ordovician. Rb/Sr whole-rock/mineral and K/Ar ages of 412-415 Ma reflect the Scandian deformation.

In the higher parts of the Upper Allochthon, in Köli (sensu Törnebohm, 32) units above those referred to earlier, fragmented ophiolites (33, 34) occur in the Støren Nappe (35) in Trøndelag; these are reported to occur in at least two stratigraphic levels. The basal Støren ophiolite fragment is overlain by mid Arenig black shales in the Hølonda area; these pass up into limestones and shales along with andesitic porphyrites and later rhyolites. In the same district, well developed ophiolites also occur at Løkken (34) and Grefstadfjell (36) and have been interpreted by the latter authors to be of late Arenig age. Extensive pillow basalts, correlative with or slightly younger than these ophiolites, also occur in the Upper Allochthon elsewhere in the Trondheim area. They are all thought to be the result of back-arc, marginal basin spreading and accretion (37). A mature magmatic arc has been recorded on Smøla and near Snåsa (38).

The development of the volcano-sedimentary edifices above the older ophiolites apparently occurred after initial obduction (32) of the latter, implying important thrusting (accompanied by greenschist facies metamorphism and folding) prior to the mid Arenig. Constraints on the timing of ophiolite obduction and thrusting of the subjacent Gula Complex are incomplete. The presence further east of Tremadoc black phyllites (39, 40) at the contact between the Gula Complex and the adjacent Meråker Nappe suggests that this deformation occurred in the late Tremadoc to mid Arenig interval.

The construction of the island-arc and back-arc basinal succession above the early Støren ophiolite was accompanied by local instability. On Smøla, an episode of folding and faulting post-dates Arenig-Llanvirn limestone deposition but pre-dates intrusion of late Ordovician diorites. A comparable event is recorded in the Løkken-Meldal district (36). Mid to late Ordovician greywackes and fanglomerates likewise testify to instability in the sedimentary basin adjacent to the volcanic islands. These subduction-related processes, apparently occurring during gradual closure of the Iapetus Ocean, were related in time to the Taconic orogeny. They did not involve major orogenic deformation, the main tectonism and metamorphism affecting the entire Støren Nappe being accomplished during the early Scandian thrusting. The lack of evidence of definite Silurian strata in the Støren Nappe may imply that the Scandian cycle started earlier in the Llandovery than has been recorded in the lower tectonic units, referred to above.

Elsewhere in the Scandinavian Caledonides, there are several fragmented ophiolites in the Upper Allochthon (33). None of the sedimentary successions overlying the igneous rocks has yielded so much biostratigraphic evidence as in Trøndelag. In many cases a pre-Ashgill age of obduction to high crustal levels can be inferred. This evidence, taken along with preliminary Rb/Sr isochrons on associated rhyolites of 436 ± 16 Ma and c. 460 Ma for a post-obduction intrusion (the West Karmøy Igneous Complex), strongly favours early Ordovician (or earlier) obduction of the ophiolites of the Bergen region, perhaps more or less coincident in time with the obduction of the Støren ophiolite in Trøndelag.

UPPERMOST ALLOCHTHON

In the Uppermost Allochthon, the tectonic units are dominated by amphibolite facies schists and gneisses; some units contain extensive marbles and amphibolites. Granite batholiths in nappes in Nordland and Trøndelag have yielded Caledonian Rb/Sr isochron ages (e.g. the Bindal granite, 415 ± 26 Ma (42)). Precambrian Rb/Sr isochron ages (42) on gneisses may date the age of the rocks or the age of the development of the penetrative fabrics and recrystallization, or both. In the Rödingsfjället Nappe of Västerbotten, schists are cut by granitic dykes yielding a 447 ± 7 Ma Rb/Sr isochron (43). The extent to which these supracrustal rocks in the Uppermost Allochthon represent Caledonian cover or pre--Caledonian basement remains to be demonstrated. However, it can be inferred that the thrusting of the Uppermost Allochthon on to the underlying tectonic units was a Scandian event.

OLD RED SANDSTONE

Emplacement of the various allochthonous units on to the Baltoscandian Platform involved deep depression of the continental margin and metamorphism of both the basement and the sedi-

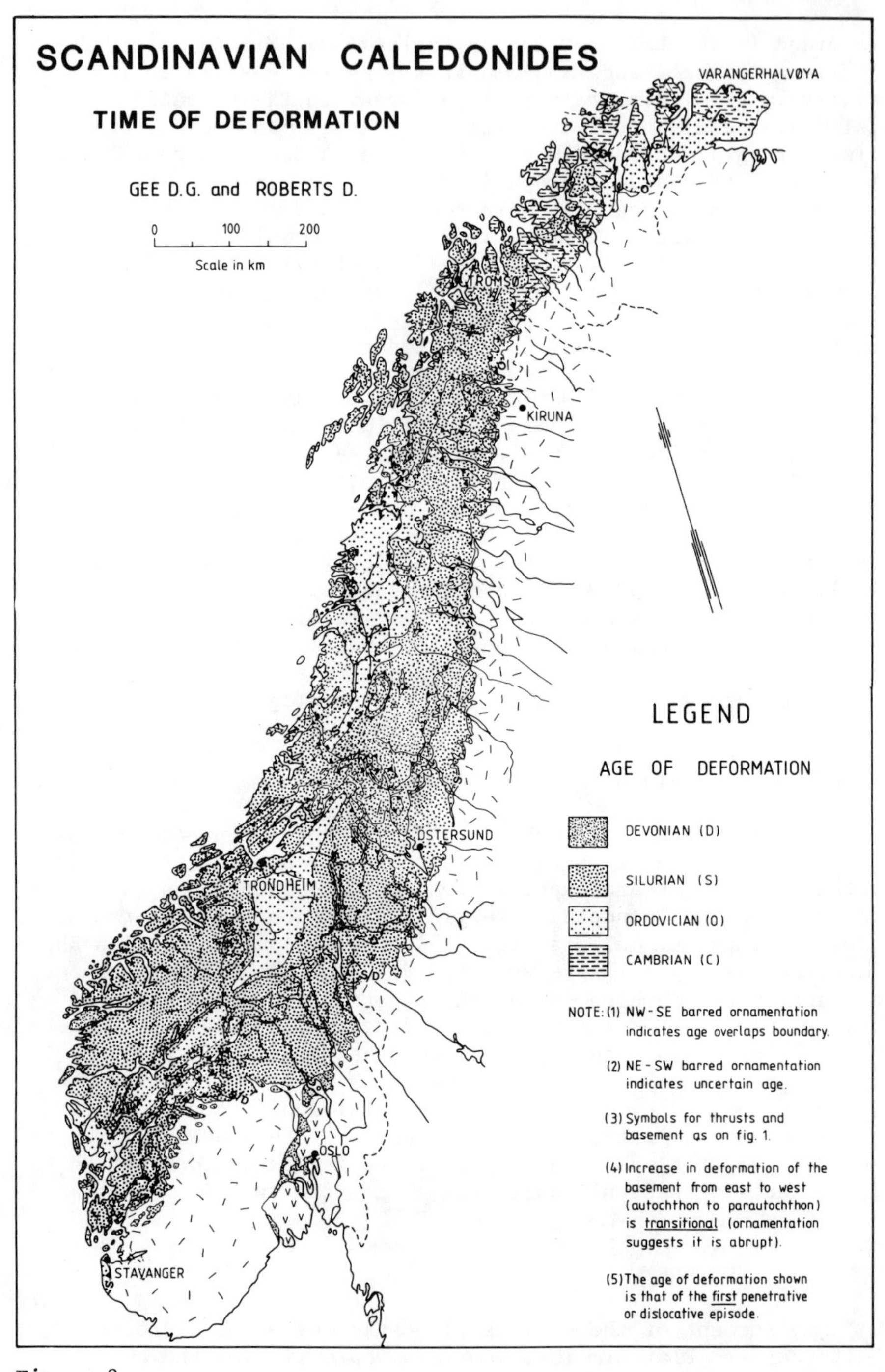
SCANDINAVIAN CALEDONIDES
TIME OF DEFORMATION
GEE D.G. and ROBERTS D.
0
100
200
Scale in km
VARANGERHALVØYA
TROMSØ
KIRUNA
ØSTERSUND
TRONDHEIM
OSLO
STAVANGER
LEGEND
AGE OF DEFORMATION
DEVONIAN (D)
SILURIAN (S)
ORDOVICIAN (O)
CAMBRIAN (C)
NOTE: (1) NW-SE barred ornamentation indicates age overlaps boundary.
(2) NE-SW barred ornamentation indicates uncertain age.
(3) Symbols for thrusts and basement as on fig. 1.
(4) Increase in deformation of the basement from east to west (autochthon to parautochthon) is transitional (ornamentation suggests it is abrupt).
(5) The age of deformation shown is that of the first penetrative or dislocative episode.

Figure 2.

mentary cover. Nd/Sm age-determinations (c. 425 Ma) of the eclogites from the basal gneisses of southwestern Norway (44) has indicated that these high-grade rocks probably obtained their pyrope-omphacite mineralogy during build-up of the Caledonian tectonic overburden.

The depression of the continental margin was followed by isostatic rebound with rapid uplift and, in the early Devonian or perhaps even latest Silurian, of deposition of an Old Red Sandstone molasse in intermontane basins (45). These fault--controlled, pull-apart basins (46, 47) accumulated very thick successions of continental sandstones and conglomerates during the early and mid Devonian. Some of these basins were subsequently subject to polyphase folding and thrusting (48, 49) involving displacements of many tens of kilometres (B.A. Sturt, pers. comm. 1982). These movements are believed to have occurred in late Devonian time. Constraints are provided by mid Carboniferous fossils in post-Caledonian sediments in the Oslo area (50) and by an Early Carboniferous weathering profile on Andøya in northern Norway (51).

CONCLUDING REMARKS

It can be concluded (Fig. 2) that the stratigraphic and age-determination available today in the Scandinavian Caledonides suggests that the Late Precambrian and Early Palaeozoic successions were subject to two principal phases of major orogenic activity; the first culminating in the Tremadoc to early Arenig and the second in the mid-late Silurian. Just as the latter extended at least into the early Devonian, along the southeastern front of the orogen, so it is possible that the former transgressed into the mid Ordovician in some areas. In parts of the Upper Allochthon of central Norway there is, however, evidence of a separate, weaker mid Ordovician deformation phase. Extensional faulting of the late-orogenic Old Red Sandstone basins accompanied deposition and was followed in the late Devonian in some areas by deformation and thrusting of the Devonian sediments and the pre-Devonian crystalline rocks.

ACKNOWLEDGEMENTS

The structural map (Fig. 2) showing time of deformation, that is presented here, is based on a 1:1 M compilation made with the assistance of Monica Beckholmen and Asbjørn Thon. As a contribution to the IGCP Project No. 27, the Caledonide Orogen, this work has been supported by the Norwegian (NAVF) and Swedish (NFR) National Science Research Councils.

REFERENCES

(1) Gee, D.G.: 1975, Am. J. Sci. 275A, pp. 468-515.

(2) Sturt, B.A., Pringle, I.R., and Ramsay, D.M.: 1978, J. geol. Soc. Lond. 135, pp. 597-610.

(3) Turner, P., and Whitaker, J.H.McD.: 1976, Sediment. Geol. 16, pp. 45-68.

(4) Worsley, D., Aarhus, N., Bassett, M.G., Howe, M.P.A., Mørk, A., and Olaussen, S.: in press, Norges geol. Unders.

(5) Bassett, M.G., Cherns, L., and Karis, L.: 1982, Sveriges geol. Unders. C 793, pp. 1-24.

(6) Reading, H.G.: 1965, Norges geol. Unders. 234, pp. 167-191.

(7) Nystuen, J.-P.: 1981, Am. J. Sci. 281, pp. 69-94.

(8) Thelander, T.: 1981, in Earth's Pre-Pleistocene glacial record, pp. 615-619, Cambridge Univ. Press.

(9) Pringle, I.R.: 1973, Geol. Mag. 109, pp. 465-472.

(10) Taylor, P.N., and Pickering, K.T.: 1981, Norges geol. Unders. 367, pp. 105-110.

(11) Gee, D.G.: 1972, Sveriges geol. Unders. C 671, pp. 1-36.

(12) Gee, D.G., Kumpulainen, R., and Thelander, T.: 1978, Sveriges geol. Unders. C 742, pp. 1-35.

(13) Nickelsen, R.P., Garton, M., and Hossack, J.R.: in press, in The Caledonide Orogen - Scandinavia and Related Areas, J. Wiley and Sons.

(14) Holland, C.H., and Sturt, B.A.: 1970, Norsk geol. Tidsskr. 50, pp. 341-355.

(15) Sturt, B.A., and Ramsay, D.M.: 1965, Norges geol. Unders. 231, pp. 1-142.

(16) Hooper, P.R.: 1971, Norges geol. Unders. 269, pp. 147-158.

(17) Robins, B., and Gardner, P.M.: 1975, Earth Planet. Sci. Lett. 26, pp. 167-178.

(18) Sturt, B.A., Miller, J.A., and Fitch, F.J.: 1967, Norsk geol. Tidsskr. 47, pp. 255-273.

(19) Sturt, B.A., Pringle, I.R., and Roberts, D.: 1975, Bull. geol. Soc. Am. 86, pp. 710-718.

(20) Ramsay, D.M., and Sturt, B.A.: 1976, Norsk geol. Tidsskr. 56, pp. 291-307.

(21) Roberts, D., and Sturt, B.A.: 1980, J. geol. Soc. London, 137, pp. 241-250.

(22) Claesson, S.: 1980, Geol. Fören. Stockholm Förhand. 102, pp. 403-420.

(23) Reymer, A.P.S., Boelrijk, N.A.I.M., Hebeda, E.H., Priem, H.N.A., Verdurmen, E.A.Th., and Verschure, R.H.: 1980, Norsk geol. Tidsskr. 60, pp. 139-147.

(24) Claesson, S.: 1981, Geol. Fören. Stockholm Förhand. 103, pp. 291-304.

(25) Stephens, M.B.: 1980, in The Caledonides of the USA, V.P.I.S.U. Mem. 2, pp. 289-298.

(26) Bruton, D.L., and Harper, O.: 1981, Norsk geol. Tidsskr. 61, pp. 153-181.

(27) Stephens, M.B.: 1982, Sveriges geol. Unders. C 786, pp. 1-111.

(28) Reymer, A.P.S.: 1979, Acad. proefschrift, pp. 1-123, Leiden.

(29) Gee, D.G., and Wilson, M.R.: 1974, Am. J. Sci. 274, pp. 1-9.

(30) Gee, D.G., and Wilson, M.R.: 1976, Am. J. Sci. 276, pp. 390-394.

(31) Klingspor, I., and Gee, D.G.: 1981, Isotopic age-determination studies of the Trøndelag trondhjemites. Terra Cognita, 1, pp. 46.

(32) Törnebohm, A.E.: 1896, Kgl. Svenska vetensk. akad. Handl. 28, pp. 1-212.

(33) Furnes, H., Roberts, D., Sturt, B.A., Thon, A., and Gale, G.H.: 1980, in Ophiolites, Proc. Int. Ophiolite Symp. Cyprus 1979, pp. 582-600.

(34) Grenne, T., Grammeltvedt, G., and Vokes, F.M.: 1980, in Ophiolites, Proc. Int. Ophiolite Symp. Cyprus 1979, pp. 727-743.

(35) Gale, G.H., and Roberts, D.: 1974, Earth Plan. Sci. Lett. 22, pp. 380-390.

(36) Ryan, P.D., Skevington, D., and Williams, D.M.: 1980, in The Caledonides of the USA, V.P.I.S.U. Mem. 2, pp. 99-103.

(37) Grenne, T., and Roberts, D.: 1980, Contrib. Mineral. Petrol., 74, pp. 374-386.

(38) Roberts, D.: 1980, Norges geol. Unders. 359, pp. 43-60.

(39) Størmer, L.: 1941, Norsk geol. Tidsskr. 20, pp. 161-170.

(40) Gee, D.G.: 1981, Norsk geol. Tidsskr. 61, pp. 93-95.

(41) Priem, H.N.A., Boelrijk, N.A.I.M., Hebeda, E.H., Verdurmen, E.A.Th., and Verschure, R.H.: 1975, Norges geol. Unders. 319, pp. 29-36.

(42) Råheim, A., and Ramberg, I.B.: in prep, Norges geol. Unders.

(43) Claesson, S.: 1980, Geol. Fören. Stockholm Förhand. 101, pp. 353-356.

(44) Griffin, W.L., and Brueckner, H.K.: 1980, Nature 285, pp. 319-321.

(45) Steel, R., Roberts, D., and Siedlecka, A.: in press, in The Caledonide Orogen - Scandinavia and Related Areas, J. Wiley & Sons.

(46) Bryhni, I.: 1964, Nature 202, pp. 384-385.

(47) Steel, R.: 1976, Tectonophysics 36, pp. 207-224.

(48) Nilsen, T.: 1968, Norges geol. Unders. 259, pp. 1-108.

(49) Roberts, D.: 1974, Norges geol. Unders. 311, pp. 89-108.

(50) Olaussen, S.: 1981, Geol. Mag. 118, pp. 281-288.

(51) Sturt, B.A., Dalland, A., and Mitchell, J.L.: 1979, Geol. Rundschau 68, pp. 523-542.

TIME OF DEFORMATION IN THE BRITISH CALEDONIDES

Derek Powell

Department of Geology, Bedford College, University of London, U.K.

The main phases of 'Caledonian' deformation in the British sector of the Caledonian Orogen vary in their nature and timing across major tectonic boundaries, namely: the Great Glen fault; the Highland Boundary fault; and the Iapetus Suture.

THE NORTHERN HIGHLANDS OF SCOTLAND

To the north-west of the Moine Thrust zone, in the Caledonian foreland (fig.1), evidence for deformation that might relate to orogenic activity within the Orthotectonic Calėdonides is given by the presence of pre-basal Cambrian, large wavelength, open folding of Upper Proterozoic (Torridonian) sediments. Further evidence is given by the possible presence of a major hiatus within the Cambro-Ordovician, Durness, shelf sequence wherein the middle and upper Cambrian are thought to be missing (fig.2).

Within the Moine Thrust zone (fig.1) both Archaean to Lower Proterozoic Lewisian gneisses, Torridonian sediments, Upper Proterozoic Moine rocks and Cambrian to Arenig (Llanvirn?) shelf sediments were strongly folded, mylonitised and thrust during post-Arenig (Llanvirn?) deformation. Late (Silurian?) thrust movement within the zone is thought to coincide with intrusion of the Loch Borrolan syenite dated at 429 $\pm$ 4 Ma (U-Pb, Zr).

South-east of the thrust zone, within the Moine metamorphic block, biostratigraphical control on the timing of deformation and metamorphism is absent, save that non-metamorphosed rocks of Lower Devonian age unconformably overlie the metamorphic rocks.

P. E. Schenk (ed.),
Regional Trends in the Geology of the Appalachian-Caledonian-Hercynian-Mauritanide Orogen, 293–299.

Information, therefore, relies on isotopic dating techniques and correlation of deformation events into the Moine Thrust zone.

Intrusion of the post-tectonic, post-metamorphic Ross of Mull granite is dated at 414 ± 3 Ma (Rb-Sr, minerals), the Grudie granite at 405 ± 15 Ma (Rb-Sr, WR), whilst intrusion of post-tectonic granites into hot country rocks is recorded at 435 ± 10 Ma (Strontian, U-Pb, Zr), c.420 Ma (Helmsdale, U-Pb, Zr), and 425 ± 4 Ma (Cluanie, Rb-Sr, Wr, M. Brook pers.comm.). The poly-phasal deformation history of the Moine rocks thus appears to have been completed by c.425 Ma as was metamorphism by c.414 Ma.

Acceptance of an age of 456 ± 5 Ma (U-Pb, Zr) for intrusion of the Glen Dessary syenite provides an important time marker since the syenite post-dates deformation and metamorphism but is itself deformed and metamorphosed. Taken with the age of intrusion of the post-tectonic granites given above, it provides time constraints bracketing late Ordovician/early Silurian deformation and metamorphism. On the other hand, an age of 467 ± 20 Ma (Rb-Sr, WR) for post-metamorphic cooling of metasediments in a nearby area of the south-western Moine appears to indicate earlier orogenic activity. Intrusion of the Carn Chuinneag granite into Moine rocks at 555 ± 10 (U-Pb, Rb-Sr, WR) might therefore suggest an additional, major phase of orogenic activity in the early Ordovician, because the granite together with its country rocks are strongly deformed and metamorphosed. Such arguments, however, presuppose non-diachroneity of orogenic activity and that the large, sub-horizontal, displacements across syn-metamorphic slide zones (ductile thrusts) have not juxtaposed crustal segments of different, early orogenic histories.

There is isotopic evidence for Precambrian elements to the orogenic history of the Moine rocks. Ages ranging from c.800 to 730 Ma from a suite of post-tectonic but deformed and metamorphosed pegmatites are held to record the age of a 'Morarian Orogeny'. An alternative, or additional, view argues for orogenic activity at 1004 ± 28 Ma. Intrusion of a late, post-tectonic, phase of the Strathalladale granite complex at 649 ± (Rb-Sr, WR) supports such arguments.

Late 'Caledonian' movements are recorded by the unconformable relationships of both the Middle and Lower ORS sediments and by the open folding of both.

THE GRAMPIAN HIGHLANDS

To the south-east of the Great Glen fault, the Riphean to at least upper, Lower Cambrian rocks of the Dalradian Supergroup

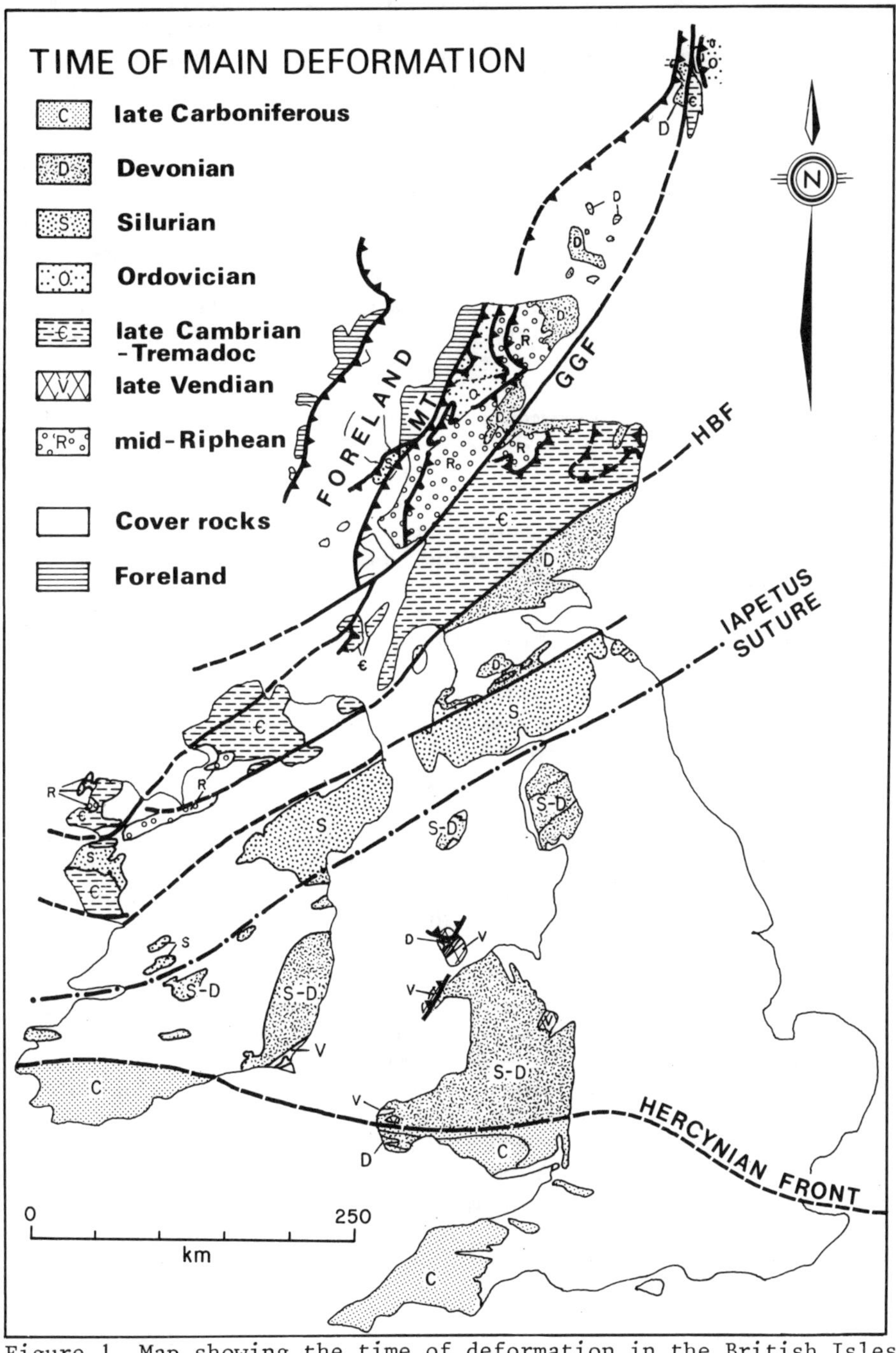

Figure 1. Map showing the time of deformation in the British Isles and Ireland.

underwent polyphase orogenic activity before deposition of Lower ORS sediments and intrusion of a suite of Newer granites at between c.415 and 400 Ma. Intrusion of an earlier suite of Ordovician granites into hot country rocks between c.475 and 460 Ma, implies the waning of metamorphic activity or re-heating, at this time. The main metamorphism of the Dalradian Supergroup overlapped and preceded intrusion of the Younger Basic complexes but was consequent upon intrusion of the Ben Vuirich granite Intrusion of the former at 489 $\pm$ 17 Ma (Rb-Sr, WR), and the latter at 514 $\pm$ 7 Ma (U-Pb, Zr), bracket major deformation and metamorphism in the early Ordovician. Intense deformation preceded intrusion of the Ben Vuirich granite, thus demonstrating late Cambrian/ Tremadocian orogenic activity. Together, these events constitute the 'Grampian Orogeny'.

Late Caledonian movements are recorded by the Lower, Middle and Upper ORS unconformities and local folding of Devonian sediments near major faults. In the Shetland Isles, cooling from regional metamorphism is recorded at 515 $\pm$ 25 Ma (Rb-Sr, WR), prograde metamorphism in the Unst nappes is suggested to be Ordovician in age, and Middle Devonian sediments are strongly folded.

THE MIDLAND VALLEY OF SCOTLAND AND THE SOUTHERN UPLANDS

Within the Midland Valley the oldest rocks, exposed at Ballantrae, are of middle Arenig age. Deformation and metamorphism, possibly related to obduction, has been dated at 483 $\pm$ 4 Ma (U-Pb, Zr) and precedes deposition of Caradocian sediments (fig.2). Subsequent major folding, probably related to fault movements, followed deposition of Lower ORS sediments but preceded deposition of Upper ORS rocks - the Middle ORS is not represented. The post-tectonic, Distinkhorn granite intrudes Lower ORS rocks and is dated at 390 $\pm$ 6 Ma (U-Pb, Zr).

Polyphase deformation of the adjacent mid-Ordovician and Silurian rocks of the Southern Uplands relates, in part at least, to subduction driven processes if the accretionary prism model for the development of these sediments is accepted. Deformation must therefore be diachronous across the Southern Uplands belt. Subsequent post-Lower/pre-Upper ORS deformation is also recorded. High level, post-tectonic granites are dated at between c.408 and 390 Ma with the Criffel-Dalbeattie granite, intruded into upper Silurian sediments giving an age of 406 $\pm$ 15 Ma (U-Pb, Zr).

THE LAKE DISTRICT

The timing of deformation in that part of the Caledonides lying immediately south-east of the Iapetus suture, the English

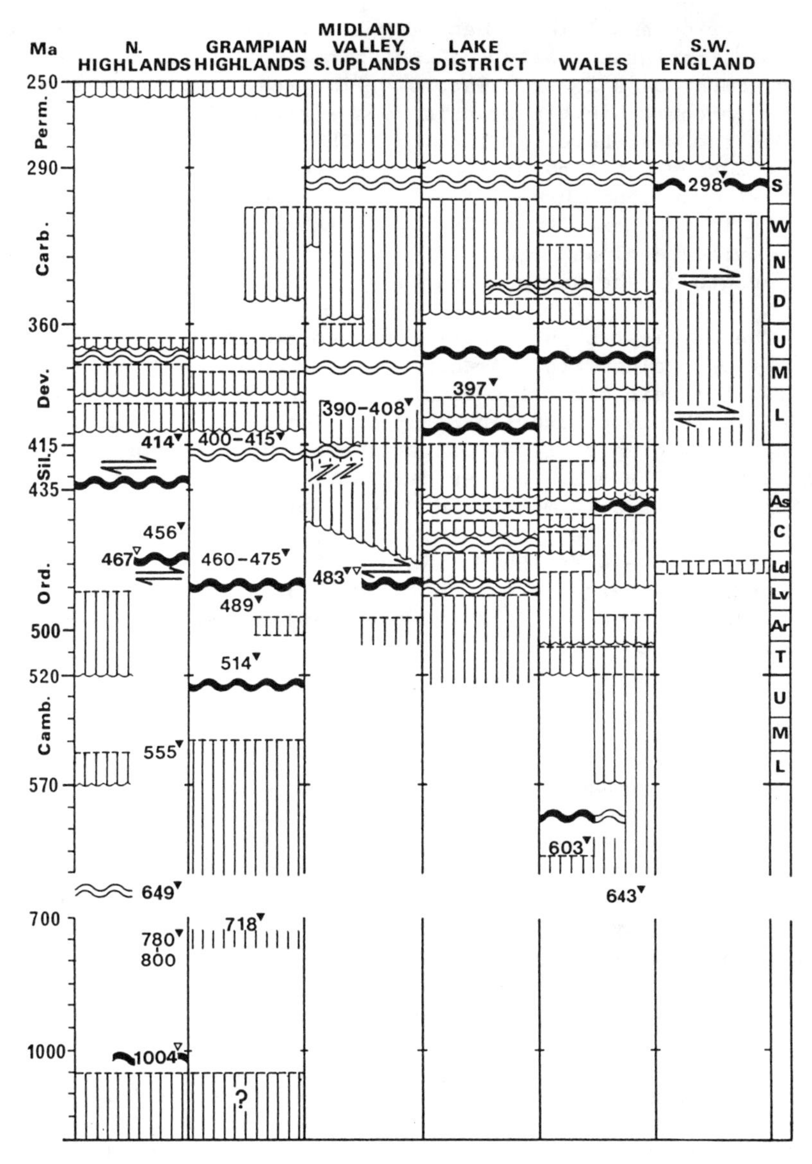

Figure 2. Time of deformation in the British Caledonides

Unconformity, open folding or tilting, intense folding, thrusting, folding and thrusting in an accretionary prism, ▼ isotopic age of intrusive event, ▽ of metamorphic event - details given in text. For the purposes of standardisation in this publication the time scale adopted is a modification of McKerrow et al. 1980, Earth planet. Sci.Lett., 51.

Lake District and Isle of Man (fig.1), is uniquely constrained both biostratigraphically and isotopically. Unfortunately, lack of agreement on a Lower Palaeozoic time scale makes correlation of the tectonic history of this area with the Orthotectonic Caledonides and elsewhere, difficult. The time scale adopted in figure 2 does not accord with that proposed locally for the Lake District. In consequence, for example, the Caradoc folding dated locally in the Lakes at c.429 $\pm$ 9 Ma (Rb-Sr, WR), might, on isotopic evidence alone, equate with the 'Silurian' tectono-metamorphic events of the Northern Highlands.

The earliest structures recognised comprise major, originally flat lying, folds within Arenig, Skiddaw Group, sediments. Their nature (tectonic or soft sediment) is uncertain but they were formed in the late Llanvirn/Llandeilo interval. Subsequent major, but open, folding preceded deposition of Caradoc (Longvillian) sediments. The most intense deformation, involving folding and regional cleavage formation, did not, however, occur until end Silurian to Lower Devonian times, before intrusion of the Shap granite at 390 $\pm$ 6 Ma (Rb-Sr, WR & U-Pb, Zr). Further regional folding and cleavage formation followed intrusion of the Skiddaw granite dated at 399 $\pm$ 4 (K-Ar, minerals) and is therefore Devonian in age.

WALES

In common with the Lake District, the Lower Palaeozoic and older rocks of Wales have suffered polyphase deformation during the 'Caledonian' Orogeny. Pre-Caledonian deformation is recorded in Anglesey and the Welsh Borders and whilst there is controversy regarding interpretation of the history of the Mona Complex in Anglesey, it appears that Cadomian events affected much of the Precambrian basement to the Welsh area (fig.2).

Within the Lower Palaeozoic rocks repeated earth movements are witnessed by unconformities and structures of varying extent and age (fig.2). Of particular note are the late Tremadoc/pre-Arenig tilting in North Wales; tilting in the Llandeilo/Caradoc interval; intra-Ashgill folding in the Welsh borders; pre- and intra-Llandovery, and post-Llandovery/pre-Wenlock folding in south-east Wales. The main phase of 'Caledonian' folding and cleavage formation would appear, however, to have been of Devonian age (possibly Middle Devonian) since, in the south-east, Silurian rocks pass conformably up into Lower Devonian and supposed Lower Devonian sediments in Anglesey are folded and cleaved.

Isotopic evidence from Wales is scant and equivocal but is held to suggest an age of c.420 to 400 $\pm$ 10 Ma for the main phase of orogenic activity.

CONCLUSIONS

Given the general lack of precision with which the timing of 'Caledonian' deformation can be dated in the British sector, and the problems that arise from lack of agreement over a Lower Palaeozoic time scale, it would appear that the only event common to all the major crustal blocks of the British Caledonides discussed here, was the intra-Devonian deformation marking the final closure of Iapetus. The pre-Devonian Caledonian deformational histories appear to differ across major tectonic boundaries but in some cases this may be due to interpretation of time scales, our lack of understanding of the behaviour of isotope systems and/or lack of sufficient data. Thus, whilst the differences in nature and timing across the Iapetus suture (fig.2) are almost certainly real, and explicable in terms of plate tectonic setting, it is less clear whether, for example, the 'Grampian Orogeny' was limited to the Grampian Highlands and did not affect Moine rocks to the north of the Great Glen fault, particularly since Northern Highland Moine-like rocks occur in the Grampian Highlands and might form a basement to the Dalradian. In addition, recognition of the role of major slides in the orogenic development of the Orthotectonic Caledonides may lead to a reappraisal of apparent differences in age. Differences may have resulted from large scale telescoping across slides which were sequentially developed over a period as long as 60 million years and which juxtaposed thrust nappes of different earlier histories.

REFERENCES

Space limitations do not permit a full reference list but useful data sources are as follows:

1. Geol.Soc.London, Special Reports Nos 1, 2, 3, 6 & 8.
2. Papers in: Harris, A.L., Holland, C.H. & Leake, B.E. (eds), *The Caledonides of the British Isles - reviewed*, 1979, Spec.Publ.geol.Soc.London, 8.
3. Gale, N.H., Beckinsale, R.D. & Wadge, A.J. 1980: Earth planet. Sci.Lett., 51, pp.9-17.
4. McKerrow, W.S., Leggett, J.K. & Eales, M.H. 1979: J.geol.Soc. London, 136, pp.755-70.
5. Pankhurst, R.J. 1982: In Sutherland, D.S. (ed), *Igneous rocks of the British Isles*, J.Wiley & Sons Ltd.
6. Piasecki, M.A.J., van Breemen, O. & Wright, A.E. 1981: Canad. Soc.Petrol.Geol, pp.57-94.
7. Powell, D., Baird, A.W., Charnley, N.R. & Jordan, P.J. 1981: J.geol.Soc.London, 138, pp.661-73.
8. Rundle, C.C. 1981: J.geol.Sco.London, 138, pp.569-72
9. Soper, N.J. & Barber, A.J. 1982: J.geol.Soc.London, 139, pp.127-38

DEFORMATION IN THE IRISH CALEDONIDES

Michael D. Max

Geological Survey of Ireland

The Caledonian orogen in Ireland exhibits Cadomian, Grampian, Taconic, Caledonian (s.s.) or Scandian, Acadian and Hercynian structural events. In addition, relics of thoroughly deformed older basement occurs both in the northwest and southeast of Ireland. Nowhere in Ireland are the margins of the Caledonian orogen seen although a major thrust on the continental shelf to the northwest marks the junction with Lewisian gneisses; identification of the margin to the southeast is more arbitrary as there is no single, sharp tectonic line. The structural grain within the orogen is NE-SW with a swing to E-W along the western seaboard.

INTRODUCTION

Virtually the whole of Ireland is underlain by rocks deformed during the Caledonian orogeny (Table 1); the margins to the orogen are not seen within mainland Ireland. The Orthotectonic, or metamorphic, and the Paratectonic, or weakly metamorphic, Caledonides[1] are the two primary tectonic divisions in Ireland (Fig. 1). These divisions reflect the onset of deformation with Grampian events largely confined to the Orthotectonic zone, where the Cadomian is absent. The youngest widespread tectonic episode in the Paratectonic Caledonides is Taconic but Taconic effects are best seen in the South Mayo Trough (Fig. 1), which is an Ordovician-Silurian basin within the Orthotectonic zone. Caledonian (s.s.) events at end Silurian-lowermost Devonian times (Table 1) are widespread. In the Orthotectonic Caledonides major metamorphic retrogression accompanying the establishment of shear belts can be related to

P. E. Schenk (ed.),
Regional Trends in the Geology of the Appalachian-Caledonian-Hercynian-Mauritanide Orogen, 301–306.

the development of the major structures and prograde metamorphism in the Paratectonic Caledonides. Widespread emplacement of granites within, and in many cases locking, the shear belts took place at about 400Ma.[2] [3] Acadian events appear to be confined to lower Devonian inliers within the Orthotectonic zone and Hercynian events are widespread in Ireland with mainly rejuvination of older structural lines in the north and progressively more important superimposed structures to the south. A zone of Hercynian cleavage folding in the south has traditionally been regarded as reflecting little or no pre-Hercynian basement control of structures.

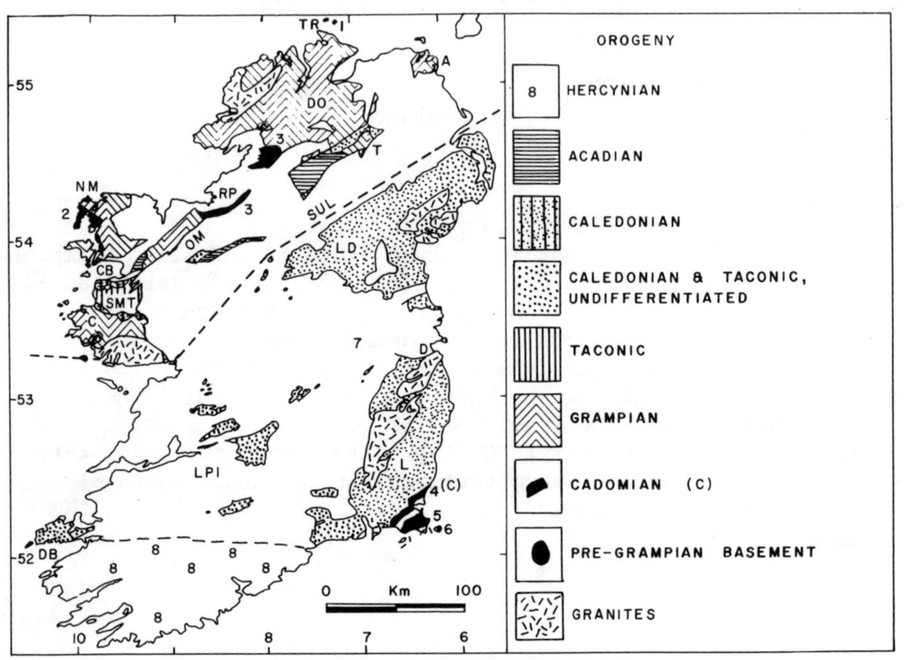

Figure 1. Structural units and location map. Orthotectonic Caledonides are north of the Southern Uplands Line, Paratectonic Caledonides to the south. A, Antrim; CB, Clew Bay; C, Connemara; D, Dublin; DB, Dingle Bay; DO, Donegal; L, Leinster; LD, Longford-Down Massif; LPI, Lower Palaeozoic inliers; NM, North Mayo; OM, Ox Mountains; RP, Rosses Point; SMT, South Mayo Trough; SUL, Southern Uplands Line; T, Tyrone Inlier; TR, Inishtrahull. Units: 1. Malin Sea Division, 2. Erris Complex, 3. Slishwood Division, 4. Cullenstown Fm., 5. Rosslare Complex, 6. Tuskar Group, 7. Hidden granulites, 8. Hercynian cleavage fold zone.

Most of the ancient pre-Caledonian basement occurs within the Orthotectonic Caledonides. The schists and gneisses of the Malin Sea Division,[4] the Erris Complex[5] and the Slishwood

Division[6] (Fig. 1) each display a unique Grampian structural style, which implies important local basement control of Grampian structural trends and major structures. In the Paratectonic Caledonides of southeast Ireland the Caledonian Cullenstown Formation[7] and the Cadomian reworked older basement[8] of the Rosslare Complex occur as imbricated tabular thrust slices along with cleaved Lower Ordovician slates. The beginning of the imbrication may be as old as Cadomian but is mainly Lower Palaeozoic and no younger than about 400Ma in age. Pre-Caledonian basement, now hidden in the vicinity of Dingle Bay, is similar to the Rosslare Complex. Lower crustal granulites seen as xenoliths in Carboniferous volcanics to the west of Dublin may not represent the character of pre-Caledonian basement here during Caledonian times. Indeed, under much of central Ireland formation of a continental crust may have taken place in the early part of the Caledonian orogeny during the closing of an Iapetus ocean.

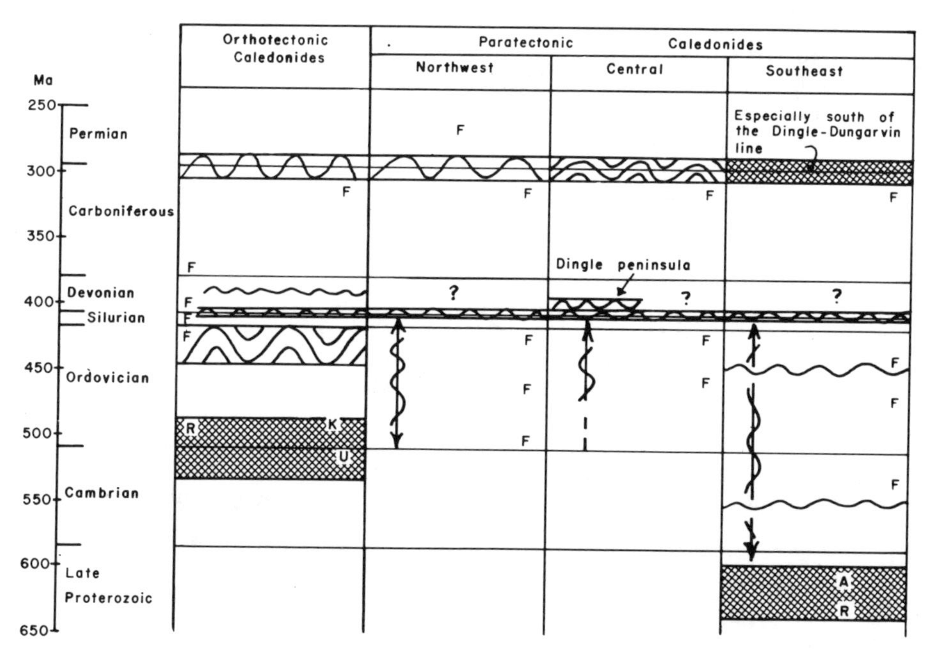

Table 1. Timing of deformation in the Irish Caledonides. Weak deformation shown by single wavy line, moderate by double and strong by cross hatching. Approximate length of deformation episode shown by height of pattern. Vertical arrows with a weak deformation symbol indicates semi-continuous deformation accompanying sedimentation. Dating method: A, Argon step heat; R, Rb-Sr whole rock; V, U-Pb; K, K-Ar fusion.

1. CADOMIAN

Cadomian rocks are confined to southeast Ireland where they are not seen in contact with the older basement. In addition to the Cullenstown formation, the Tuskar Group, which is another unit of late Precambrian to Cambrian rocks[9], may be generally comparable with the Gwna Group of Anglesey[10]. The contacts are unexposed but are almost certainly major faults or thrusts. The contact with the Lower Palaeozoic rocks to the northeast in Leinster is also a fault, but with evidence of late major, horizontal movement. The history of deformation here is imprecise because of the lack of exposure and structural telescoping in the imbrication of the various structural units. The structural trend is about NE-SW but as no major structures are seen, this may reflect subsequent reorientation.

2. GRAMPIAN

The Orthotectonic Caledonides occur in a series of inliers from Antrim, Tyrone and Donegal along the Ox Mountains to North Mayo and Connemara (Fig. 1). The characteristic sediments of the Grampian zone are Dalradian[11], which now includes what had been regarded as the upper part of the Moine succession. These were deposited from about 750 to 530Ma on a cratonic platform whose fragmentation is seen as a prelude to the onset of the Grampian episode[12]. The structural trends within the Grampian zone can be generally related to structural level. In North Mayo at the deepest Grampian structural level, the trend of the basement-cored structures is NW-SE. This trend turns to a more easterly trace in North Mayo at higher structural levels and the turn to a more normal NE trace is seen in Donegal[13]. In southwest Donegal and the northeast Ox Mountains basement and cover are in a thrust relationship wherein the original thrust surface may have been close to horizontal[13]. The highest level structures are true nappes, which are seen in SE Donegal and Connemara.

There is some contention about the Ox Mountains Succession in that one school of interpretation regards it as pre-Caledonian in age[14] while another regards it as part of the Dalradian succession[15,16]. A pre-Grampian history has not been demonstrated. It has been subjected to Grampian and subsequent events and is regarded here as part of the Dalradian-related sedimentary sequence. Structures are upright and trend NE-SW except in highly faulted areas such as the SW Ox Mountains and the Clew Bay region. Post-580, pre-400Ma shearing and retrogression is common.

3. TACONIC

The Taconic phase is not seen distinctly in Ireland except in the South Mayo Trough (Fig. 1) where shallow marine Silurian sediments rest unconformably upon a tectonic unit of deformed and weakly metamorphosed Arenig, Llanvirn and Llandeilo rocks. Elsewhere in the Paratectonic Caledonides, the structural situation is more mixed and probably reflects the accumulation of Ordovician and Silurian sediments in an oceanic environment which was undergoing semi-continuous deformation. The mechanism of continental accretion suggested for the Southern Uplands of Scotland, the Longford-Down massif and the northern part of the Paratectonic inliers of central Ireland demands individual deformation of each tectono-stratigraphic slice of the imbricating wedge periodically and on a local basis from Arenig to end-Silurian times. This is clearly a most complex framework and a unique Taconic event has not been separately identified in the northern Paratectonic zone.

In Leinster and the central Paratectonics about the supposed Iapetus suture, no widespread Taconic event has been noted other than a tilting of older rocks and the development of a widespread sub-Caradoc unconformity. Important structures may have been overlooked or mis-interpreted, however, and even if no discrete Taconic event exists, an alternate structural framework more closely resembling that suggested for the Longford-Down area could also be suggested at least for the Leinster area.

4. CALEDONIAN (sensu strictu) SCANDIAN

This is the most widespread and well defined structural phase in the Paratectonics and is also important in the Orthotectonic Caledonides. Emplacement of most of the major igneous bodies took place at the end of this episode. Structures are largely upright, trend NE-SW and were accompanied by low-grade metamorphism. It is clearly a late episode which is superimposed upon more important regional and local structures. Few major structural dislocations appear to have been generated at this time and the structural importance of this episode is being reevaluated as the complexity of the earlier structural history is recognized.

5. ACADIAN

This is a minor structural phase. Inliers of Lower and Middle Devonian rocks along southeast and west central part of the Orthotectonic zone show a weak cleavage accompanying folding and very slight metamorphism. Structural trends may be generally

E-W but are often unclear or conflicting even within individual inliers. There are one or two areas in the South Mayo Trough and along the SW flank of the Leinster massif where deformation of this age is suspect but these are as yet unproven.

6. HERCYNIAN

Hercynian faulting and folding is seen throughout Ireland usually concentrated near older structures in the Caledonian basement, and the structures are most intense to the south. The nature of the basement is not known beneath the zone of cleavage folding in the Variscan belt (Fig. 1), although it is probably Paratectonic Caledonides underlain by or interleaved with Cadomian and older basement. Restricted shear belts are often associated with some of the major faults and thrusts in southern Ireland. The greater part of Ireland is not overlaid by pervasive Hercynian deformation and the older structural history is not obscured.

REFERENCES

1. Dewey, J.F. 1969. *Nature* 222, pp. 124-129.
2. O'Connor, P.J., Long, C.B., Max, M.D., Kennan, P.S., Halliday, A.N. and Roddick, J.C. *Geol. J.* (in press).
3. Hutton, D.H.W. 1982. *J. Geol. Soc. London* 139, pp. 615-631.
4. Max, M.D. 1981. *Prog. Underwater Sci.* 6, pp. 65-69.
5. Max, M.D. and Sonet, J. 1979. *J. geol. Soc. London* 136, pp. 379-382.
6. Max, M.D. and Long, C.B. (in preparation).
7. Max, M.D. and Dhonau, N.B. 1974. *Geol. Surv. Ireland Bull.* 1, pp. 447-458.
8. Winchester, J.A. and Max, M.D. 1982. *J. geol. Soc. London*, 139, pp. 309-319.
9. Max, M.D., Grant, P. and Ryan, P.D. 1981. *Jl. Struct. Geol.*
10. Barber, A.J. and Max, M.D. 1979. *J. geol. Soc. London*, 136, pp. 407-432.
11. Harris, A.L. and Pitcher, W.S. 1975. *Geol. Soc. Lond. Spec. Rep.* 6, pp. 52-75.
12. Harris, A.L., Baldwin, C.T., Bradbury, H.J., Johnson, H.D. and Smith, R.A. 1978. *Geol. Soc. London Spec. Issue* 10, pp. 115-138.
13. Pitcher, W.S. and Berger, A.R. 1972. *The Geology of Donegal*, 435 pp.
14. Phillips, W.E.A. 1978. *Geol. Surv. Canada Pap.* 78-13, pp. 97-103.
15. Yardley, B.W.D., Long, C.B. and Max, M.D. *Geol. Soc. London Spec. Publ.* 8, pp. 369-374.
16. Taylor, W.E.G. 1968. *Proc. R. Irish Acad.* 67, pp. 63-82.

TIMES OF DEFORMATION IN THE CANADIAN APPALACHIANS

KEPPIE, J. D., Dept. of Mines and Energy, Halifax, Nova Scotia, Canada; ST. JULIEN, P., Dept. of Geology, Laval University, Quebec, P.Q., Canada; HUBERT, C. & BELAND, J., Dept. of Geology, University of Montreal, Montreal, Quebec, Canada; SKIDMORE, B., Dept. of Natural Resources, Quebec, P.Q., Canada; FYFFE, L.R., Geological Survey Branch, Dept. of Natural Resources, Fredericton, New Brunswick, Canada; RUITENBERG, A.A., & McCUTCHEON, S.R., Geological Survey Branch, Dept. of Natural Resources, Sussex, New Brunswick, Canada; WILLIAMS, H., Dept. of Geology, Memorial University, St. John's, Newfoundland, Canada; BURSNALL, J., Dept. of Geology, St. Lawrence University, Canton, New York, U.S.A.

This short paper summarizes the constraints on the times of deformation upon which the Structural Map of the Appalachian Orogen in Canada (1) was based. Space does not permit a detailed description of the structural styles, which may be found in (2) and references therein.

The times of deformation fall into several periods (Table 1): Mid Proterozoic or earlier (e.g. Grenvillian, Micmacian), Late Proterozoic (Cadomian = Avalonian), isolated Early Ordovician (Penobscotian), Mid-late Ordovician (Taconian), Devonian and possibly latest Silurian (Acadian) and Permo-Carboniferous (Hercynian).

MID PROTEROZOIC OR EARLIER

Rocks of the Grenville Province form the cratonic margin of ancient North America and generally crop out along the western side of the Appalachian Orogen (3). Elsewhere in the orogen, Mid Proterozoic rocks crop out in isolated basement blocks such as in the >1600 Ma gneisses in the Chain Lakes block (4), in the >1000 Ma Gneisses in Cape Breton Island (5), and in the >1180 Ma gneisses in the Meguma Zone (6). Since these rocks formed prior

P. E. Schenk (ed.),
Regional Trends in the Geology of the Appalachian-Caledonian-Hercynian-Mauritanide Orogen, 307–313.

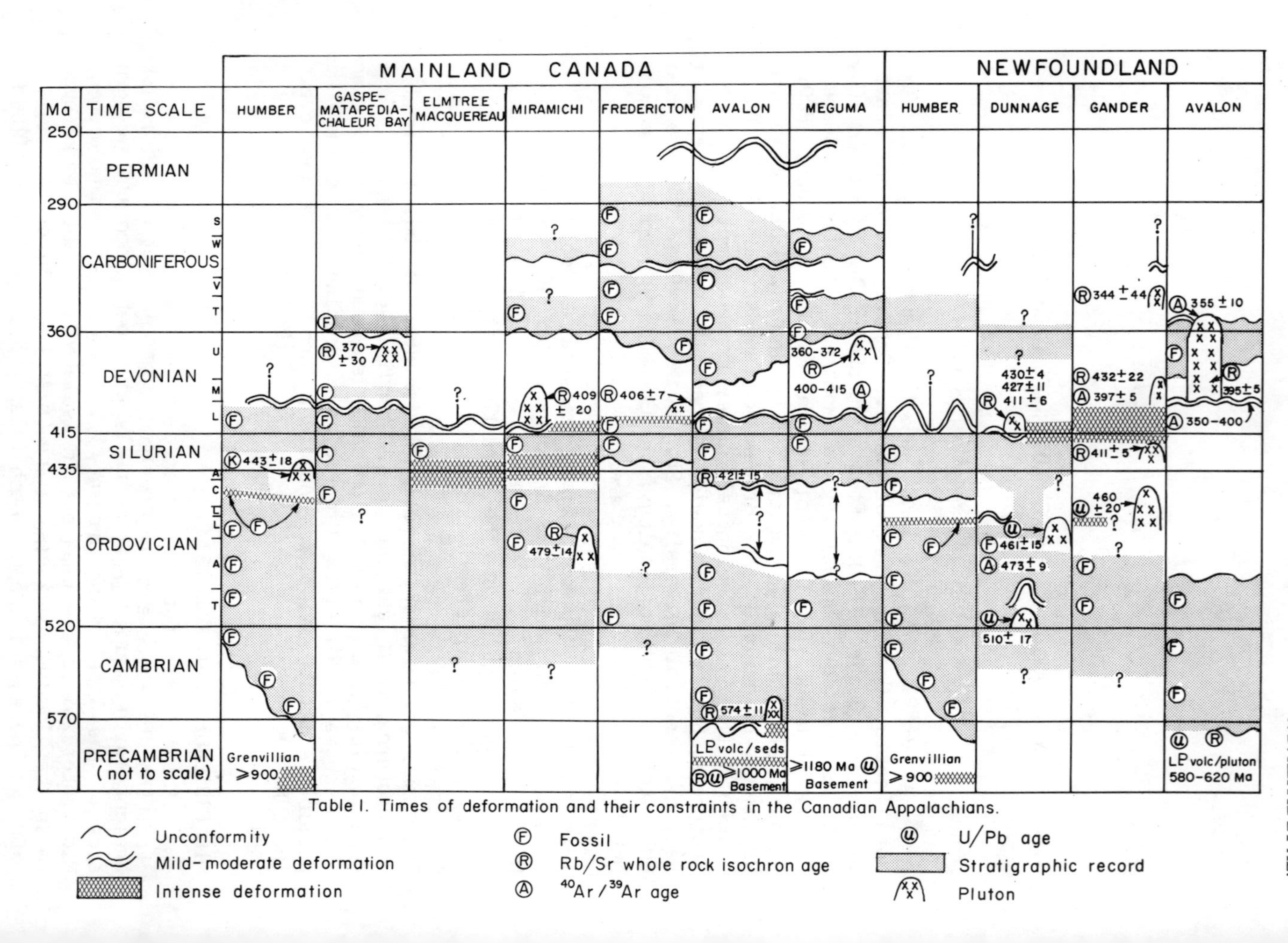

Table I. Times of deformation and their constraints in the Canadian Appalachians.

to the Appalachian cycle, they are beyond the scope of this paper.

LATE PROTEROZOIC

The Late Proterozoic Cadomian (Avalonian) deformation is confined to the Avalon Zone (Table 1). The intensity of deformation varies from weak to absent in Newfoundland and the central to eastern Caledonian Highlands of New Brunswick, where it is represented by disconformities, nonconformities, and unconformities, to penetrative polyphase deformation in Nova Scotia and the western Caledonian Highlands (7). Lower limits on the time of deformation are only available in Newfoundland where volcanics of the Marystown and Love Cove Groups have yielded U-Pb zircon dates of 608 ± 25 Ma and 590 ± 30 Ma respectively (8), and an U-Pb age on the Holyrood Granite of 620 ± 2 Ma (9). Dating of the volcanic rocks in Nova Scotia and southern New Brunswick by Rb-Sr has yielded anomalously young ages. Upper limits on the time of deformation are provided by Early Cambrian fossiliferous strata lying unconformably to disconformably upon older rocks, and post-tectonic plutons the oldest of which is the Shunacadie pluton at 574 ± 11 Ma (Rb-Sr whole rock isochron 10) and the 593 ± 28 Ma K-Ar age on hornblende from the Coxheath Hills pluton (11) both in Cape Breton Island, Nova Scotia. If the base of the Cambrian is at 570 Ma then these data bracket the time of deformation between 570 and 620 Ma.

EARLY ORDOVICIAN

Deformation during this time interval is restricted to Twillingate in Newfoundland (12). Here, the Twillingate granite with an U-Pb intrusive age of 510 ± 17 Ma is affected by a weak to penetrative foliation, which in turn is cut by post-tectonic mafic dykes. The oldest of these dykes yielded a 473 ± 9 Ma ^{40}Ar-^{39}Ar age on hornblende.

MID-LATE ORDOVICIAN

The effects of the Taconian deformation are recorded mainly from the Humber Zone along the northwestern flank of the orogen (Table 1), although local effects also occur in the Dunnage, Avalon and Meguma Zones (1). In the Humber Zone, the structures are predominantly nappes and thrusts which become refolded towards the southeast. The earliest record of regional Taconian deformation in the Humber Zone is given by the age of genetically related development of olistostrome. In Quebec, this olistostrome is Caradocian becoming younger across the zone from the internal nappes to the external nappes (13). It is also diachronous along the Humber Zone because the same olistostrome is Llanvirn in Newfoundland. An upper limit on the time of deformation is given by the Caradocian age of the Long Point Group that unconformably overlies the transported rocks (14, 15). In Quebec it is provided by cooling ages on hornblende produced during associated regional metamorphism, the oldest of which is a K-Ar age of 443 ± 18 Ma on amphibolites from Mont Albert (13).

In the Dunnage Zone of Newfoundland, the rocks are not generally polydeformed and they are everywhere less intensely deformed than those of the orthotectonic part of the Humber Zone. In central Notre Dame Bay, Ordovician and Silurian sequences are conformable and were first penetratively deformed by the Acadian (16). The time of deformation on the Baie Verte Peninsula is bracketed by the Arenigian Snooks Arm Group, the U-Pb zircon age of 461 ± 15 Ma on the pre-Taconian deformed Burlington granodiorite (8), and the oldest, post-tectonic, unconformably overlying, unfossiliferous Cape St. John and MicMac Lake Groups of presumed Silurian age. Tight constraints on the times of deformation in the Gander Zone in Newfoundland are absent and all could be attributed to the Acadian orogeny.

Taconian polyphase deformation appears to be youngest in the central part of the orogen in Quebec and northern New Brunswick: Macquereau, Elmtree and Miramichi blocks (17) (Table 1). The time of deformation is best defined in the Miramichi block where the stratigraphic record is most complete. Here the time of deformation is bracketed by the mid-Caradocian age of the youngest deformed rocks (17) and the Ludlovian age of the unconformably overlying strata. In the Elmtree block, the oldest post-Taconian rocks are Llandovery C3-C4 (18). It is interesting to note that the central part of the Matapedia Basin was the site of continuous deposition from Caradocian through the Silurian, i.e. synchronous with Taconian deformation both to the north and south (17).

In the Avalon and Meguma Zones of Nova Scotia and southern New Brunswick, an unconformity is present between Cambrian-Arenigian and earliest Silurian strata (Table 1) (19). Structures associated with this regional uplift are only rarely recorded, e.g. adjacent to the Hollow Fault in the northern Antigonish Highlands Llandoverian rocks rest unconformably upon recumbently folded Cambrian-early Ordovician rocks (20).

DEVONIAN (AND POSSIBLY SILURIAN)

The effects of the Acadian Orogeny are widespread through the Canadian Appalachians varying from upright folds over rigid basement blocks to polyphase deformation in some troughs. Examination of the Structural Map of the Appalachian Orogen in Canada (1) and Table 1 suggests that the time of Acadian deformation ranges from Silurian to Devonian. However, while this may be the case, it may also be an artifact of the dating method and the placing of the Siluro-Devonian boundary at 415 Ma. Thus, where the stratigraphic record is used, the youngest pre-Acadian, penetratively deformed rocks are Emsian (Early Devonian) in age (21, 22) and the oldest post-tectonic strata range from Middle Devonian to Early Carboniferous. Although locally, the Mid-Late Devonian rocks are also tilted and folded they are generally not penetratively deformed. However, where stratigraphic relationships are limited or absent as in the Gander Zone and parts of the Dunnage Zone in Newfoundland, the oldest post-tectonic plutons are the Hodges Hill pluton at 411 ± 6 Ma and the Middle Brook pluton at 432 ± 22 Ma (both Rb-Sr whole rock isochrons) (23). At face value these data imply a Silurian or Siluro-Devonian time of deformation. However, in the Fredericton trough, deformed Gedinnian Eastport Formation is intruded by the post-tectonic St. George granite dated at 406 ± 7 Ma by Rb-Sr whole rock isochron (24). Direct measurements of the age of the Acadian deformation by ^{40}Ar-^{39}Ar total gas ages on slates in the Meguma Group range from 400 to 415 Ma (25) and phyllites in the Love Cove Group from the Newfoundland Avalon Zone range from 350-400 Ma (8). Considering the data as a whole it appears that either the 415 Ma age on the Siluro-Devonian boundary is too young, or that the isotopic data are inaccurate. If the Siluro-Devonian boundary is older, then all the deformation could be accommodated in the Devonian.

PERMO-CARBONIFEROUS

Permo-Carboniferous Hercynian deformation was most intense around the margins of the Magdalen Basin (19) dying out onto the surrounding platforms and towards the centre of the basin. Structures are generally upright folds with and without attendant cleavage; however, refolded recumbent folds and thrusts are developed in certain narrow zones and in the Visean rocks where gravity sliding on evaporites occurs. In and around the Magdalen Basin, where the stratigraphic record is most complete, there appear to be three main times of deformation during Late Tournaisian, Westphalian B-C and Late Permian times. However, there are also many other local unconformities as would be expected in a predominantly terrestrial sequence. The Westphalian time of deformation is documented by the sub-Pictou

Group unconformity which cuts across rocks ranging in age from early Paleozoic to Westphalian B. The Pictou Group spans the Westphalian C to Early Permian, and was deformed during the break in the stratigraphic record between Early Permian and Late Triassic times.

REFERENCES

(1) Keppie, J.D., Ruitenberg, A.A., Fyffe, L.R., McCutcheon, S.R., St. Julien, P., Skidmore, B., Beland, J., Hubert, C., Williams, H. and Bursnall, J.: 1982, Memorial University of Newfoundland Map No. 4.

(2) St. Julien, P. and Beland, J. (Eds.): 1982, Geol. Assoc. Can. Spec. Pap. 24.

(3) Williams, H.: 1978, Memorial University of Newfoundland Map No. 1.

(4) Naylor, R.S., Boone, G.M., Boudette, E.L., Ashenden, D.O. and Robinson, P.: 1973, EOS, Amer. Geophys. Union Trans. 54(4), pp. 495.

(5) Olszewski, W.J., Jr., Gaudette, H.E., Keppie, J.D. and Donohoe, H.V.: 1981, Geol. Soc. Amer. Abstr. 13(3), pp. 169.

(6) Keppie, J.D., Odom, A.L. and Cormier, R.F.: 1983, Geol. Soc. Amer. Abstr.

(7) Ruitenberg, A.A., Giles, P.S., Venugopal, D.V., Buttimer, S.M., McCutcheon, S.R. and Chandra, J.J.: 1979, New Brunswick Dept. Nat. Res. Min. Res. Br. Mem. 1, 213 p.

(8) Dallmeyer, R.D. and Odom, A.L.: 1980, Newfoundland Dept. of Mines and Energy Report 80-1, pp. 143-146.

(9) Krogh, T.E. and Papezik, V.S.: 1983, Geol. Soc. Amer. Abstr.

(10) Poole, W.H.: 1980, Geol. Surv. Canada Paper 80-1C, pp. 165-169.

(11) Wanless, R.K., Stevens, R.D., Lechance, G.R. and Edmonds, C.M.: 1968, Geol. Surv. Canada Paper 67-2A, pp. 141.

(12) Williams, H., Dallmeyer, R.D. and Wanless, R.K.: 1976, Can. J. Earth Sci. 13, pp. 1591-1601.

(13) St. Julien, P. and Hubert C.: 1975, Amer. J. Sci. 275A, pp 337-362.

(14) Rodgers, J.: 1965, Geol. Assoc. Can. Proc. 16, pp. 83-94.

(15) Bergstrom, S.M., Riva, J. and Kay, M.: 1974, Can. J. Earth Sci., 11, pp. 1625-1660.

(16) Van der Pluijm, B., Karlstrom, K. and Williams, P.F.: 1982, Marit. Seds. and Atlantic Geol. 18, pp. 47.

(17) Fyffe, L.R.: 1982, Geol. Assoc. Can. Spec. Pap. 24, pp. 117-130.

(18) Noble, J.P.A.: 1976, Can. J. Earth Sci. 13, pp. 537-546.

(19) Keppie, J.D.: 1982, Geol. Assoc. Can. Spec. Pap. 24, pp. 263-280.

(20) Murphy, J.B., Keppie, J.D. and Hynes, A.: 1982, Nova Scotia Department of Mines and Energy Map No. 82-5.
(21) St. Peter, C. and Boucot, A.J.: 1981, Marit. Seds. and Atlantic Geol. 17, pp. 88-95.
(22) Boucot, A.J.: 1960, 21st Int. Geol. Cong. Rept. 12, pp. 129-137.
(23) Bell, K., Blenkinsop, J. and Strong, D.F.: 1977, Can. J. Earth Sci. 14, pp. 456-476.
(24) Fyffe, L.R., Pajari, G.E. and Cormier, R.F.: 1981, Geol. Soc. Amer. Abstr 13(3), pp. 133.
(25) Reynolds, P.H. and Muecke, G.K.: 1978, Earth and Planet. Sci. Lett. 40, pp. 111-118.

TIMING OF OROGENIC EVENTS IN THE U.S. APPALACHIANS

Philip H. Osberg

University of Maine at Orono, Orono, Maine, USA 04469

ABSTRACT

In the New England Appalachians, where the cover sequences are reasonably complete and where many plutons exist, time brackets for deformational features are well established. Deformational events occur over a short duration of time and appear as distinct events. These events are more-or-less synchronous across the orogen. To the south in the Middle and Southern Appalachians, where the cover sequences to a considerable extent have been removed by erosion, deformations are less well bracketed in time, and ambiquity exists over large tracts as to the age of the deformational features observed in the rocks. These difficulties inhibit any attempt to test synchroniety along the orogen.

Thermal effects, manifest by plutons and metamorphism, correlate with the deformations, but the thermal effects are longer lived than the deformational effects.

The Appalachian orogen is composite, having been constructed over a period of ~350.my. The effects of these events are not uniformly distributed over the fold-thrust belt.

GENERAL STATEMENT

An orogenic event consists of deformational and thermal features produced in a single orogenic cycle. Deformational features include multiple foliations, fold forms, thrusts and high-angle faults. Thermal features include several generations

P. E. Schenk (ed.),
Regional Trends in the Geology of the Appalachian-Caledonian-Hercynian-Mauritanide Orogen, 315–337.

of plutons and polymetamorphism. Our main interest here is to date the deformational and thermal events.

Recognition of a deformational event is most convincingly made where rock bodies, either unconformable sequences or undeformed plutons, cut across structural features included in an older sequence of rocks. If subsequently the post-deformational bodies are themselves deformed, the cross-cutting relations between them and the older structural features are commonly preserved. However, where the younger deformational event is intense, the distinction between older and younger events may be blurred.

Deformational features cannot be dated directly, and their age is established by setting lower and upper age bounds. The lower bound is the youngest age of a deformed rock body, and the upper bound is the oldest age of rocks that cross-cut the previously deformed sequence. There is always a range of uncertainty in age of deformation between the lower and upper bounds.

The ages of plutonic rocks are established by radiometric methods. Only Pb-U and Rb-Sr methods have been utilized in this study; K-Ar dates are thought to give argon-closure dates that may be far removed in time from dates of intrusion.

The ages of the climax of metamorphism cannot be measured directly and can only be inferred from regional studies which measure the diachroniety in mineral pairs, such as, biotite and and hornblende. Diachronous ages can be projected back in time to approximate the age of metamorphism, but because of the uncertainty of assumptions involved, the oldest hornblende age is used as the minimum age of metamorphism. Single K-Ar ages, in general, fall within the age range of metamorphism, but by themselves may be misleading. One exception to this is the age of muscovite in rocks that have recrystallized at temperatures below the Argon-closure temperature for muscovite.

Taking an extended view of the orogen of which the Appalachains is only a segment, the following orogenic events have been identified; A Helikian event, > 1.0 b.y. ago; a Cadomian event in late Precambrian time; a Penobscot (Grampian) event in late Cambrian-early Ordovician time; a Taconic event in Middle to Upper Ordovician time; an Acadian event in Early Devonian time; and an Alleghanian (Hercynian) event in Permian time. The Helikian event in the Appalachian segment is restricted to the Grenville basement and the deformational and thermal features of this event are present in the basement far beyond the western limit of the Appalachian orogen. Clearly these features were inherited and were not formed in response to the building of the Appalachian orogen; therefore, they will not be considered further in this paper.

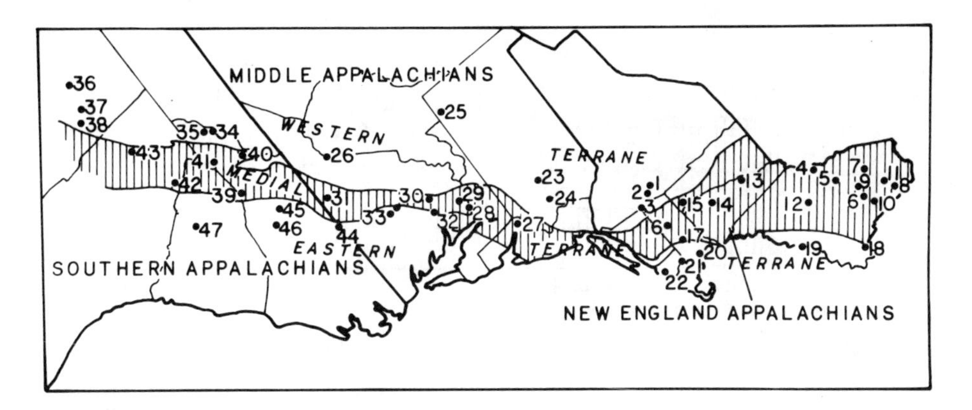

Figure 1. U.S. Appalachians showing New England, Middle and Southern regions as well as western, medial and eastern terranes. Numbers refer to localities at which information was obtained on the timing of deformations.

OROGENIC EVENTS IN THE APPALACHIANS

For the purposes of discussion the Appalachian orogen is divided into the New England Appalachians, the Middle Appalachians and the Southern Appalachians (Fig. 1). In addition, the Broughton-Blue Ridge suture (1) and the west limit of "Avalon" define western, central and eastern terranes within each segment. Localities where data exist that bear on the timing of deformations are shown in Figure 1 as well, and descriptions of these localities are given in the Appendix. Figure 2 shows the timing of deformational events for each terrane and segment, and Figure 3 shows the distribution of ages of dated plutons for each segment, and where data are available, the duration of metamorphic events.

New England Appalachians

Elements of Cadomian, Penobscot (Grampian), Taconic, Acadian and Alleghanian (Hercynian) deformational effects have been recognized in New England, and they are distributed among the terranes as shown in Figure 2. The Cadomian event has been identified only in the eastern terrane. The Penobscot (Grampian) has been documented only in northern Maine near the east margin of a possible small continent (Chain Lakes). Effects of the Penobscot (Grampian) could occur to the south along the Bronson Hill antiform, but requisite bounds by which to identify them are not present. The Taconic deformational effects are found in all three terranes, although they are most intense adjacent to the boundary between the western and medial terranes. A small

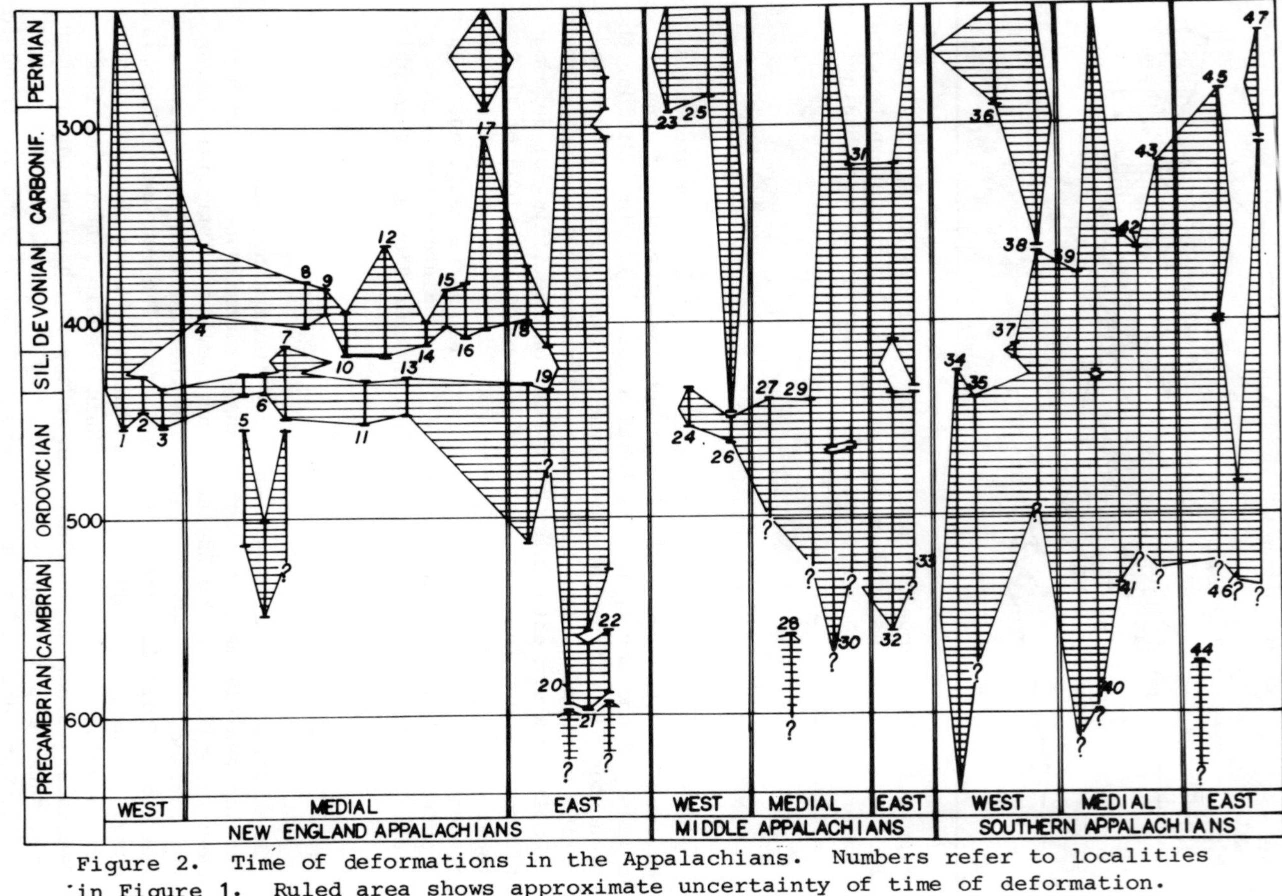

Figure 2. Time of deformations in the Appalachians. Numbers refer to localities in Figure 1. Ruled area shows approximate uncertainty of time of deformation.

hiatus occurs in the medial terrane (locality 7) that is locally known as the Salinic Disturbance (21); this hiatus may be epierogenic, connected to the growth of an Early Devonian island arc. Acadian deformation spreads across all terranes. It is weak and cannot be dated in the western terrane, but it is reasonably dated in the other terranes. Alleghanian (Hercynian) deformations appear in the southern part of the western and medial terranes and in the eastern terrane. Both lower and upper bounds have been recorded only in the eastern terrane, however.

Deformation appears as discrete events in Figure 2, with intervals in which deposition took place separating the periods of deformation. However, locally, a younger deformational event bites down through an older event, thereby removing the evidence for the older deformation. Within the uncertainty of the geologic constraints the deformations are nearly synchronous across the orogen.

The times of deformation are compared with the ages of plutons in Figure 3. A time line indicates the absolute age, and the frequency of dated plutons (3-17) in ten million year increments is shown as an histogram. The approximate duration of deformational activity is also shown. Ages of metamorphism (4, 18) are also shown where such information is available. The duration of metamorphism extends from the closure data of hornblende to that of biotite. Each deformational event has an associated cluster of plutons and in some cases a metamorphic event, but in each case the thermal activity far outlived the deformational activity.

Middle Appalachians

Cadomian, possible Penobscot (Grampian), Taconic, Acadian and Alleghanian (Hercynian) orogenic elements are recognized in the Middle Appalachians. Cadomian deformation has been reasonably dated only in the medial terrane. The possibility of the Penobscot (Grampian) deformation depends on the validity of the Rb-Sr dates at localities 30 and 31. These date a part of a thermal event, and, if the thermal event postdates the deformation as the geologic relations suggest (19), the dates are more consistent with Penobscot (Grampian) deformation than with Taconic. Pre-Late Ordovician or Silurian deformation in the eastern terrane could be either Penobscot (Grampian) or Taconic. There is no resolution of this matter. Taconic deformation is dated in the western terrane, and dates on igneous rocks in the medial terrane (20, 21) indicate a Taconic thermal event, but not necessarily a deformational event. Taconic deformation could be present across the entire orogen in the Middle Appalachians, but the geologic relations as presently understood do

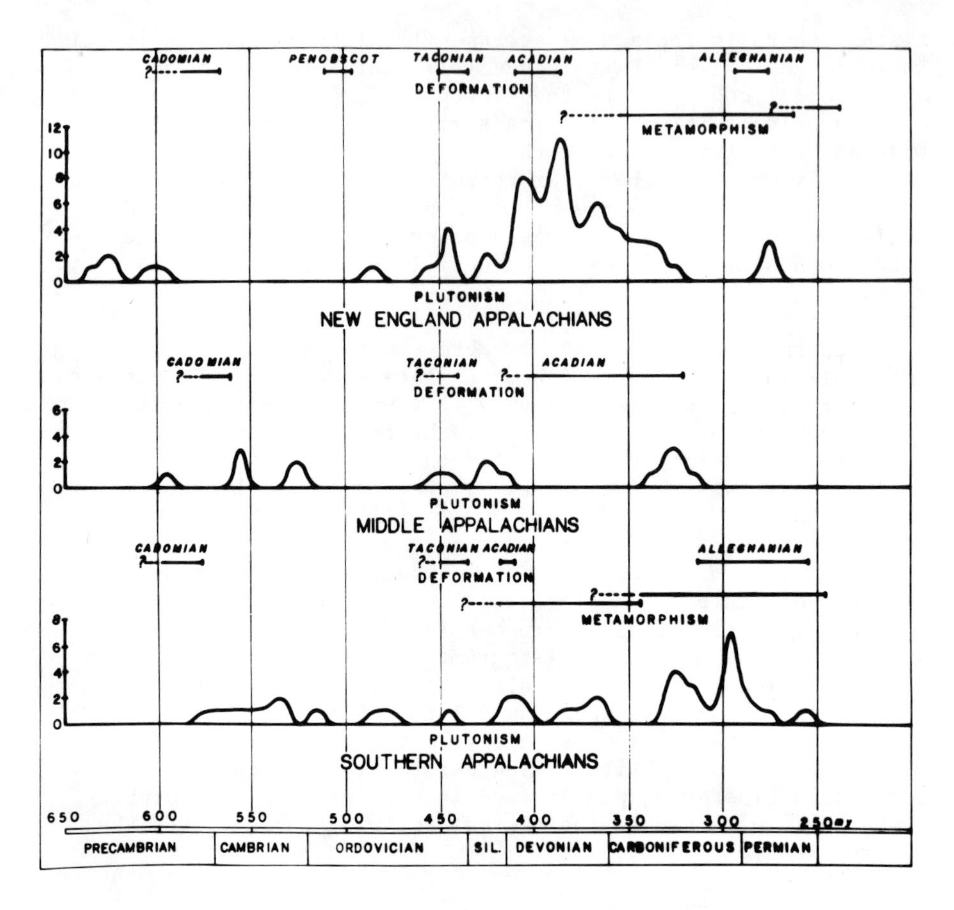

Figure 3. Frequency of plutonism with time in the Appalachians. Frequency is numbers of plutons per 10 million year increment. Durations of deformations and metamorphism are shown where data are available.

not require this to be the case. Acadian deformation has been indicated only in the eastern terrane (locality 32). The premise of Acadian deformation here depends substantially on the assumption that the Falmouth plutons (~320 my) (22) predate the Alleghanian (Hercynian) deformation, as is the case for plutons of that age in New England. Similary, the ~320 my age of pegmatites at locality 31 may imply but do not demand Acadian thrusting. Alleghanian (Hercynian) deformation has been positively identified only in the western terrane, where Carboniferour and lowest Permian strata are deformed. If the Alleghanian (Hercynian) deformation affected the medial and eastern terranes, its intensity was such as to leave little recognizable fabric in the Carboniferous plutons.

The cover sequence in the medial and eastern terranes of the Middle Appalachians has been deeply eroded exposing large tracts of Cambrian(?) and Precambrian rocks but much smaller tracts of younger rocks. This deep level of erosion has removed much of the evidence that would constrain the times of deformation. Consequently, we do not know that deformation occurred as discrete events in this region or whether or not these events were synchronous across the orogen.

Plutons (19-26) in the Middle Appalachians are not numerous, and the histogram in Figure 3 is far from statistical. In any event their distribution is consistent with the distribution in New England, and in a similar way, these are thought to be the thermal effects of orogenic cycles. Obvious differences in the two regions are the presence of 500-600 my plutons in the Middle Appalachians, and the absence of a strong Acadian thermal event.

Southern Appalachians

Effects of Cadomian, possible Penobscot (Grampian), Taconic, Acadian and Alleghanian (Hercynian) are present in the southern Appalachians (Fig. 2). Cadomian deformation has been indicated only in the eastern terrane. Possible Penobscot (Grampian) deformation is based on muscovite ages from rocks of low greenschist facies at locality 46. Assumptions of no argon loss and no excess argon in a complex terrane such as this may make these data suspect. Conceivably, Penobscot (Grampian) deformation could be present across the orogen in the southern Appalachians, but the geologic constraints for its identification are not present. Taconic thermal effects are recorded at localities 34, 35 and 40. At locality 40 deformation and the thermal effects were in part simultaneous, and at the other localities the thermal effects imply but do not demand a Taconic deformation. Taconic deformation could have spread across the orogen, but evidence for its definition is for the most part lacking. An Acadian deformation appears tightly bracketed in the western terrane (locality 37), but uncertainties concerning the age of the deformed strata and questions about the validity of the metamorphic ages make the bracket less convincing than it appears. Devonian and Carboniferous thermal effects (localities 39, 41, 42, 43, and 45) define an Acadian thermal event, and imply Acadian deformations across the region. Unfortunately, geologic relations that could uniquely identify these effects are mostly absent. Folded Carboniferous strata in the western terrane and the imposition of deformation fabrics on older than ~300 my old plutons in the eastern zone are evidence of

Alleghanian (Hercynian) deformation. Faint deformation fabrics in carboniferous plutons in the medial terrane may be due to Alleghanian (Hercynian) deformation as well.

The preservation of cover sequences are such in the Southern Appalachians that distinct deformational events cannot generally be defined. This being the case, it is impossible to comment on the synchroneity of these deformations across the orogen.

The frequency distribution with time of plutons (27-35) is similar to the distribution in the Middle Appalachians (Fig. 3). Plutons having ages between 500-700 my are numerous. A thermal pulse represented by plutons with ages between 330 and 290 my is the strongest in the Southern Appalachians. Whether the 300-330 my group of plutons represents a waning of an Acadian thermal event as is suggested in New England, or whether this group of plutons was intruded at the initial stages of an Alleghanian (Hercynian) thermal event is not clear. At any rate, these plutons, except for the 250 my pluton are pre-Alleghanian (Hercynian) deformation. The Southern Appalachians do not possess a marked Acadian thermal event like that in New England, however.

DISTRIBUTION OF DEFORMATIONAL FEATURES

The distribution and nature of deformational features are shown in Figure 4. Special symbols indicate the character of the structural features, and the background shades indicate the age of first deformation in the rocks currently exposed at the surface. Considerable uncertainty is involved in the application of shades, and they must be considered as generalities. However, where the uncertainties are sufficiently great, the area is separately shaded. Plutons are unshaded where they are reported to be undeformed.

Helikian deformation is confined to antiforms and slices of North American continent caught within the orogen. Cadomian deformation is confined to the eastern parts of New England and the medial terrane of the Middle Appalachians. The Penobscot (Grampian) is present in northern Maine, but it could have a large distribution in the Piedmont of the Middle and Southern Appalachians. Part of the reason for the uncertainty in the Piedmont is the difficulty of distinguishing Penobscot (Grampian) and Taconic deformations. Taconic deformation is well established in western New England. Ordovician rocks deformed in the Taconic are also exposed in Northern Maine, along the Bronson Hill antiform in Connecticut, Massachusetts and New Hampshire, and in the Miramachi antiform in eastern Maine. Another belt of Ordovician rocks lies adjacent to the coast of Maine, but,

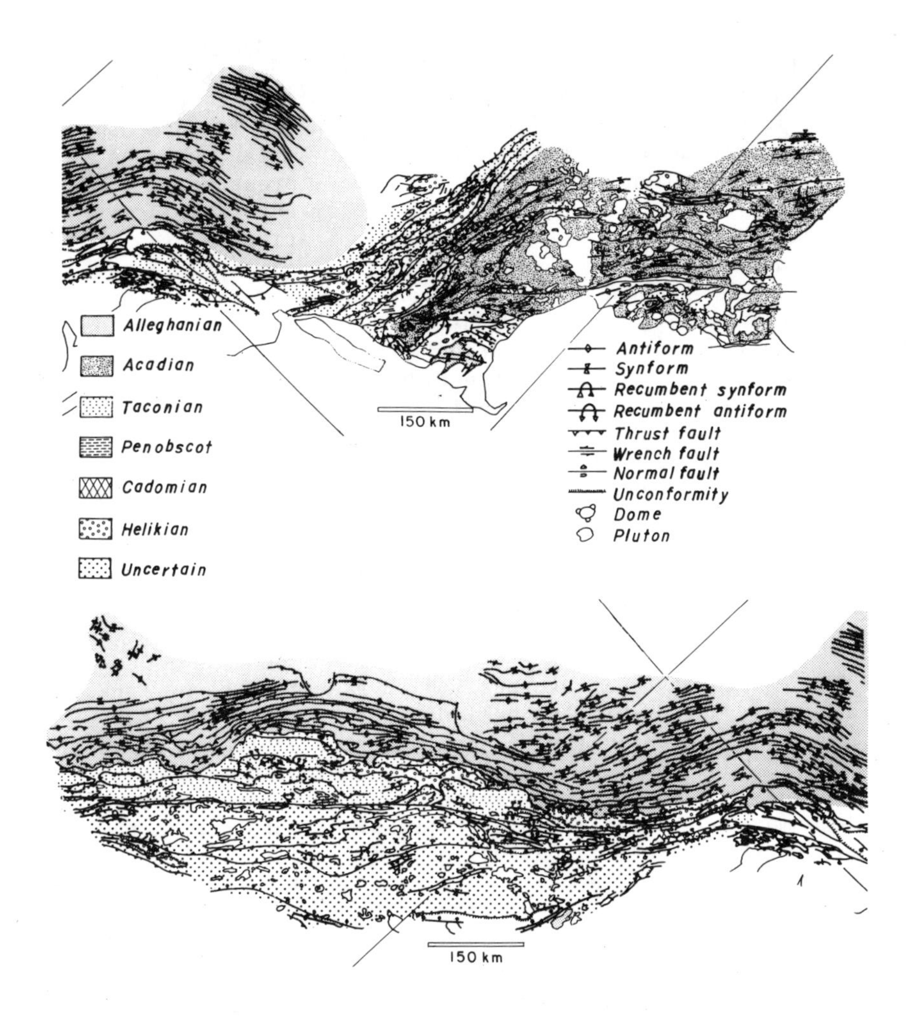

FIGURE 4. DISTRIBUTION OF DEFORMATIONAL EVENTS IN THE U.S. APPALACHIANS

because it is not clear that this section is more deformed than adjacent Siluro-Devonian rocks, they are shaded as to indicate uncertainty in their age of deformation. Well defined Taconic deformation is recognized adjacent to the Hudson Highlands and the Reading Prong in the Middle Appalachians, and the Blue Ridge is interpreted to be first deformed in the Taconic as well. The Acadian is present mainly in New England where a large Siluro-Devonian sedimentary section is preserved. It is also present in the Talladega Belt (36) over an area that exposes the Jemison Chert. A small spot indicating Acadian deformation is located at Fredericksburg, Virginia (22), and the Acadian shade might be more common to the south in the Piedmont had Devonian strata withstood the ravages of erosion. Alleghanian (Hercynian) deformation is mostly confined to the western edge of the orogen in the Middle and Southern Appalachians. A small basin of carboniferous rocks near Worcester, Massachusetts (37) and north of Narragansett Bay in Rhode Island (38) were deformed in the Alleghanian (Hercynian) for the first time. Presumably, the Alleghanian (Hercynian) front crosses southern New England to connect the Pocono Plateau with the deformed carboniferous in New Brunswick.

CONCLUSIONS

The relatively complete section of Paleozoic cover rocks and numerous plutons provide well defined brackets for the deformational events in the New England Appalachians. The cover sequence is less complete in the Middle and Southern Appalachians, and consequently the ranges of uncertainty that bracket the deformational events become progressively broader to the south. In the Southern Appalachians, particularly, there is considerable uncertainty in the identification of deformational events.

Where the brackets for deformational events are well established (New England and parts of the Middle Appalachians), they are produced over a short range of time and appear as discrete events. Furthermore, the deformational events are more or less synchronous across the orogen.

Plutonic activity, as measured by the frequency of plutons, is cyclic, and, where sufficiently well studied, metamorphic events correlate in time to the plutonism. These thermal effects correlate with times of deformation, although the thermal effects are much longer lived than the deformational events.

The Appalachian orogen is composite, having been constructed over a span of ~350 my by several distinct orogenic events. The effects of these events are not uniformly distributed over the orogen. Cadomian effects are found east of the Broughton-Blue Ridge suture (Fig. 1 and 4), and may be due to accretions of small continents to North America, although some may have been inherited with "Avalon". The Penobscot (Grampian) effects are striking in New England, but elsewhere their presence is clouded by the ambiguity of the geologic data. In general, effects of the Taconic orogeny extend farther west than effects of the Acadian all along the orogen except in Alabama in the Talladega Belt. The western front of the Alleghanian orogeny lies east of those of the Taconic and Acadian orogenies in a large part of New England, but in southern New England the Alleghanian (Hercynian) effects reach progressively westward, and in the Middle and Southern Appalachians the Alleghanian (Hercynian) front lies west of the deformational effects of older orogenies.

ACKNOWLEDGEMENTS

Figure 4 is taken from an unpublished time-of-deformation map compiled largely by James F. Tull, Robert D. Hatcher, Jr., Avery A. Drake, Jr. and Leo M. Hall. This author wishes to acknowledge with appreciation the kind consent of these compilers for the use of their information. Obviously, any errors that have crept into this version are mine. The color overlay was not part of the original map.

REFERENCES

1. Williams, Harold: 1978, Lithotectonic Map of the Appalachians: Memorial University, Newfoundland, Map No. 1.
2. Boucot, A.J.: 1962, in Coe, Kenneth (ed.) Some aspects of the Variscan fold-belt, 9th Intern.-Univ. Geol. Congress., Manchester Univ. Press, pp. 155-163.
3. Brookins, D.G.: 1976, in Page, L.R. (ed.) Contributions to the stratigraphy of New England: Geol. Soc. America Mem. 148, pp. 129-145.
4. Dallmeyer, R.D.: 1979, in Skehan, J.W., S.J., and Osberg, P.H. (eds.) The Caledonides in the U.S.A.: Geological excursions in the northeast Appalachians: Weston, Weston Observatory, pp. 63-71.
5. Fairbairne, H.W., Moorbath, Stephen, Ramo, A.D., Pinson, W.H., Jr., and Hurly, P.M.: 1967, Earth and Planetary Sci. Letters, 2, pp. 321-328.
6. Faul, Henry, Stern, T.W., Thomas, H.H. and Elmore, P.L.D.: 1963, Am. Jour. Sci., 261, pp. 1-19.
7. Hermes, D.D., Gromet, L.P. and Zartman, R.E.: 1981, in Boothroyd, J.C. and Hermes, O.D. (eds.) NEIGC, 73rd Ann. Meeting: Kingston, R.I., pp. 315-338.

8. Lyons, J.B.: 1979, in Skehan, J.W., S.J. and Osberg, P.H. (eds.) The Caledonides in the U.S.A.: Geological excursions in the northeast Appalachians: Weston, Weston Observatory, pp. 73-92.
9. Lyons, J.B. and Faul, Henry: 1968, in Zen, E-an and others (eds.) Studies of Appalachian geology - Northern and maritime. New York, Intersci. Pubs., pp. 305-318.
10. Moench, R.H. and Zartman, R.E.: 1976, in Lyons, P.C. and Brownlow, A.H. (eds.) Studies in New England Geology: Geol. Soc. America Mem. 146, pp. 203-238.
11. Naylor, R.S.: 1969, Geol. Soc. America Bull., 80, pp. 405-428.
12. Naylor, R.S.: 1971, Sci., 172, pp. 558-560.
13. Ratcliffe, N.M. and Zartman, R.E.: 1976, in Page, L.R. (ed.) Contributions to the stratigraphy of New England: Geol. Soc. America Mem. 148, pp. 373-412.
14. Spooner, C.M. and Fairbairne, H.W.: 1970, Geol. Soc. America Bull., 81, pp. 3663-3670.
15. Zartman, R.E., Snyder, George, Stern, T.W., Marvin, R.F. and Buckman, R.C.: 1965, U.S. Geol. Survey Prof. Paper 525-D, pp. D1-D10.
16. Zartman, R.E. and Marvin, R.F.: 1971, Geol. Soc. America Bull., 82.
17. Zartman, R.E. and Naylor, R.S.: 1972, Geol. Soc. America Abstracts with Programs, 4, pp. 54.
18. Dallmeyer, R.D.: 1981, Geol. Soc. America Abstracts with Programs, 13, pp. 127-128.
19. Muth, K.G., Arth, J.G. and Reed, J.C., Jr.: 1979, Geol., 7, pp. 349-350.
20. Grauert, B. and Wagner, M.E.: 1975, Amer. Jour. Sci., 275, pp. 683-691.
21. Pavlides, Louis, Arth, J.C., Daniels, D.L. and Stern, T.W.: 1982, Geol. Soc. America Abstracts with Programs, 14, pp. 584.
22. Pavlides, Louis: 1980, U.S. Geol. Survey Prof. Paper 1146, pp. 1-29.
23. Higgins, M.W., Sinha, A.K., Zartman, R.E. and Kirk, W.S.: 1977, Geol. Soc. America Bull., 88, pp. 125-132.
24. Seiders, V.M., Mixon, R.B., Stern, T.W., Newell, M.F. and Thomas, C.B., Jr.: 1975, Amer. Jour. Sci., 275, pp. 481-511.
25. Sinha, A.K., Hanon, B.B., Sans, J.R. and Hall, S.T.: 1980, in Wones, D.R. (ed.) Proceedings on the Caledonides in the U.S.A., I.G.C.P. Project 27: Blacksburg, Virginia Polytechnic Institute and State University Mem. 2, pp. 131-135.
26. Tilton, G.R., Doe, B.R. and Hopson, C.A.: 1970, in Fisher, G.W. and others (eds.) Studies of Appalachian geology: Central and southern: New York, Intersci. Pub., pp. 429-434.

27. Butler, J.R. and Fullagar, P.D.: 1978, Geol. Soc. America Bull., 89, pp. 460-466.
28. Ellwood, B.B., Wenner, D.B., Mose, D. and Amerigian, C.: 1980, Jour. Geophys. Res., 85, pp. 6521-6533.
29. Fullagar, P.D.: 1971, Geol. Soc. America Bull., 82, pp. 2845-2862.
30. Fullagar, P.D. and Butler, J.R.: 1979, Amer. Jour. Sci., 279, pp. 161-185.
31. Glover, Lynn, III and Sinha, A.K.: 1973, Amer. Jour. Sci., Cooper Volume, pp. 234-251.
32. Odom, A.L. and Fullagar, P.D.: 1973, Amer. Jour. Sci., Cooper Volume, pp. 133-149.
33. Russell, G.S.: 1978, Unpublished Ph.D. thesis, Fla. State University.
34. Snoke, A.W., Kish, S.A. and Secor, D.T., Jr.: 1980, Amer. Jour. Sci., 280, pp. 1018-1034.
35. Whitney, J.A., Jones, L.M. and Walker, R.L.: 1976, Geol. Soc. America Bull., 87, pp. 1067-1077.
36. Tull, J.F.: 1980, _in_ Wones, D.R. (ed.) Proceedings for the Caledonides in the U.S.A., I.G.C.P., Project 27, Blacksburg, Virginia Polytechnic Institute and State University Mem. 2, pp. 167-177.
37. Grew, E.S.: 1973, Amer. Jour. Sci., 273, pp. 113-129.
38. Murray, D.P. and Skehan, J.W., S.J.: 1979, _in_ Skehan, J.W., S.J. and Osberg, P.H. (eds.) The Caledonides in the U.S.A.: Geological excursions in the northeast Appalachians: Weston, Weston Observatory, pp. 1-35.
39. Zen, E-an: 1967, Geol. Soc. America Spec. Paper 97, 107 p.
40. Berry, W.B.N.: 1962, Geol. Soc. America Bull., 73, pp. 696-718.
41. Ratcliffe, N.M., Bird, J.M. and Bahrami, Beshid: 1975, _in_ Ratcliffe, N.M. (ed.) N.E.I.G.C. 67th Ann. Meeting Guidebook, New York, City College of New York, pp. 55-86.
42. Albee, A.L. and Boudette, E.L.: 1972, U.S. Geol. Survey Bull. 1297, 110 p.
43. Boucot, A.J.: 1961, U.S. Geol. Survey Bull. 111-E, 188 p.
44. Neuman, R.B.: 1980, Geol. Soc. America Abst. with Programs, 12, pp. 75.
45. Eisenberg, R.A.: 1981, Geol. Soc. America Abstr. with Programs, 13, pp. 131.
46. Neuman, R.B.: 1967, U.S. Geol. Survey Prof. Paper 524-I, 37 p.
47. Hall, B.A.: 1970, Maine Geol. Survey Bull. 22, 63 p.
48. Boucot, A.J., Field, M.T., Fletcher, Raymond, Forbes, W.H., Naylor, R.S. and Pavlides, Louis: 1964, Maine Geol. Survey Quad. Map Ser. no. 2, 123 p.
49. Rankin, D.W.: 1969, _in_ Zen, E-an and others (eds.) Studies of Appalachian geology: Northern and maritime: New York, Intersci. Pub., pp. 355-369.

50. Pavlides, Louis: 1972, U.S. Geol. Survey Geol. Quad. Map GQ-1094.
51. Roy, D.C. and Mencher: 1976, in Page, L.R. (ed.) Contribution to the stratigraphy of New England: Geol. Soc. America Mem. 148, pp. 25-52.
52. Pankiwskyj, K.A., Ludman, Allan, Griffin, J.R. and Berry, W.B.N.: 1976, in Lyons, P.C. and Brownlow, A.H. (eds.) Studies in New England Geology: Geol. Soc. America Mem. 146, pp. 263-280.
53. Dallmeyer, R.D., Van Breeman, Otto and Whitney, J.A.: 1982, Amer. Jour. Sci., 282, pp. 79-93.
54. Billings, M.P.: 1935, Geology of the Littleton and Moosilauke quadrangles, New Hampshire: New Hampshire Planning and Development Commission, Concord, N.H., 51 p.
55. Harwood, D.S.: 1973, U.S. Geol. Survey Bull. 1346, 90 p.
56. Hepburne, J.C.: 1975, in Harwood, D.S. (ed.) Tectonic studies of the Berkshire massif, western Massachusetts, Connecticut, and Vermont. U.S. Geol. Survey Prof. Paper 888, pp. 33-49.
57. Boucot, A.J.: 1969, in Zen, E-an and others (eds.) Studies of Appalachian geology: Northern and maritime: New York, Intersci. Pub., pp. 83-94.
58. Ashwal, L.D., Leo, G.W., Robinson, Peter, Zartman, R.E. and Hall, D.J.: 1979, Amer. Jour. Sci., 279, pp. 936-969.
59. Ruitenburg, A.A.: 1968, Geology and mineral deposits of Passamaquoddy Bay area: New Brunswick Dept. Nat. Res. and Min. Res. Br., Rept. of Invest. 7, 47 p.
60. Abbott, R.N., Jr.: 1978, in Ludman, Allan (ed.) N.E.I.G.C., 70th Ann. Meeting, Guidebook, Fleishing, Queens College Press, pp. 17-37.
61. Gates, Olcott: 1975, Maine Geol. Survey, Geol. Map Ser. 3.
62. Schluger, P.R.: 1973, Geol. Soc. America Bull., 84, pp. 2533-2548.
63. Smith, G.O., Bastin, E.S. and Brown, C.W.: 1907, U.S. Geol. Survey Geol. Atlas 149, 14 p.
64. Brookins, D.G., Berdan, J.M. and Stewart, D.B.: 1973, Geol. Soc. America Bull., 84, pp. 1619-1628.
65. Smith, B.M.: 1978, M.Sc. thesis, Brown University, Providence, R.I., 94 p.
66. Wood, G.H., Jr. and Bergin, M.J.: 1970, in Fisher, G.W. and others (eds.) Studies of Appalachian geology: Central and southern: New York, Intersci. Pub., pp. 147-160.
67. Drake, A.A., Jr., Davis, R.E. and Alvord, D.C.: 1960, U.S. Geol. Survey Prof. Paper 400B, pp. 3180-3181.
68. Stose, G.W. and Ljungstedt, O.A.: 1931, Geologic Map of Pennsylvania: Penn. State Topographic and Geol. Survey.
69. Cooper, B.N.: 1964, in Lowry, W.D. (ed.) Tectonics of the southern Appalachians: Blacksburg, Virginia Polytechnic Institute, Dept. Geol. Sci. Mem. 1, pp. 81-114.

70. Drake, A.A., Jr. and Lyttle, P.T.: 1981, U.S. Geol. Survey Prof. Paper 1205, 16 p.
71. Fullager, P.D. and Dietrick, R.V.: 1976, Amer. Jour. Sci., 276, pp. 347-365.
72. Deuser, W.G. and Herzog, L.F.: 1962, Jour. Geophys. Res., 67, pp. 1997-2004.
73. Brown, W.R.: 1969, Geology of the Dillwyn quadrangle, Virginia: Virginia Div. of Min. Res. Rept. of Inv. 10, 77 p.
74. Dallmeyer, R.D.: 1975, Amer. Jour. Sci., 275, pp. 444-460.
75. Kish, S.A., Fullagar, P.D. and Dabbagh, A.E.: 1976, Geol. Soc. America Abstr. with Programs, 8, pp. 211-212.
76. Thomas, W.A., Tull, J.F., Bearce, D.N., Russell, Gail and Odom, A.L.: 1980, in Wones, D.R. (ed.) Proceedings for the Caledonides in the U.S.A., I.G.C.P. Project 27: Blacksburg, Virginia Polytechnic Institute, Dept. Geol. Sci. Mem. 2, pp. 91-97.
77. Kish, S.A. and Harper, C.T.: 1973, Geol. Soc. America Abst. with Programs, 5, pp. 409.
78. Fullagar, P.D.: 1981, South Carolina Geol., 25, pp. 29-32.
79. Butler, J.R.: 1973, Amer. Jour. Sci., 273-A, pp. 72-88.
80. Dallmeyer, R.D.: 1978, Amer. Jour. Sci., 278, pp. 124-149.
81. Ross, C.R. and Bickford, M.E.: 1980, in Stormer, J.C. and Whitney, J.W. (eds.) Geological, geochemical, and geophysical studies of the Elberton Batholith, eastern Georgia: Atlanta, Georgia Dept. Nat. Res. Guidebook 19, pp. 52-62.
82. Sundeluis, H.W.: 1970, in Fisher, G.W. and others (eds.) Studies of Appalachian geology: Central and southern: New York, Intersci. Pub., pp. 351-368.
83. Kish, S.A., Butler, J.R. and Fullagar, P.D.: 1979, Geol. Soc. America Abstr. with Programs, 11, pp. 184-185.

APPENDIX

1. Castleton, N.Y. Zen (39) has described diamictites at the west margin of the Giddings Brook slice that he interprets as wildflysch formed at the front of an advancing, thrust plate. Berry (40) has described graptolites of Caradocian age from the matrix of the wildflysch. It has been assumed that these fossils date the emplacement of the lower Taconic slices.

2. Hudson, N.Y. Ratcliffe and others (41) have described the geology of the Mount Ida Quarry. Dolostones belonging to the Manlius Limestone overlie slate of the Germantown Formation. A limestone conglomerate containing clasts of green slate forms the basal beds of the Manlius Limestone. The slate fragments are cut by a penetrative foliation that in some clasts is misaligned to the fracture cleavage that

cuts the Manlius Limestone. Because the Germantown Formation lies within the Giddings Brook slice which was emplaced in Caradocian time (See 1) and the Manlius Limestone lies unconformably on the slice, the age of the deformation is within the range Caradocian to upper Llandoverian.

3. South Sandesfield, MA. Ratcliffe and Zartman (13) describe thrust zones in the crystalline rocks of the Berkshire massif that carry the crystallines westward over younger autochthonous and allochthonous rocks. Recumbent folds associated with the thrusts deform all rock sequences. The Caradocian Walloomsac Slate is the youngest unit in the autochthonous sequence. The thrusts are cross-cut locally by granite that has a preliminary Late Ordovician age by Pb-U methods.

   The thrust related features post-date the emplacement of the Taconic slices, at least two folding episodes, and a metamorphic episode. These features are considered to have been produced sequentially in the range between Caradocian and Ashgillian time.

4. Jackman, ME. Albee (42) has described the geology in the northern part of the Chain Lake Massif. He reports folds with "axial plane cleavage" in the Seboomook Formation of Siegenian age. The folded Seboomook Formation is cut by the Hog Island Granite that has been dated as 340±28 my by Rb-Sr methods.

5. Greenville, ME. Boucot (43) has described the geology of the Moose River synclinorium and in that report suggested the correlation of the Lobster Mountain volcanics and the volcanic part of the Kennebec Formation at Cornish Farm. At Lobster Mountain the volcanics are Ashgillian (44) and at Cornish Farm the volcanics are Caradocian (43). Boone (1982, personal communication) indicates that the Lobster Mountain Volcanics and Kennebec Formation cut across more highly deformed units of the underlying stratigraphic section. The oldest units in the underlying section are intruded by trondjhemitic dikes that have been dated as 515 my (45), so that the age of the younger units in the underlying section must be uppermost Cambrian to Lower Ordovician. An intense deformational event occurred between Upper Cambrian/Lower Ordovician and Caradocian time.

   The rocks of the Lobster Mountain Volcanics and Kennebec Formation are deformed with similar style and intensity as the Siluro-Devonian section that unconformably overlies them. This unconformity, between Ashgillian and Lower Silurian rocks, represents only a mild deformational event or even a diastrophic event.

6. Shin Pond, ME. Neuman (46) described the Shin Brook Formation containing Arenigian fossils that he reports to be unconformably on the Grand Pitch Formation which contains Oldhamia. Units of the Grand Pitch Formation are more highly deformed than the overlying Shin Brook Formation and the character of the deformation in the Shin Brook Formation is similar to that in nearby Siluro-Devonian rocks.

   A few miles west of Shin Brook, Neuman (46) has described Ashgillian polymictic conglomerates and sandstones that lie between Grand Pitch Formation (Cambrian?) and Lower Silurian conglomerates. The Grand Pitch Formation is more highly deformed than the conglomerates of Ashgillian age, but there is little different in deformational character between the Ashgillian rocks and the overlying Siluro-Devonian section.

7. Spider Lake, ME. Hall (47) has described a section in which highly deformed rocks of Cambrian(?) age are overlain unconformably by much less deformed rocks of Caradocian age.

   A second unconformity is recorded between Caradocian rocks and Gedinnian rocks, but the character and intensity of deformation in both Caradocian and Gedinnian sections is similar.

8. Presque Isle, ME. Boucot and others (48) described the Mapleton Sandstone of Eifelian age resting unconformably on rocks as young as Gedinnian. Rocks of the Silurian and lowest Devonian section are deformed into tight to open folds that are truncated by the nearly flat-lying Mapleton Sandstone.

9. Trout Brook Farm, ME. Rankin (49) has described a section of sandstones and slates and ash-flow tuffs that are as young as Emsian. This section is overlain unconformably by the Trout Brook Formation that contains plants of Middle Devonian age. The Lower Devonian sequence is deformed by open folds, whereas the Trout Brook Formation is nearly flat-lying being preserved as downfaulted blocks.

10. Smyrna Mills, ME. Pavlides (50) has described the Smyrna Mills Formation which contains fossils as young as Wenlockian to Ludlovian age. The Smyrna Mills Formation is deformed by open to tight folds. These folds are cut by plutons dated by K-Ar method as 384-397 my.

11. Ashland, ME. Roy and Mencher (51) indicate that the Frenchville Conglomerate of Llandoverian age rests with angular unconformity on the Winterville Formation of Caradocian age. The state of deformation in the two sequences is not markedly different.

12. Hartland, ME. The Hartland Granite cuts a major syncline in the Silurian section of south-central Maine. The Silurian section contains rocks as young as Ludlovian and possibly Pridolian age (52), and the Hartland pluton has been dated as 360±8 my (53).

13. Littleton, NH. Billings (54) described foliated granite (Highlandcroft) that intrudes the Ammonoosac Volcanics. The Ammonoosac Volcanics have been traced into western Maine where they interfinger with slates containing Caradocian fossils (55). The Highlandcroft plutons where dated give Rb-Sr ages of 450-420 my (11). Both the granite and the volcanic section are overlain unconformably by a fossiliferous Siluro-Devonian section as old as Llandoverian.

14. Hopkington, NH. Lyons (8) indicates that the Spaulding Quartz Diorite cuts structures that deform the Kearsarge Member of the Littleton Formation. The Kearsarge Member is multiply deformed whereas the Spaulding Quartz Diorite is "structureless" except near its margin. The Kearsarge Member has been traced (Hatch, 1981, personal communication) into the Carrabasett Formation in western Maine, which is either Pridolian or Gedinnian in age. The Spaulding Quartz Diorite has been dated as 402±5 my by Rb-Sr whole rock methods (8).

15. Guilford, VT. Hepburn (56) in mapping the Guilford Dome has demonstrated that the section including the Standing Pond Amphibolite and the Gile Mountain Formation are folded by recumbent isoclinal folds and subsequently domed. The Gile Mountain Formation is thought to be Lower Devonian in age (57). The recumbent folds and the dome are cut by the Black Mountain Granite, which Naylor (12) has dated as 383±7 my by Rb-Sr methods.

16. Belchertown, MA. Ashwal and others (58) have described the Belchertown pluton as cutting across recumbent folds that deform the Siluro-Devonian section of western Massachusetts. The Siluro-Devonian section is possibly as young as Gedinnian and the emplacement age of the Betchertown pluton is 380±5 my. It should be noted that the outer parts of the Belchertown pluton are deformed suggesting that either it is in part syntectonic or it was deformed by a later event.

17. Worcester, MA. Grew (37) has described the geology in the vicinity of Worcester. Conglomerates and fossiliferous carbonaceous phyllites of Upper Carboniferous age occur along the Clinton-Newbury fault. These Carboniferous rocks are deformed and metamorphosed to garnet grade. Grew did not observe a metamorphic discontinuity between

the Carboniferous and older rocks. Close by, the Harvard Conglomerate rests with unconformity on the Ayer Granite, which has an ill-defined radiometric age of 520-410 my (15).

18. Calais, ME. Rutenburg (59) has described an unconformity between the Lower Silurian(?) Oak Bay Conglomerate and the Cookson Formation of Tremadocian age. Differences in structural features below and above the unconformity are negligible.

    The Red Beach Granite (60) intrudes the Siluro-Devonian section south of Calais. The granite has post-orogenic textures and is dated at 400 my based on whole rock and mineral Rb-Sr ratios (14). The youngest stratified rocks cut by the Red Beach Granite belong to the Eastport Formation of Gedinnian age (61). The Red Beach Granite and the Silurian-Lower Devonian section are unconformably overlain by the Perry Formation of Upper Devonian age (62). The Perry Formation occupies a gently dipping basin relative to the more highly deformed rocks beneath the unconformity.

19. Penobscot Bay, ME. The Ames Knob Formation (63) lies unconformably on the North Haven Greenstone of unknown age. Clasts of greenstone contain a foliation that predates the time of incorporation into the conglomerate at the base of the Ames Knob Formation. Siluro-Devonian volcanics of the Castine Formation are broadly folded and faulted (64) and are cut by the non-foliated Vinyl Haven Granite dated at 361±6 my by Rb-Sr whole rock methods (3).

20. Boston, MA. The Dedham Granodiorite, Milbridge Granite and Northbridge Granite Gneiss are dated in the range 600-650 my on the basis of U-Th-Pb systems in zircon (17). These rocks intrude foliated Marboro Formation and high grade metamorphic rocks of uncertain age.

21. North Attleboro, MA. Fossiliferous Lower Cambrian sandstone rests unconformably on Dedham Granodiorite ($\simeq$600 my) at Hoppen Hill (5). The Cambrian rocks are only moderately folded.

22. Narragansett Bay, RI. The Newport Granite, dated at 595±12 my (65), intrudes deformed volcanics (38). Polydeformed Cambrian sections are in fault contact with the older strata.

    Rocks dated palynalogically as Westphahm B are multiply deformed and subsequently intruded by the undeformed Narragansett Pier Granite, which is dated at 272 my by U-Pb techniques (7).

23. Pocono, PA. Wood and Bergin (66) have described folds and thrusts in rocks as young as the Pottsville Formation, dated on the basis of plant fossils as Westphalen. No meaningful upper limit of the time of deformation is present.

24. Allentown, PA. Drake and others (67) compared the intensity of deformation in Siluro-Devonian sections to that in Cambro-Ordovician sections. In this vicinity the Silurian Shawangunk conglomerate overlies the Martinsburg Shale (Caradocian) unconformably. The older sections are profoundly more deformed than the sections above the unconformity.

25. Uniontown, PA. Rocks of latest Carboniferous and earliest Permian are present in the core of the Latrobe Syncline (68). No upper bound for the deformation is present.

26. Giles County, VA. Cooper (69) described local unconformities in anticlinal crests at which Blackford beds (Middle Ordovician) cut down into the Knox Dolomite (Lower Ordovician). In adjacent synclines the relations are conformable with no hiatus. The rocks of both sequences are folded.

27. Willmington, DE. Zircons from the Willmington Complex have yielded Pb-U ratios that plot near the lower intercept of the concordia (20). The lower intercept age is 441 my. The Willmington complex is thought to be of volcanic origin and to be equivalent to the James River Formation, which has been dated at ~500 my.

28. Annadale, VA. Drake and Lyttle (70) have mapped a stack of thrust plates that carry the Eastern Fairfax Sequence onto the Sykesville Formation and the Piney Branch Allochthon onto the Peters Creek Schist. Both slices are unconformably overlain by the Popes Head Formation and all sequences are intruded by the Occoquan Granite, which has been dated at 560 my (24).

29. Ellisville, VA. Pavlides and others (21) have described schistose melanges involved in imbricated thrust slices and intruded by the post-deformational Ellisville pluton dated at 440±8 my by a Rb-Sr whole rock isochron.

30. Great Falls, VA. Muth and others (19) have described multiply deformed rocks in the Potamac River that are intruded by pegmatites and small bodies of granite, postdating F_1 and F_2 but prior to F_3. Muscovite from the pegmatites and granites gives a Rb-Sr mineral age of 469±20 my.

31. Lynchburg, VA. Fullagar and Dietrick (71) have described a gneiss-schist terrane cut by the Leatherwood Granite which has been dated as 464±20 my by Rb-Sr whole rock methods. These rocks are interpreted to have been subsequently transported westward as a thrust slice. This thrust slice is cut by pegmatites from which mica separates have yielded a Rb-Sr mineral date of 321±17 my. (72).

32. Fredericksburg, VA. Pavlides (22) has mapped Quantico Formation dated as late Ordovician to Silurian on the basis of fossils as unconformably overlying Chopawamsic Formation and its metamorphic equivalents. The Chopawamsic Formation contains volcanics that are intruded by the Occoquan Granite dated at 560 my (24). Rocks above and below the unconformity are intensely deformed, but the Chapowamsic Formation contains an additional phase of folding relative to the Quantico Formation. The stratigraphic section is intruded by the deformed Falls Run Granite dated at 410 my and the undeformed Falmouth plutonic suite dated at 320± my (22).

33. Arvonia, VA. Brown (73) has described an unconformity at the base of the Arvonia Slate, which has been dated as late Ordovician on the basis of fragmentary fossils. The Evington Group, underlying the unconformity consists of a volcanic-sedimentary section that is considered to be Cambrian in age by correlation with the Chopawamsic Formation near Fredericksburg. The Hatcher complex is intrusive into the Evington Group. The Arvonia Formation is intensely folded and cleaved.

34. Cherokee, NC. Dallmeyer (74) determined $^{40}Ar/^{39}Ar$ incremental gas-released ages in biotite (343-350 + 15 my) and on hornblende (415-427 + 15 my) from Grenville aged rocks. The rocks studied are in the kyanite zone of Paleozoic metamorphism. Based on the difference in age between biotite and hornblende and on assumptions about the intensive parameters of kyanite metamorphism, average uplift rates were calculated. Using this uplift rate, Dallmeyer believes that appropriate depths for Kyanite metamorphism were attained at ~480 my.

35. Bryson City, NC. Kish and others (75) described undeformed pegmatites intruding polydeformed and metamorphosed rocks of late Precambrian age. Whole rock Rb-Sr age on the pegmatite intrusives gives a 440±13 my age.

36. Birmingham, AL. Thomas (76) states that the highest stratigraphic unit involved in folding is the Pottsville Formation of Westphalen B age.

37. Jemison, AL. Tull (36) has described the geology within the Talladega Belt. A section that includes the fossiliferous Lower to Middle Devonian Jemison Chert is variously metamorphosed, polydeformed and involved in large thrust slices. Whole-rock K-Ar ages on low grade phyllites that contain micas crystallized below their argon-retention temperatures give a range of ages from 417 to 329 my (77).

38. Ashland, AL. Russell (33) has indicated that the Bluff Spring Granite is locally discordant to the regional schistosity in the Wedowee Group of uncertain age. The Bluff Springs Granite has a Rb-Sr whole rock age of 366±9 my. It is itself moderately well foliated.

39. Gray Court, SC. The Gray Court Granite (78) intrudes polydeformed and metamorphosed Precambrian(?) rocks. The Gray Court Granite is undeformed and has a 378±24 my age.

40. Spruce Pine, NC. Butler (79) has summarized the geology near Spruce Pine. Deformed and metamorphosed rocks of lare Precambrian(?) age are intruded by pegmatites and alaskites that are dated by Pb-U methods at 430±20 my. The pegmatites and alaskites are synchronous with a second fold sequence and a second phase of metamorphism. A third fold phase postdates the intrusion of the pegmatites and alaskites.

41. Rosman, NC. Odom and Fullagar (32) examined the blastomylonites found within the Henderson Gneiss along the Brevard zone. The Henderson Gneiss has an uncertain protolith, and outside of the Brevard zone a Rb-Sr whole-rock isochron age of 535±27 my is obtained. A Pb-U concordia intercept age of 538 my is in good agreement. Within the Brevard zone, blastomylonites give an age of 356±8 my. This age must be a minimum age for motion within the Brevard zone.

42. Atlanta, GA. Deformed and metamorphosed gneisses and schists have been studied by Dallmeyer (80) using $^{40}Ar/^{39}Ar$ methods. The rocks are of uncertain age, but the regional metamorphism is ~365 my on the basis of diachronous post-metamorphic cooling ages of biotite and hornblende.

43. Elberton, GA. The Elberton Granite is undeformed and intrudes Precambrian(?) polydeformed gneisses and amphibolites (28). Ross and Bickford (81) have determined the age of the Elberton Granite to be 320±20 my.

44. Roxboro, NC. Glover and Sinha (31) indicate that late Precambrian to early Cambrian(?) volcanic and epiclastic rocks were folded and faulted before the intrusion of a granodiorite dated at 575±20 my.

45. Salisbury, NC. Butler and Fullagar (27) describe the Southmont pluton as intruding well foliated and metamorphosed rock that they deduce to be older than ~520 my because of intrusive relations elsewhere in the Charlotte belt. The Southmont pluton is itself slightly foliated and has Rb-Sr whole-rock ages of 402±2 my and 396±4 my. Nearby the Churchlane pluton is clearly undeformed and yields a Rb-Sr whole-rock age of 282±6 my.

46. Abermarle, NC. Sundelius (82) has described the geology in the vicinity of Abermarle as consisting of folded and slightly metamorphosed volcaniclastics, phyllites and sandstones having ages possibly as young as Middle Cambrian. Kish and others (83) have repoted an average whole-rock K-Ar age of 483±15 my for samples taken from this section. Phengitic mica is thought to be the only K-bearing phase in the samples. The samples are of low grade phyllites that were recrystallized below temperatures at which argon would be released.

47. Batesville, SC. Snoke and others (34) have described the geologic relations in part of the Keokee belt. Multiple deformed and polymetamorphosed rocks thought to be in part equivalent to those of the adjacent Slate Belt were intruded by the Lake Murray pluton dated at 313±24 my by Rb-Sr whole-rock methods. The Lake Murray pluton has a gneissic fabric that parallels schistosity in the country rocks. Aplitic dikes associated with the Lake Murray pluton are isoclinally folded with axial planar cleavage parallel to the main schistosity of the country rocks and pegmatites associated with the pluton cut the main schistosity but are themselves folded. In the vicinity of Batesville, the Edgefield Granite cuts the younger folds, but it possesses a weak foliation that is deformed by still younger folds. The Edgefield Granite is 254±11 my old as dated by Rb-Sr whole rock methods.

STRUCTURAL DOMAINS OF THE HERCYNIAN BELT IN MOROCCO

Alain PIQUÉ

Institut de Géologie, Université Louis Pasteur,
1, rue Blessig 67084 Strasbourg Cédex (France)

ABSTRACT - The hercynian belt of Morocco is divided into several domains on the basis of the style and the timing of the main Hercynian orogenic event.

By its geographic position, at the northwestern corner of Africa as well by the wide range, from Precambrian to Tertiary, of the represented deposits, Morocco offers a particular interest for geologic correlations, espacially for the reconstruction of the circum-Atlantic Paleozoic belts. Several structural domains can be described, each of them characterized by its own geologic history and by the timing and the intensity of the main deformation.

THE STRUCTURAL DOMAINS OF MOROCCO

Schematically, these domains are disposed from the South, for the oldest, to the North for the most recent, suggesting a progressive building of continental Morocco upon the African Shield.

- *The african domain* is represented only in the southern provinces. The oldest series are known in the "Dorsale reguibate", the 2 000 Ma West-African Shield. This basement is overlain by undeformed series of Paleozoic age in the North (Tindouf basin) and Meso- to Cenozoic age in the West (Coastal plain).

- North of the Tindouf basin, in the *Anti-Atlas domain*, this old basement is overprinted by the Panafrican orogeny (680-570 Ma) which comprised locally a stage of ocean opening, followed by ophiolite obduction and active margin deformation. The Upper Pro-

P. E. Schenk (ed.),
Regional Trends in the Geology of the Appalachian-Caledonian-Hercynian-Mauritanide Orogen, 339–345.

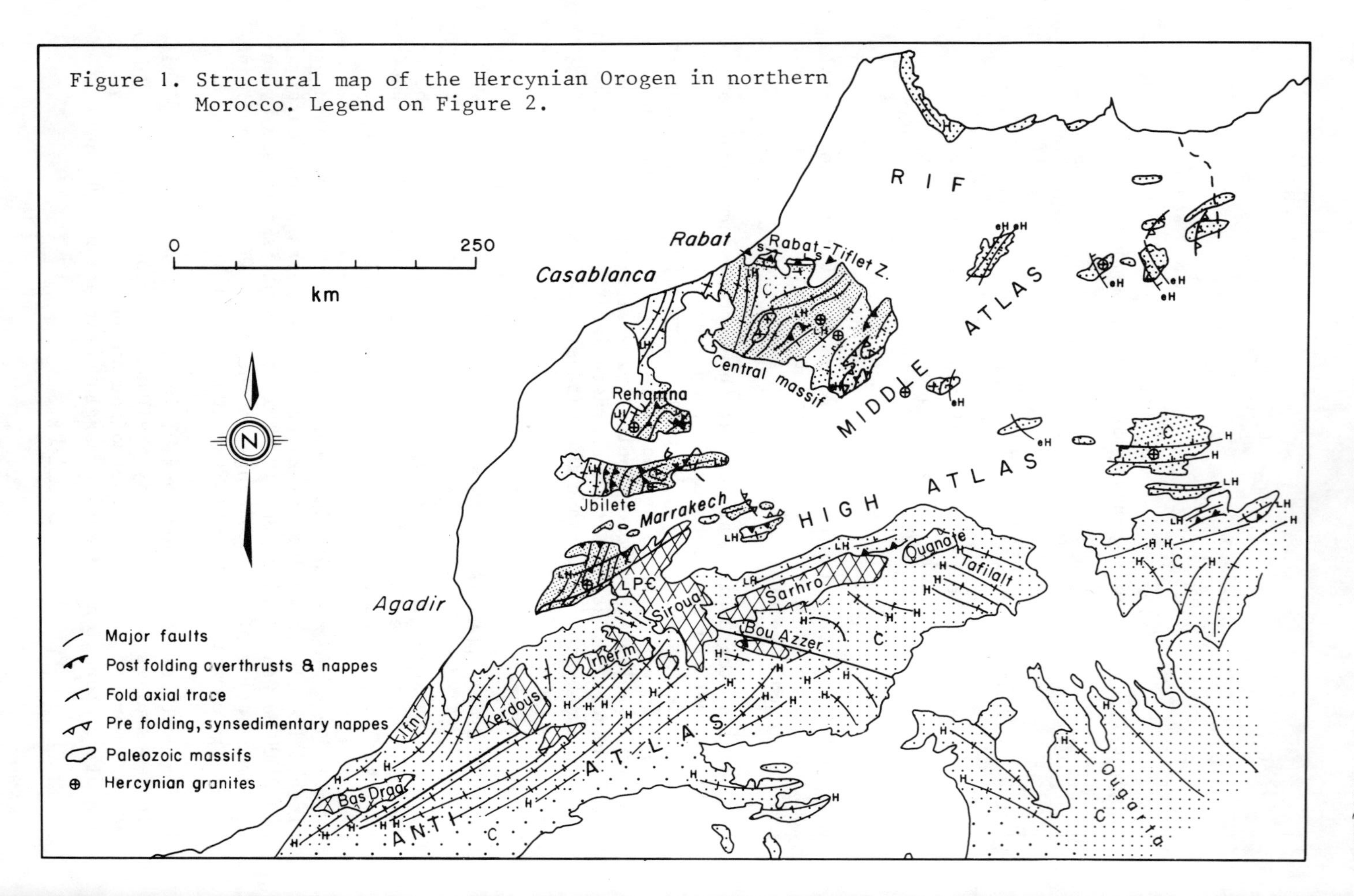

Figure 1. Structural map of the Hercynian Orogen in northern Morocco. Legend on Figure 2.

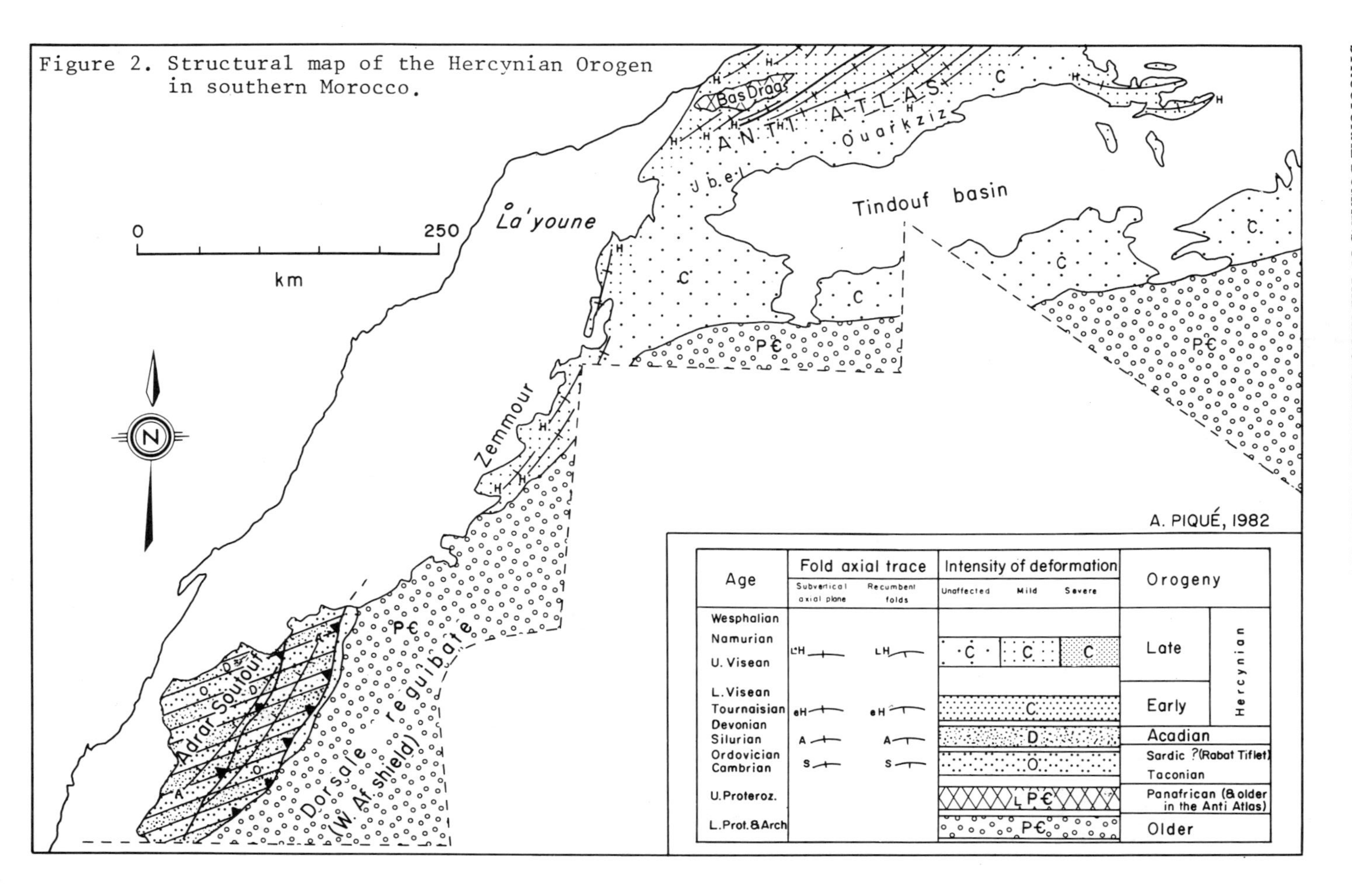

Figure 2. Structural map of the Hercynian Orogen in southern Morocco.

terozoic and Paleozoic cover of this panafrican belt was mildly deformed during the Hercynian orogeny : the hercynian regional folds are open and upright and their axial trends vary from NE-SW in the western part of the domain toward a NW-SE direction in the Ougarta belt, East of the Anti-Atlas.

- *The Mesetan domain* extends North of the Anti-Atlas. The Meseta represents the folded hercynian belt, eroded and overlain unconformably by mesozoic and cenozoic series which remained undeformed. The paleozoic blocks which constitute the atlasic basement can be studied together with the mesetan domain as they represent disrupted blocks of an initially continuous domain.

- *The atlasic domain* represents the meso- and cenozoic ensialic belts of High Atlas and Middle Atlas, extending through the hercynian belt.

- *The Rif*, the northernmost domain, is a fragment of the alpine belt of Western Mediterranean. As the Betic and Tellian (Algerian) parts of the orogen, it includes Paleozoic blocks where traces of the Hercynian orogeny are recorded.

THE HERCYNIAN BELT IN MOROCCO

Inside the mesetan domain, the paleozoic blocks included in the atlasic belt and the Anti-Atlas domain, *i.e.* the entire region affected by the Hercynian deformation, several belts can be distinguished on the basis of their stratigraphic history and overall of the age and the style of their main deformation event. Everywhere, except in the Rabat-Tiflet zone, the main paleozoic deformation postdated the Devonian - Carboniferous boundary. In other words, neither Taconic nor Acadian phases are known in the Mesetan domain of Hercynian Morocco.

- *The Rabat-Tiflet zone* is the northernmost part of the Mesetan domain. It is a narrow zone, trending E-W, always in fault contact with the other zones of Meseta. Its series, mostly greywackes and shales, are considered to represent Cambrian, because their strong facies analogies with paleontologically dated Cambrian in Meseta. These series have been folded and metamorphosed prior to the emplacement of the Rabat-Tiflet granite, at 430 $\pm$ 2 Ma (Rb/Sr method). Therefore, this folding occurred during Upper Cambrian or Ordovician. During the Hercynian orogeny, this zone was thrusted southward upon the devonian strata of the Rabat anticlinorium.

- *The Eastern Meseta* is characterized by overturned and recumbent folds which appeared often under greenschists metamorphic conditions. Almost everywhere this folding predates the Upper Visean. At Midelt, the metamorphism, contemporaneous with the main deformation, has been dated at 367 ± 7 Ma (Rb/Sr). Elsewhere, Upper Visean transgressive strata overlie unconformably folded series as young as Lower Visean. It is thus probable that this pre-Upper Visean folding occurred at various times inside the region. Another caracteristic feature of the Eastern belt is the importance of the chaotic facies of Upper Visean age, principally at its western and southern rims. Whether or not these olistostromes and wildflyschs were related to gravity nappes emplacement, they originated from eastern parts of the Meseta, which were surelevated with regard to Western Meseta and Anti-Atlas.

In the northern half of this Eastern belt, the Upper Visean was marked, often with chaotic deposits, by an important volcanism. The products, mostly breccias and tuffs, accumulated on several hundreds of meters. Their petrographic nature is andesitic and, at a lesser extent, dacitic or rhyolitic.

- *The Western Meseta* distinguishes itself from the Eastern belt by the age and the style of its deformation. Here, the pre-Upper Visean deformation event did not occur. It was represented, at the most, by the coming of coarse detrital sediments of Upper Visean age in the sedimentary basin. Unlike to Eastern Meseta, too, the magmatism, which is here Late Devonian to Lower Visean, is bimodal, represented by mafic plutonic and effusive rocks (gabbros, dolerites and spilites) and acid tuffs. In the western Meseta sedimentary basins, the Paleozoic sedimentation remained continuous and marine up to, at least, Namurian. The Paleozoic series were folded before Middle Westphalian which is represented by continental redbeds. The folds have generally steeply dipping axial planes and metamorphism was weak, except in some narrow and linear zones, which acted as shear zones. Westward, both deformation and metamorphism intensities decrease and the coastal block, the westernmost part of this belt, is almost free of any metamorphic evolution and penetrative deformation. Locally, espacially in the southern part of the belt, the Namuro-Westphalian main folding was followed by more of less important flat thrustings.

- *The Anti-Atlas domain* constitutes the southern limit of the Hercynian belt of Morocco. The regional fold trends vary from West to East. Their wave length is generally kilometric or larger. They are broad and upright and cleavage and low metamorphism occur only in the deepest pelitic facies. Often, the folds are genetically related to shear zones which evolve into strike-slip faults. These strike slip faults affect Lower Visean stratas, giving a lower limit for the age of the deformation. The Upper limit cannot the easily determined because Middle and Upper Carboniferous is not represented in the properly so-called Anti-Atlas, but only in the Tindouf basin farther South where it is undeformed.

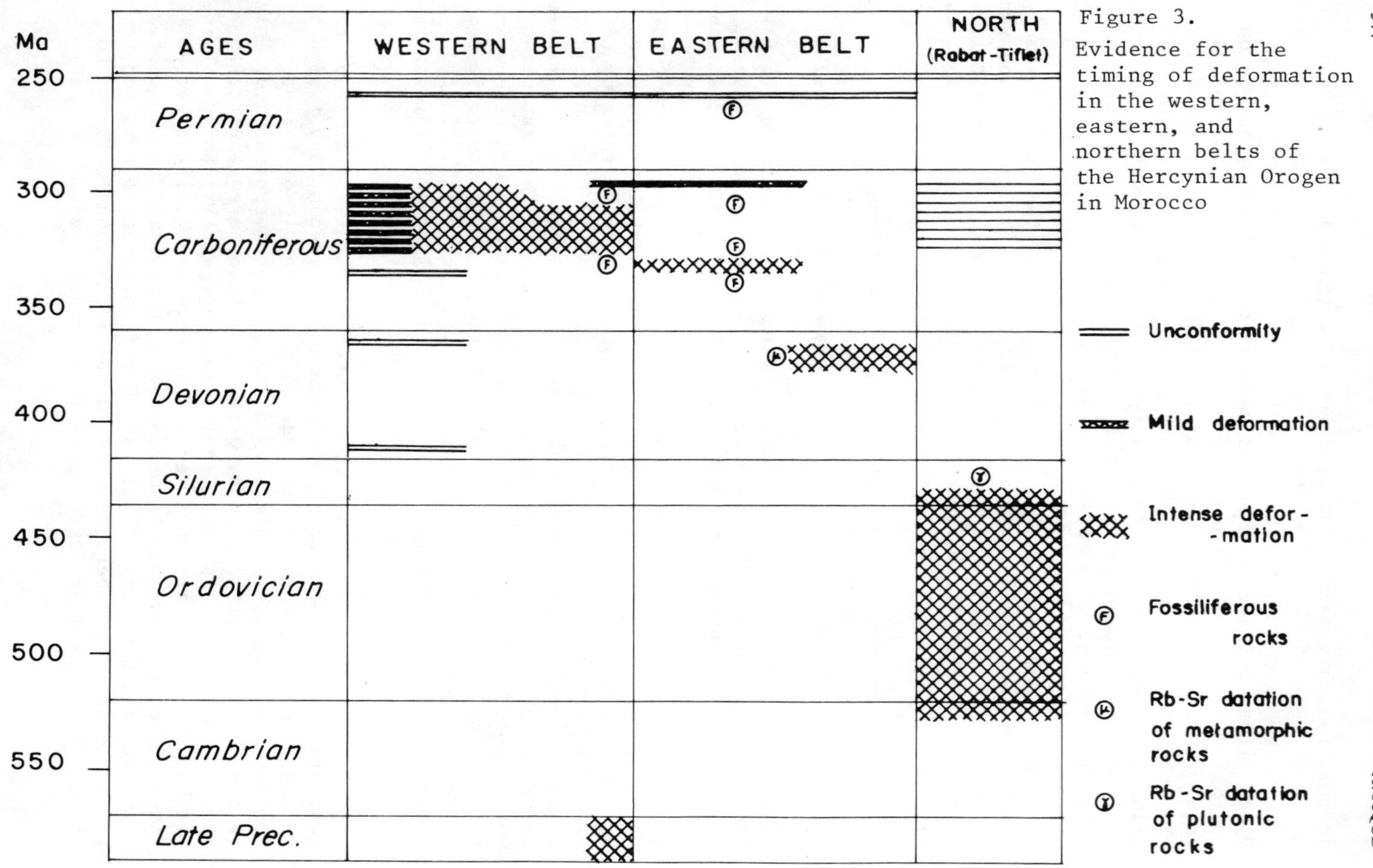

Figure 3.
Evidence for the timing of deformation in the western, eastern, and northern belts of the Hercynian Orogen in Morocco

DISCUSSION. THE GEOTECTONIC EVOLUTION OF HERCYNIAN MOROCCO

The above distinction between Eastern and Western Meseta could be interpreted as the distinction between the inner and outer belts of the chain. In the inner belt, the early folds are often recumbent and metamorphism is relatively accentuated. In the outer belt, on the contrary, the folds are late, generally upright and metamorphism is often low. At first sight, this interpretation fits with the results given by the study of the Dinantian magmatism : indeed, recent studies showed that the rocks, principally andesitic, of Eastern Meseta, represent a calk-alkaline series and the trace and rare earth elements show their similarities with modern volcanic series related to subduction processes. At the contrary, the spilites, keratophyres and dolerites of Dinantian age in Northwestern Meseta are alkaline to transitionnal types similar of those found in present continental rifts or backarc basins. So, this tectonic polarity : Eastern internal zone and Western external zone, could represent a geotectonic polarity during middle to upper Paleozoic times : Oceanic crust was plunging westward under the active margin of the Eastern Meseta, whereas a marginal basin developed in Western Meseta. Later, this subduction stage was followed by the collision between the margin (Eastern Meseta) and a more eastern continental block. This collisional event resulted in the main deformation of first the Eastern Meseta and then the Western Meseta. The continental suture, presently hidden under the recents deposits of Algeria, could be followed farther North.

Yet, this model does not account for the totality of the facts observed. Firstly, the andesitic volcanic episod of Upper Visean age did not affect the entire Eastern Meseta, but only its northern part. Consequently, the subduction which seems to be indicated by this volcanism was active only in that part of the eastern Meseta. Secondly, a difficulty arises if one considers the age of this volcanism with respect to the deformation : inside Northeastern Meseta, the volcaniclastic upper Visean series unconformably overlie folded and metamorphosed series. In other words, the volcanism, which is associated to the subduction, postdates the folding and consequently the subduction was probably inactive and the collision achieved when the magma reached the surface. Thirdly, the vergence of the pre-Upper Visean folds is not the same everywhere : in the northeastern part of the Meseta, the folds are overturned and recumbent to the NW. In the southeastern part, at the contrary, the vergence is toward the East. Doubtless, this feature is not totally explained by the collision mechanism.

STRUCTURE OF THE MAURITANIDES

LECORCHE, Jean-Paul

LA CNRS n°132 "Etudes geologiques ouest-africaines" Faculte des Sciences St-Jerome 13397 MARSEILLE CEDEX 13 FRANCE

The polyphase Mauritanide belt has been gradually developed from south to north during three major orogenic events, Panafrican, Taconian and Early Hercynian. The present aspect of the belt is due to superposition of Hercynian elements (internal nappes) on both Taconian and Panafrican elements (external nappes). These nappes cover the northern two-thirds of the belt and overthrust the West African shield and its Upper Proterozoic to Palaeozoic cover to the east. Taconian and Panafrican elements in parautochthonous, locally allochthonous position occupy the southern third of the belt. A discussion of the arguments for the present interpretation emphasizes the problems being solved.

The main Mauritanides orogenic belt (1) extends N-S for about 2000 km from Southern Morocco to Liberia. It looks like a narrow folded and metamorphosed ribbon, stretching between the Upper Proterozoic to Palaeozoic Tindouf and Taoudeni basins in the east, and the Mesozoic-Cenozoic coastal plains of Morocco, Mauritania and Senegal in the west. Both Tindouf basin and the northernmost quarter of the belt are isolated from Taoudeni basin and from the rest of the chain by a NE-SW uplift of old basement known as Dorsale Reguibat or Reguibat Uplift. The latter is itself overlain to the west by sediments of coastal plain. South of Senegal River, a part of the belt runs southwestward but is rapidly hidden by Senegal basin. The other part runs southward across Eastern Senegal and Guinea where it is partly overlain by Late Ordovician to Upper Devonian sediments of Bove Basin, Sierra Leone and Liberia.

P. E. Schenk (ed.),
Regional Trends in the Geology of the Appalachian-Caledonian-Hercynian-Mauritanide Orogen, 347–353.

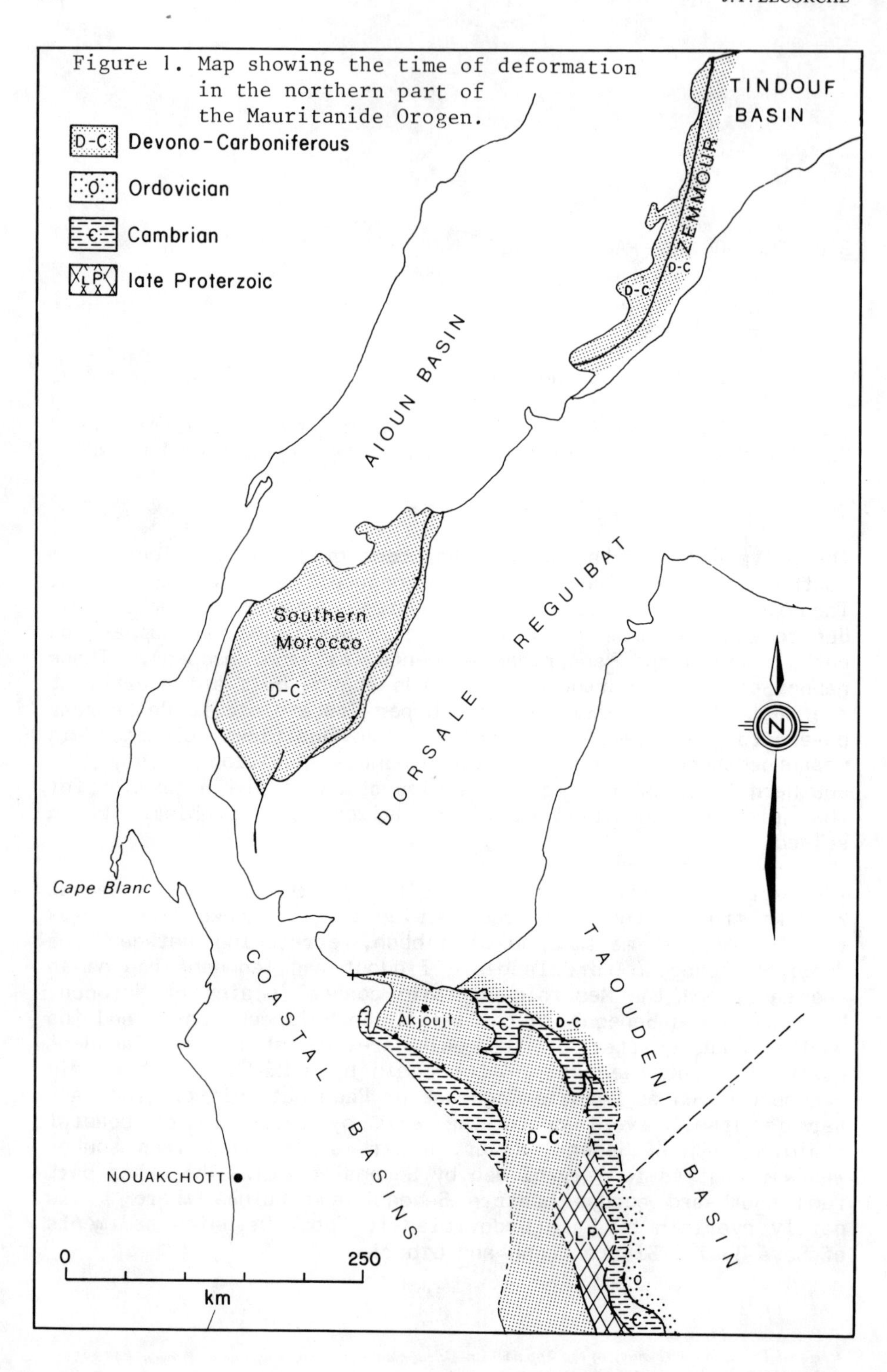

Figure 1. Map showing the time of deformation in the northern part of the Mauritanide Orogen.

THE FORELAND

The West African craton and the Taoudeni and Tindouf basins together form the foreland of the belt. The shield outcrops in the Dorsale Reguibat in the north, southwards in the Kayes and Kenieba inliers, and in the Dorsale de Leo in the south. The ages of the basement range from 2700 Ma to 1600 Ma and those of the cover from 1000 Ma to 350 Ma (Frasnian). In the case of the cover, where lithostratigraphy is well known (2), chronostratigraphy is roughly defined between Upper Proterozoic geochronological data (3) and fossiliferous Silurian. The Cambrian base is still undefined and the Cambrian-Ordovician boundary is only presumed (4).

As we possess few geochronological data in the belt, the timing of events is hypothetical.

THE BELT

The structure of the belt is characterized by an evolution of the style and of the timing of deformations from the south to the north.

In the Rokelides of Sierra Leone (5), the belt is composed of granitic, basic and ultrabasic rocks, strongly metamorphosed, folded and thrust over the folded and locally para-autochthonous Rockel River Group. This slightly metamorphosed clastic group lies unconformably on the craton (Dorsale de Leo) and corresponds to the Upper Proterozoic cover. A late Panafrican K/Ar age of 550 Ma would indicate the end of the orogeny. But rocks of the western overthrust formations are refolded and could be of older possibly Eburnean (2000 Ma) origin.

In Guinea and Southeastern Senegal, Villeneuve (6) distinguishes three main events. The oldest, assumed to be Panafrican, affects Upper Proterozoic sedimentary cover and prefolded older rocks which are refolded in turn and lightly thrust with their presumed Cambrian-Ordovician cover before the sedimentation of Silurian-Devonian Bove Basin. A post-Devonian event moderately affects the north of this area.

In Northeastern Senegal, the region of the Senegal River (7) is a key zone. The former terranes, of which we see here the extension, are largely overthrust by nappes of various (volcanic, volcani-clastic and clastic) materials. Old geochronological data (8) indicate Taconian and Early Hercynian events. Nevertheless it is probable that an earlier late Panafrican event has affected the lowest parautochthonous polydeformed

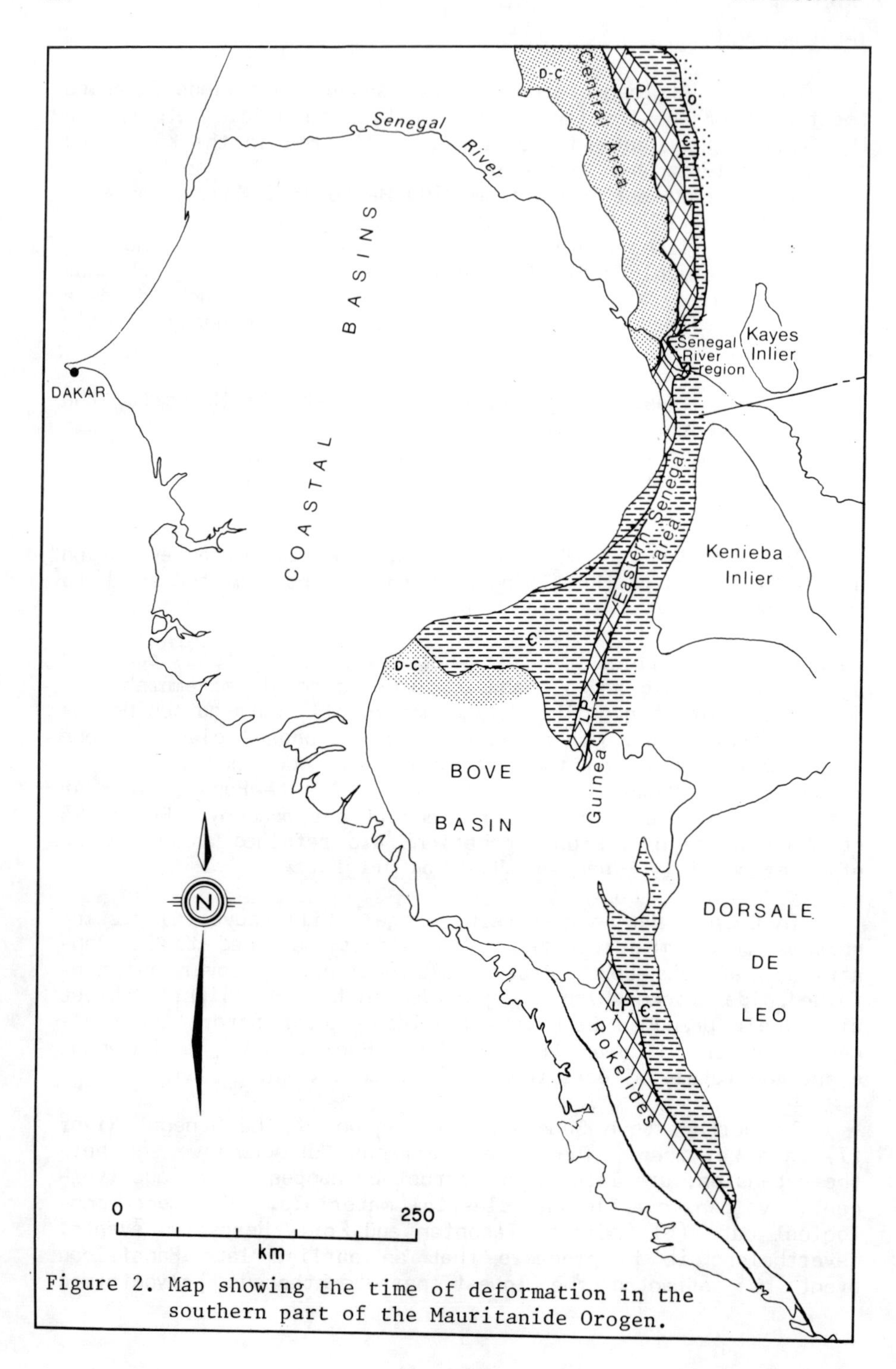

Figure 2. Map showing the time of deformation in the southern part of the Mauritanide Orogen.

formations before Taconic orogeny. The allochthonous formations, set during or immediately after Late Devonian times, would represent metamorphosed and more strongly folded equivalents of the lowest formations, deposited in deeper conditions with volcanic and ultra-basic rocks (9).

In the central area of the belt, Dia (10) has shown that the outcropping belt results from thrusting of several clastic, volcanic, ultra-basic units with different types and times of deformations, on a late Proterozoic to Palaeozoic foreland. Dia (11) distinguishes old, partly ophiolitic, polydeformed series assumed to be of Panafrican origin. These series are refolded with metamorphosed mixtites and flyshoid sequences in which two phases occur. The first phase would be Cambrian (Late Panafrican), the second Middle Ordovician. A third group of formations overthrusts the former units and corresponds to late movements of post-Devonian age. Locally folded, clastic rocks, dated Devonian, have been observed in these formations.

Northwards, in the Akjoujt region, Lecorche (12) describes i) external (or lower) nappes, composed of volcanics, mixtites (tilloids) and associated formations characterized by lower greenschist metamorphism and affected by two phases of folding, the first one undated, the second related to Taconian event; ii) internal (or upper) nappes of clastic to mainly volcanic material and basement-like terranes, independently metamorphosed and folded before being thrust over on the lower formations and finally, together with them, pushed on the craton itself (Dorsale Reguibat) or on its Palaeozoic cover. This last event is related to post-Devonian orogeny.

On the northern flank of the Dorsale Reguibat, in Southern Morocco, the external nappes are missing and the internal nappes, with ultra-basic soles, lie directly on the craton or on a thin Upper Ordovician to Devonian cover (13). For Marchand and Bronner (14), these nappes, mainly studied on aerial photographs, form a very large klippe the northern part of which is hidden below the Cretaceous sediments of Aioun Basin. The emplacement of these nappes is also post-Devonian.

Outcrops of the belt do not exist farther in the north, but the entire western margin of the Tindouf Basin or Zemmour (15) is progressively folded westward. Strongest deformation is reached in the Dhlou belt where limited thrusting is dominant and affects Devonian formations.

DISCUSSION

The post-Devonian overthrusting appears to be everywhere from Southern Morocco as far as Senegal River area. It is related, in the foreland to folding and limited thrusting of Devonian (up to Frasnian) rocks. The lack of younger formations do not permit an upper limit for this event to be determined. It is generally inferred to be early Hercynian, but without evidence.

The terranes implicated in the belt are of various origins. The internal nappes of Southern Morocco, Akjoujt region and central area of the belt are assumed to be Hercynian. In fact, the age of material and the age(s) of deformations are unknown in most cases. Some evidence for such an interpretation only exists in the central area of the belt where folded clastic Devonian rocks are known in the allochthonous formation.

The external nappes are known from Akjoujt region to the Senegal River area. Composed of only clastic to volcaniclastic formations in the north, they become gradually rich in prefolded basic volcanics and ultrabasic rocks towards the south. The metasedimentary units are generally similar mixtites and flyschoid green series, polydeformed but not intensively deformed, and metamorphosed in the greenschist facies. Timing of deformation is deduced from the angular unconformities observed in the foreland. Consequently the two main proposed periods of orogeny, Late Panafrican and Taconian are not directly related with the belt. In the oldest basic and ultrabasic rocks, timing of the first deformations is inferred to be early Panafrican but could be Eburnean (2000 Ma).

Southwards, in Eastern Senegal, Guinea and Sierra Leone, Hercynian tectonism is very lightly imprinted or missing. Taconian ages have been measured (8) in Eastern Senegal and might be confirmed in the other regions. Panafrican orogeny seems to be more developed than in the north. However confusion between early Panafrican and Eburnean is possible.

It would therefore be of major interest to intensify the geochronological program which is just starting. Nevertheless the main obstacle remains: the absence of plutonism in the entire outcropping belt.

CONCLUSION

The Mauritanide belt is still far from being known as well as its Caledonian-Appalachian sisters. But now it must be integrated in the models of evolution of the Palaeozoic belts.

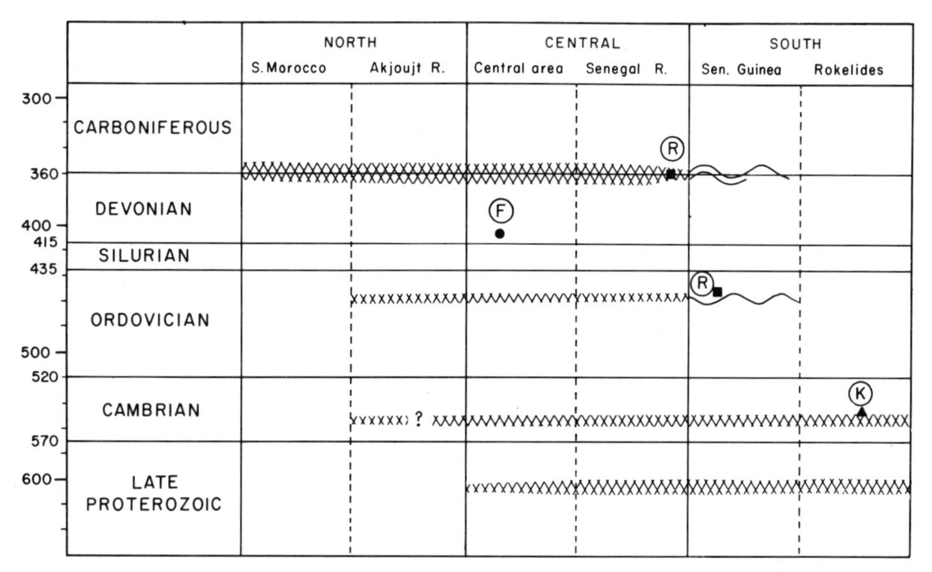

Table I : Times of deformation in the Mauritanides.

LEGEND

~ unconformity

≈ mild – moderate deformation

XXX intense deformation

● Ⓕ fossil

■ Ⓡ Rb/Sr whole rock isochron age

▲ Ⓚ K/Ar age

REFERENCES

(1) Sougy, J.: 1962, Geol. Soc. Amer. Bull., 73, pp. 871-876.
(2) Trompette, R.: 1973, Trav. Lab. Sci. Terre St-Jerome, Marseille, B7, 702 p.
(3) Clauer, N.: 1976, These Doctor. Etat, Strasbourg, 227 p.
(4) Legrand, P.: 1969, Bull. Soc. geol. Fr., 11, pp.251-256.
(5) Allen, P.M.: 1969, Geol. Rundschau, 58, pp. 588-620.
(6) Villeneuve, M.: 1980, C. R. somm. Soc. geol. Fr., 2, pp. 54-57.
(7) Le Page, A.: 1978, C. R. Acad. Sci., Paris, 286, D, pp. 1853-1856.
(8) Bassot, J. P., Bonhomme, M., Roques, M. and Vachette, M.: 1963, Bull. Soc. geol. Fr., 5, pp. 401-405.
(9) LePage, A.: 1982, person. commun.
(10) Dia, O., Lecorche, J. P. and Le Page, A.: 1979, Rev. Geol. dyn. Geogr. phys., Paris, 21, pp. 403-409.
(11) Dia, O.: 1982, person. commun.
(12) Lecorche, J. P.: 1980, These Doctor. Etat, Marseille, 445 p.
(13) Sougy, J.: 1962, Bull. Soc. geol. Fr., 4, pp. 436-445.
(14) Marchand, J. and Bronner, G.: 1982, person. commun.
(15) Sougy, J.: 1964, Ann. Fac. Sci. Univ. Dakar, 12, pp. 1-695.

COTICULES - A KEY TO CORRELATION ALONG THE APPALACHIAN- CALEDONIAN OROGEN?

P.S.Kennan, M.J.Kennedy

Department of Geology, University College, Belfield, Dublin 4.

ABSTRACT.

Marginal belts of coticules (spessartine quartzites) and tourmalinites can be recognised in the Caledonian-Appalachian orogen. These distinctive rocks occur both in clastic sequences and in pelagic sediments deposited on oceanic crust. The host sequences are almost invariably unfossiliferous and, hence, poorly dated. The occurrence of coticule-tourmalinite is so wide-spread and often so continuous within mappable units along strike that long range correlation of the sequences containing these special lithologies may be possible.

INTRODUCTION.

Coticules are quartzites rich in spessartine garnet first defined as such by Renard (1). Since Emerson (2) first noted coticules in the Hawley Formation in Massachusetts, these distinctive rocks have been reported from various parts of the Caledonian-Appalachian orogen. However, it is only in New England that any overview of their significance in space and time within the orogen has emerged (3).

Coticules are typically found as thin beds in black, often sulphide-bearing, pelitic host rocks. Less often, the host rocks are coarser clastics. Garnet-bearing amphibolites and carbonates are

P. E. Schenk (ed.),
Regional Trends in the Geology of the Appalachian-Caledonian-Hercynian-Mauritanide Orogen, 355–361.

common associates. Thin-sections usually show multitudes of small equidimensional garnets embedded in a quartz matrix. Of the various accessory minerals present, tourmaline may be considered almost essential (1). Coticules are often closely associated with quartz-tourmaline rocks (tourmalinites); the two form a single sedimentary association.

The origin of coticules is problematical. Their composition and lack of any clastic textures has suggested that, prior to metamorphism, they were manganiferous sedimentary ironstones (4), manganese-rich sands (5), chert (6) and siliceous sediments rich in montmorillonite clay (7). Significantly, identical spessartine quartzites are recognised as genetically associated with iron-stones, quartz-tourmaline rocks and stratabound sulphide deposits of volcanoganic origin - notably at Broken Hill, New South Wales (8, 9). Significantly also, coticules are identical to the manganiferous cherts that cap many ophiolites, e.g. the Caledonian ophiolite at Karmoy, Norway (10). Merguerian (11) considers the New England coticules to belong to the sedimentary cover of the pre-Taconic ocean crust.

Coticules compare with oceanic or ophiolitic cherts on the one hand and with certain types of iron formation on the other. A common feature of both sediment types is an extended lateral continuity. In the Caledonian-Appalachian orogen, they - and the related tourmalinites - occur in two continuous marginal belts (Figure 1).

DISTRIBUTION.

In New England, coticules occur widely within the Bronson Hill Anticlinorium and on the eastern side of the Green Mountain Anticlinorium; they belong to the Middle Ordovician Partridge Formation and its lateral equivalents (3). Northeastwards along the regional strike, coticules are a part of the pre-Middle Ordovician Dead River Formation in western Maine (12). A close association with tourmaline is recognised (3, 5). To the south-east, the second marginal belt is marked by occurrences in the Marlboro and Nashoba formations of eastern Massachusetts (13). The age of these formations is very uncertain. They may be Middle Ordovician (14) and they may relate to the tourmalinites spatially associated with base-metal deposits at Penobscot Bay in southern Maine (15).

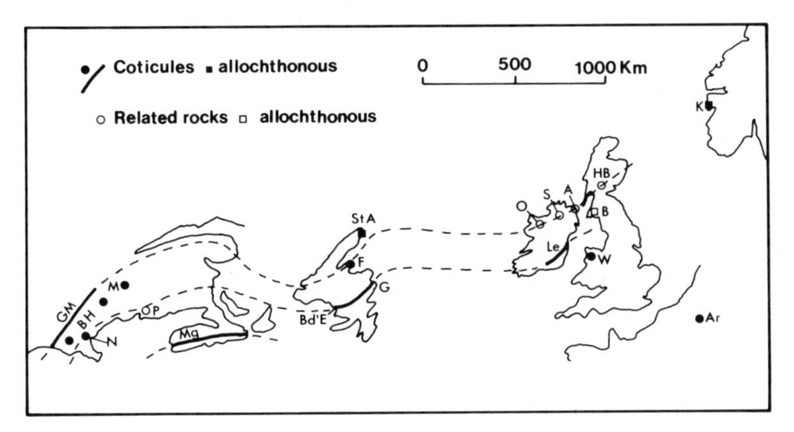

Figure 1.

Marginal belts of coticules, tourmalinites and related rocks in the northern Appalachians and southwestern Caledonides. Continental fit is based on Scrutton (30). Locations mentioned are as follows: GM: Green Mountain Anticlinorium; BH: Bronson Hill Anticlinorium; M: western Maine; N: Nashoba-Marlboro Formations; P: Penobscot Bay; Mg: Meguma Group; St.A: St. Anthony Complex; F: Fleur de Lys Supergroup; G: Gander Group; B.d'E: Baie d'Espoir Group; O: Ox Mountains; S: Sperrin Mountains; B: Ballantrae; Le: Leinster; W: Wales; K: Karmoy; Ar: Ardennes.

In Newfoundland, the northeast marginal belt is identified by the occurrence of coticules in the Birchy Schist of the Fleur de Lys Supergroup (16). Similar garnetiferous metacherts occur, not very far away, within the dynamothermal aureole of the White Hills Peridotite (17, 18). Within the southeastern belt, coticules occur in metasediments belonging to the Gander Group in the vicinity of Aspen Cove. These sediments are pre-Llanvirnian and probably Aranigian and older (19). Coticules also occur in the correlative early Middle Ordovician or older Riches Island Formation of the Baie d'Espoir Group of south-central Newfoundland (20).

The Gander Zone occurrences correlate well with similar rocks - the Garnetiferous Beds of Leinster (4) - in southeast Ireland. In both places the coticules are, as is typical, thinly-bedded, display complex plastic styles of folding and are enclosed in aluminous, sometimes graphitic and often limonite-stained schists. In southeast Ireland, coticules and associated tourmalinites define a narrow unbroken belt that extends for some 130 km along strike; the host successions are of Cambro-Ordovician age (21). Though true coticules have not been recognised in Fleur de Lys equivalents (Dalradian Supergroup) in Ireland, rocks showing exceptional enrichments in tourmaline mark the trace of the northwestern belt through the Ox Mountains in Mayo (22), the Sperrin Mountains in Tyrone (23) and in Antrim (24). A possible though problematic continuation northeastwards may be found in the manganese cherts of the partly ophiolitic Highland Border Complex of Arenig age in Scotland (25, 26). The cherts in the nearby Ballantrae ophiolite, which are of the same age (27), may also be related.

The coticules of the type locality in the Ardennes are Lower Ordovician (7, 28). They and those of the Meguma Group of Nova Scotia (6) and of the Manganese Group of the Harlech Dome in Wales (29) lie outside the main marginal belts. The Welsh examples are unique in their Middle Cambrian age.

DISCUSSION.

Coticules may prove useful in long range stratigraphic correlation within the Caledonian-Appalachian orogen. The chert-ironstone nature of the coticule protolith suggests so. So also does the fact that coticules are demonstrably confined to individual formations the continuity of which can be demonstrated by mapping over considerable distances along strike, e.g. 350 km in New England, 130 km in Ireland, 240 km in eastern Newfoundland. Coticules are a distinctive lithology which appear in many places, whether in an oceanic or continental margin setting, to mark what was essentially a single depositional interval. Even though individual coticule-bearing sequences are very difficult to date accurately, the overall pattern of the age ranges (Figure 2) suggests that the significant presence of coticules indicates a Lower Ordovician age widely in the orogen.

Coticule-bearing sequences on the northwestern side of the lapetus may be slightly older than on the opposite margin.

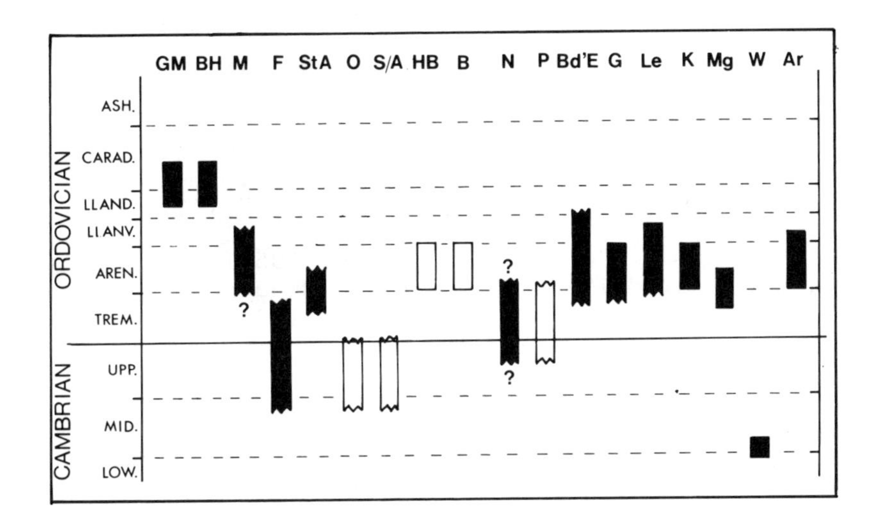

Figure 2.

Ages of coticules and related rocks shown on Figure 1. Ages are based on references quoted in the text. Key to localities is as in Figure 1.

It is on the southeastern margin of lapetus that the continuity that coticules provide may prove especially useful. Any continuity on this margin is difficult if not impossible to recognise upon stratigraphy alone. Here, coticules, tourmalinites and related lithologies are invaluable. They may be equally valuable wherever similar unfossiliferous successions occur, e.g. in pre-Cambrian terrains.

REFERENCES.

1. Renard,A. : 1879, Mem.Cour.Mem.Sav.Etr.Acad.Roy. Belg.XLI.
2. Emerson,B.K. : 1898, U.S.Geol.Surv.Monograph 29, 790p.
3. Kim,S.W. : 1975, Jour.geol.Soc. Korea 11, pp. 36-68.
4. Brindley,J.C. : 1954, Sci.Proc.R.Dubl.Soc. 26, pp. 245-262.
5. Clifford,T.N. : 1960, Neues Jahrh.Mineral. Abhandlungen 94, pp.1369-1400.
6. Schiller,E.A., Taylor,F.C. : 1965, Am.Mineral. 50, pp. 1477-1481.
7. Kramm,U. : 1976, Contrib.Mineral.Petrol. 56, pp. 135-155.
8. Stanton,R.L. : 1976, Trans.Instn Ming Metall. B.85, pp.132-141.
9. Plimer,I.R. : 1980, Mineralium Deposita 15, pp. 275-289.
10. Sturt,B.A., Thon,A., Furnes,H. : 1979, Geology 7, pp. 316-320.
11. Merguerian,C. : 1980, Geol.Soc.Amer. Abstracts with Programes 12, p. 73.
12. Boone,G.McG. : 1973, Maine Geol.Surv.Bull. 24, 136 p.
13. Skehan,J.W., Abu-Mostafa,A.A. : 1976, Geol.Soc. Amer.Mem. 148, pp. 217-240.
14. Hall,L.M., Robinson,P. : 1982, "in" St. Julien,P., Beland,J., Eds. Geol.Assoc.Canada Special Paper 24, pp. 15-41.
15. Slack,J.F. : 1980, Maine Geol.Surv.Special Economic Studies Series No. 8, 25p.
16. Kennedy,M.J. : 1971, Geol.Assoc.Canada Proc. 24, pp. 59-71.
17. Talkington,R.W., Jamieson,R.A. 1979 "in" Malpas, J.,Talkington, R.W. Eds. Memorial University of Newfoundland Department of Geology Report No. 8. pp. 43-52.
18. Jamieson,R.A. : 1981, J.Petrol. 22, pp. 397-449.
19. Kennedy,M.J.: Bazinet,J.P. Can.J.Earth Sci.(in prep.)
20. Colman-Sadd,S.P. : 1980, Am.J.Sci. 280, pp. 991-1017.
21. Bruck,P.M., Potter,T.L., Downie,C. : 1974, Proc. R.Irish Acad. 74B, pp. 75-84.
22. Taylor,W.E.G. : 1968, Proc.R. Irish Acad. 67B, pp. 63-82.
23. Hartley,J.J. : 1938, Proc.R. Irish Acad. 44B, pp. 141-171.

24. Bailey,E.B., McCallien,W.J. : 1934, Trans.R.Soc. Edinburgh LVIII, pp. 163-177.
25. Henderson,W.G., Robertson,H.F. : 1982, J.Geol. Soc. London 139, pp. 433-450.
26. Curry,G.B., Ingham,J.K., Williams, A. : 1982, J. Geol.Soc. London 139, pp. 453-456.
27. Bluck,B.J., Halliday,A.N., Aftalion,M., MacIntyre,R.M. : 1980, Geol 8, pp. 492-495.
28. Michot,P. : 1980, Geologie des pays europeens : France, Belgique, Luxembourg. Dunod. pp. 486-594.
29. Binstock,J.L.H. : 1977, Unpublished Ph.D. thesis, Harvard University, 325 p.

METALLIC MINERAL ZONATION RELATED TO TECTONIC EVOLUTION OF THE NEW BRUNSWICK APPALACHIANS

A. A. Ruitenberg and L. R. Fyffe

Department of Natural Resources, Geological Surveys Branch, respectively Sussex and Fredericton, N.B. Canada

Abstract

A great variety of stratiform and granitoid related mineral deposits occur in the New Brunswick Appalachians. The distribution of these deposits is discussed in relationship to the tectonic evolution of the province. Special emphasis is placed on the disrupted margins of and the deformed zones crossing the Avalon Microcontinent. A brief comparison is made with deposits that are found in a similar tectonic setting elsewhere in the Appalachian-Caledonian Orogen.

Introduction

The distribution of metallic mineral deposits in New Brunswick shows a distinct relationship to major tectonic elements and/or structural zones (Fig. 1). Low grade base metal sulphide deposits occur in an Upper Precambrian volcanic and sedimentary sequence along the southeastern margin of the Caledonia Zone (Avalon Zone) (1). Gold-bearing quartz veins also occur in this zone.

The large Bathurst base metal sulphide deposits occur in an Ordovician felsic volcanic sequence in the northern Miramichi Zone (2). Numerous small base-metal sulphide deposits occur in Silurian mafic and felsic volcanic and interbedded sedimentary rocks, along the southeastern margin of the Magaguadavic Zone (Fredericton Trough) (3, 12).

P. E. Schenk (ed.),
Regional Trends in the Geology of the Appalachian-Caledonian-Hercynian-Mauritanide Orogen, 363–373.

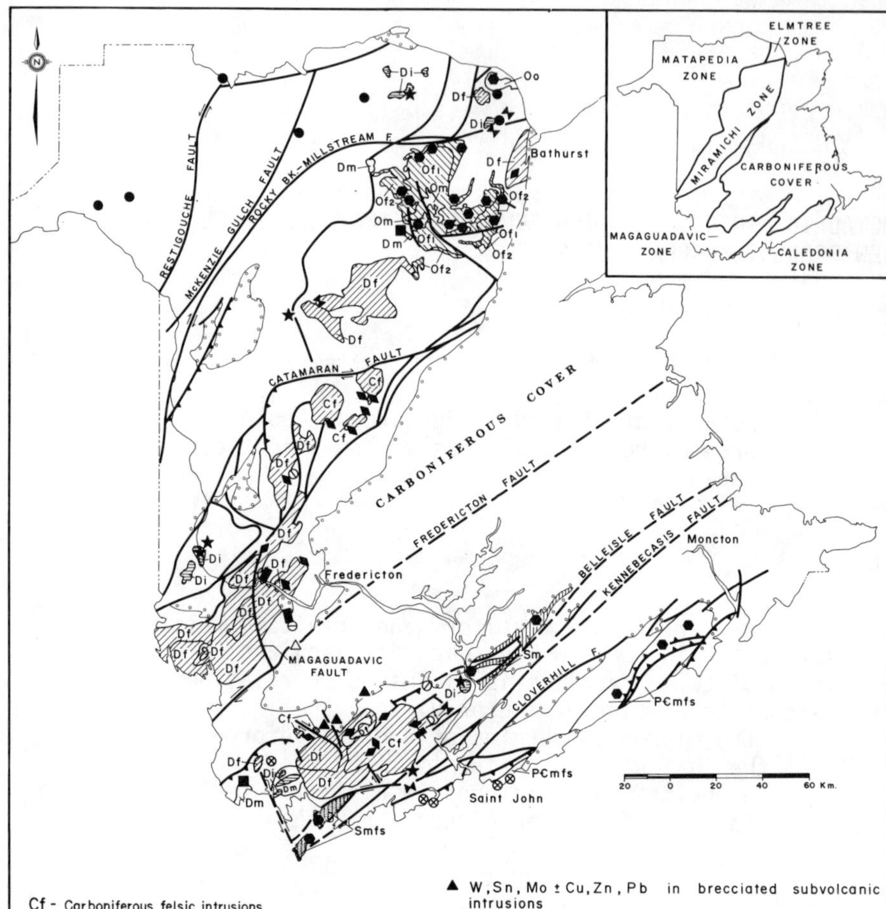

Cf - Carboniferous felsic intrusions

Df - Devonian felsic intrusions

Di - Devonian intermediate intrusions

Dm - Devonian mafic intrusions

Smfs - Silurian mafic volcanic, felsic volcanic and sedimentary rocks

Om - Ordovician mafic volcanic rocks

Of2 - Ordovician porphyritic metarhyolithic tuff

Of1 - Ordovician metarhyolitic rocks

Oo - Ordovician ophiolitic complex

PЄmfs - Upper Precambrian mafic volcanic, felsic volcanic and sedimentary rocks

▲ W, Sn, Mo ± Cu, Zn, Pb in brecciated subvolcanic intrusions

♦ W, Sn, Mo, F, Be ± Cu, Pb, Zn in quartz and greisen veins

⊗ Au ± Pb, Zn, Cu in quartz veins

⊖ Sb ± Cu, Pb, Zn, U in quartz veins

▮ W ± Mo in quartz-carbonate veins and skarns

╋ Sn, Cu, Zn, Pb in skarns

★ Fe, Cu ± Mo, Pb, Zn in porphyry stocks

⧗ Zn, Pb, Cu, Ag in carbonate-quartz veins

● Fe, Cu ± Mo, Pb, Zn in contact aureoles

■ Massive and disseminated Fe, Ni, Cu in mafic intrusions

● Volcanogenic stratiform Pb, Zn, Cu deposits

FIG. 1. Map showing distribution of metallic mineral deposits in New Brunswick

Contact metasomatic and vein type base metal sulphide and porphyry copper deposits are associated with Devonian I-type igneous complexes in the Matapedia Zone and the northwest margin of the Miramichi Zone. In contrast, tin-tungsten molybdenum deposits are associated with S-type granitoid complexes in the southwestern Magaguadavic and central Miramichi Zones.

A description of the major tectonostratigraphic zones in New Brunswick is given by Ruitenberg *et al* (2). The structural evolution of the province is described by Fyffe (4) and Ruitenberg and McCutcheon (5). The Precambrian rocks are summarized by Ruitenberg *et al* (6) and Giles and Ruitenberg (7). Only the geology of the major mineral belts is briefly described here. The reader is referred to the bibliography for more detailed descriptions.

Stratiform Base Metal Sulphide Deposits

Upper Precambrian. Upper Precambrian stratiform base metal sulphide deposits occur in the eastern part of the Caledonia Zone, which is composed mainly of basaltic and rhyolitic tuffs and intercalated marine sedimentary rocks of the Coldbrook Group. No deposits are known in the terrestrial basaltic and rhyolitic flows and tuffs that occupy most of the central and western parts of this zone (6).

The Coldbrook volcanic sequence formed in an intracratonic rift zone (7) that preceded the Iapetus Oceanic Cycle.

The Upper Precambrian deposits and their host rocks are intensely deformed; they are all low grade, but some are quite extensive. The main metallic minerals are sphalerite, chalcopyrite, galena, minor tennantite and brecciated pyrite in a carbonate-quartz-talc matrix. The sulphide deposits are enveloped by a chloritic zone which is in turn surrounded by a zone of intense silicification and micaceous alteration (6).

Ordovician. Large Ordovician stratiform base metal sulphide deposits occur in metasedimentary rocks that are intercalated with a thick sequence of metamorphosed rhyolitic rocks and porphyritic felsic tuff (crystal tuff) (8) of the Tetagouche ensialic arc sequence in the northern Miramichi Zone (Fig. 1) (2). No deposits are known in the overlying basaltic pillow lavas and tuffs. The Tetagouche felsic volcanic sequence conformably overlies quartzite and pelitic rocks which are considered to represent a continental rise prism deposited largely on Avalonian basement (2, 4). The Tetagouche Arc is believed to have formed during closing of the Iapetus Ocean by subduction to the south (beneath the Avalon Zone).

The sulphide deposits in the northern Miramichi Zone are massive and disseminated pyrite, sphalerite, galena and chalcopyrite. Iron formation, in the form of oxide, chlorite and carbonate facies, overlies the base metal sulphide ores. The deposits and their host rocks were intensely polydeformed during the Ordovician Taconian orogeny (8, 9).

A small massive zinc-copper sulphide deposit occurs in silicified mafic volcanic rocks associated with the Ordovician ophiolitic complex in the Elmtree Zone, immediately north of the Miramichi Zone. The ore is believed to occur in pipe-shaped bodies (8).

Silurian. Several small stratiform base metal sulphide deposits occur in a Silurian marine volcanic belt (Mascarene-Nerepis Belt) along the southern margin of the Magaguadavic Zone (Fredericton Trough) and in its southwestern extension in the State of Maine. This bimodal volcanic sequence probably formed along a short lived rift zone (10). This rifting environment could be comparable to the one that resulted in the bimodal submarine volcanic sequence that formed in the Japanese island arc about 9 million years ago (11).

Two types of deposits can be distinguished in this Silurian volcanic belt. In the northeastern part of the belt, deposits composed of massive and disseminated pyrite, chalcopyrite and minor amounts of silver minerals occur in mafic tuff intercalated with marine slate and sandstone (9, 12). In the southwestern part of the belt, small deposits composed of pyrite, sphalerite, galena and chalcopyrite occur in rhyolite tuff and intercalated pelitic sedimentary rocks. Associated volcanic breccias indicate proximity to volcanic centres. In the State of Maine, the quite extensive Black Hawk deposit occurs in sedimentary rocks that are probably distal equivalents of the Silurian volcanics.

Mineral Deposits Associated With Granitoid And Gabbroid Intrusions And Their Subvolcanic Equivalents

A comprehensive synthesis of these deposits was given by Ruitenberg and Fyffe (13). Emphasis in this discussion is placed upon distribution of the deposits as related to tectonic evolution.

Deposits in Gabbroid Intrusions. Deposits composed of massive and disseminated nickeliferous pyrrhotite, pentlandite and chalcopyrite occur in Devonian norite and gabbro stocks in the southwestern Magaguadavic Zone and in a prominent gabbro dyke in the Miramichi Zone (Fig. 1). The nickeliferous intrusions in the Magaguadavic Zone were emplaced in a dilatant zone immediately west of a prominent sinistral fault that cuts across Ordovician pelitic metasedimentary rocks. These sedimentary rocks were penetratively deformed during the Early-Middle Devonian Acadian orogeny (14).

Deposits Associated with I-Type Granitoid Intrusions. Numerous lead, zinc and copper sulphide deposits are associated with high level, intermediate to felsic I-type intrusions in the Matapedia Zone and along the northwest margin of the Miramichi Zone. A few have been formed along the southern margin of the Magaguadavic Zone. Petrological and petrochemical studies (13) have shown that these I-type intrusions have a highly variable SiO_2 content and contain large amounts of intermediate rocks with hornblende rather than mica as the characteristic mineral. K_2O/Na_2O and Rb/Sr are also relatively low.

The Matapedia Zone has been interpreted to be underlain by a remnant of Iapetus Oceanic crust that was not destroyed during the Taconian orogeny (2, 15). The oldest rocks exposed in the zone are Middle Ordovician clastic turbidites containing abundant volcanic detritus, which are conformably overlain by Upper Ordovician to Lower Silurian limestone and calcareous slate of the Matapedia Group (16). The relative abundance of the mantle derived (I-type) intrusions in the Matapedia Zone can be related to the relatively thin continental crust in this area. In areas of thicker continental crust, such as along the southeastern margin of the Magaguadavic Zone, intrusion of mantle derived magmas was confined to major fault zones. The intrusions were emplaced immediately after the Acadian orogeny.

Three types of deposits associated with I-type intrusions can be distinguished. These include: contact metasomatic base metal sulphide, vein-type base-metal sulphide and porphyry-copper deposits.

A large number of contact metasomatic deposits are associated with small Devonian stocks of mainly intermediate composition that cut the calcareous rocks of the Matapedia Group. The largest known of these deposits are along prominent east- or northeast-striking wrench faults. In addition to pyrite, pyrrhotite, sphalerite, galena, and chalcopyrite, small amounts of scheelite and molybdenite are locally associated with these deposits.

Carbonate-quartz veins (Fig. 1) containing pyrite, pyrrhotite, galena, sphalerite and chalcopyrite are commonly associated with the contact metasomatic deposits and are therefore probably genetically related. The largest known of these deposits are associated with intermediate to felsic stocks that were intruded along the Rocky Brook-Millstream dextral wrench fault (Fig. 1).

Porphyry copper deposits occur mainly in the Matapedia and northwestern Miramichi Zones, and in a few localities along the southeastern margin of the Magaguadavic Zone. The largest known of these deposits occur along the northwestern margin of the Miramichi Zone. Pyrite, pyrrhotite, chalcopyrite and minor molybdenite occur as disseminations and veinlets in intensely fractured granodiorite porphyry. Argillic and phyllic alteration are associated with the mineralized zones.

Deposits Associated With S-Type Intrusions. Numerous tin-tungsten-molybdenum and antimony deposits with or without base metal sulphide minerals are associated with late phases of granitic stocks and related subvolcanic intrusions, ranging in age from Late Devonian to Early Carboniferous. These occur in a broad north trending arc-shaped zone traversing the southwestern Magaguadavic and central Miramichi Zones (Fig. 1). These intrusions are late phases of S-type igneous complexes that were emplaced along prominent dilatant fault zones within thick continental crust. Petrological and petrochemical investigations (13) have demonstrated that the S-type intrusions in the province, in contrast with the I-type, have high levels of SiO_2 that range within narrow limits, and K_2O/Na_2O and Rb/Sr are also relatively high.

The intrusions associated with the lithophile mineral deposits were emplaced at relatively shallow depth and they cut across Ordovician pelitic metasediments. Regional geochemical surveys indicated that these Ordovician metasediments generally have a relatively high background in various trace elements including tin, tungsten and molybdenum.

Greisen and micaceous quartz veins containing varying amounts of cassiterite, stannite, wolframite, molybdenite, arsenopyrite and base metal sulphide minerals occur within and along the contacts of highly fractionated, late phases of S-type intrusions. Fluorite and topaz are the most characteristic nonmetallic minerals in these veins. Tin-tungsten-and molybdenum-bearing greisens associated with Carboniferous subvolcanic complexes occupy broad zones within and along the margins of the youngest intrusive phases. In addition to the greisen deposits, low grade scheelite-bearing granitic intrusions and associated skarns were found in the northwestern part of the Magaguadavic Zone (Fig. 1).

Extensive stibnite and native antimony-bearing quartz veins have been mined in the northwestern part of the Magaguadavic Zone. Smaller occurrences have been found along a prominent fault in the southern Magaguadavic Zone and in the central Miramichi Zone (13).

Gold Deposits. Economically interesting amounts of gold have been found recently in quartz veins that cut intensely deformed sedimentary rocks of probable Eocambrian age in the coastal area of southern New Brunswick. The sedimentary rocks are intruded by granitic stocks of possible Devonian or younger age. The gold-bearing veins are associated with major thrust and wrench faults (17). Other gold-bearing quartz veins, in southern New Brunswick, cut across Ordovician pelitic metasedimentary rocks that have been subjected to intense cross folding during the Acadian orogeny (12, 14).

Summary And Conclusions

The distribution of metallic mineral deposits in New Brunswick shows a distinct relationship to the margins of the Avalon Microcontinent.

Precambrian stratiform base metal sulphide deposits occur in mafic and felsic tuffs and intercalated marine sedimentary rocks of the Coldbrook Group along the southeastern margin of the Avalon Zone. The Coldbrook volcanic rocks formed in an intracratonic rift zone that preceded the Iapetus Oceanic Cycle. Similar base metal sulphide deposits occur in other fragments of this volcanic belt in southern Cape Breton, Nova Scotia (Forchu Group) (18), and the "Carolina Slate Belt" (19). It is notable that gold-bearing quartz veins are also common in all these areas.

The Ordovician stratiform base metal sulphide deposits are confined to metasedimentary rocks intercalated with metamorphosed rhyolitic rocks and porphyritic felsic tuff (crystal tuff) of the Tetagouche Group in the northern Miramichi Zone. The Tetagouche Group represents an ensialic arc sequence that formed along the northern margin of the Avalon Zone (continental basement) during the closing of the Iapetus Ocean by subduction to the south. Similar stratiform base metal sulphide deposits occur in Ordovician metamorphosed felsic volcanic and sedimentary rocks in Wicklow County southeastern Ireland (Avoca Deposits) (20). These occur along the northern margin of Avalon equivalent rocks (21).

One small base metal sulphide deposit has been found in the Ordovician mafic volcanic rocks of the Elmtree ophiolitic complex, north of the Tetagouche island arc sequence. Similar deposits occur in mafic volcanic rocks associated with the ophiolitic complexes in the Notre Dame Bay area (Dunnage Zone) of central Newfoundland and the Serpentine and Sutton-Bennett Belts in the Eastern Townships of Quebec (22, 9). In these areas, base metal sulphide deposits are also associated with island arc sequences overlying oceanic crust (9, 18, 22). It is possible that similar deposits occur at depth in the Matapedia Zone in northern New Brunswick, but none have been discovered.

Small copper and lead-zinc-copper sulphide deposits occur respectively in Silurian mafic tuffs and rhyolitic tuffs with intercalated metasediments along the southern margin of the Magaguadavic Zone (Fredericton Trough) in southern New Brunswick and its southwestern extension in the State of Maine. The bimodal volcanic host rocks probably formed along a rift zone in the Avalonian continental basement. These deposits are comparable to the slightly younger (Gedinnian) base metal sulphide deposits in the western Armorican Massif in France (23).

Contact metasomatic and vein type base metal sulphide and porphyry copper deposits are associated with Devonian I-type intrusions mainly in the Matapedia Zone and along the northwestern margin of the Miramichi Zone and locally (along major faults) in the southern Magaguadavic Zone. The relative abundance of the mantle-derived (I-type) intrusions in the Matapedia Zone and the northwestern margin of the Miramichi Zone is probably related to relatively thin continental crust in this area. It is also possible that the deposits associated with the I-type intrusions represent at least in part remobilized older stratiform deposits.

Tin-tungsten-molybdenum deposits are associated with highly fractionated, late stages of S-type intrusive complexes and their subvolcanic equivalents. These intrusions occur along dilatant fault zones in a broad arc-shaped belt that cuts across the southwestern Magaguadavic and central Miramichi Zones (both on Avalonian continental basement). The S-type intrusive complexes probably derived their elements from Ordovician sedimentary rocks. Comparible lithophile deposits occur in southern Maine (24, 25), southern Nova Scotia (26), southern Newfoundland (27), Cornwall (U.K.) (28), and the Armorican and Central Massifs of France (Cartes des Gites Minéraux de la France, Feuilles Nantes et Lyon, Bureau de Recherche Géologiques et Minières 1978, 1979).

Acknowledgements

The authors would like to thank Barbara W. M. Carroll, John B. Hamilton, and Steven R. McCutcheon for critical reading of the manuscript. Maurice Mazerolle drafted the map.

References

1. Williams, H. 1978. Tectonic lithofacies map of the Appalachian Orogen. Memorial University, St. John's, Newfoundland.

2. Ruitenberg; A. A., Fyffe, L. R., McCutcheon, S. R., St. Peter, C. J., Irrinki, R. R. and Venugopal, D. V. 1977. Evolution of pre-Carboniferous tectonostratigraphic zones in the New Brunswick Appalachians. Geoscience Canada 4, pp. 171-181.

3. McKerrow, W. S. and Ziegler, A. M. 1971. The Lower Silurian paleogeography of New Brunswick and adjacent areas. Jour. Geol. 1, pp. 635-646.

4. Fyffe, L. R. 1982. Taconian and Acadian structural trends in central and northern New Brunswick in St. Julien, P. and Béland, J. editors, Major structural zones and faults of the Northern Appalachians. Geol. Assoc. Canada Special Paper 24, pp. 117-130.

5. Ruitenberg, A. A. and McCutcheon, S. R. 1982. Acadian and Hercynian structural evolution of southern New Brunswick in St. Julien, P. and Béland, J., editors, Major structural zones and faults of the Northern Appalachians. Geol. Assoc. Canada Special Paper 24, pp. 131-148.

6. Ruitenberg, A. A., Giles, P. S., Venugopal, D. V., Buttimer, S. M., McCutcheon, S. R. and Chandra, J. 1979. Geology and mineral deposits of the Caledonia Area. Mineral Resources Branch, New Brunswick Department of Natural Resources, Memoir 1, 213 p.

7. Giles, P. S. and Ruitenberg, A. A. 1977. Stratigraphy, paleogeography and tectonic setting of the Coldbrook Group in the Caledonia Highlands of southern New Brunswick. Can. Jour. of Earth Sc. 14, pp. 1263-1275.

8. Fyffe, L. R. and Davies, J. L. 1982. Sulphide deposits of the Bathurst area, New Brunswick. Field Trip Guidebook B, IGCP Project 27, NATO Advanced Study Institute, Atlantic Canada, August 1982.

9. Ruitenberg, A. A. 1976. Comparison of volcanogenic mineral deposits in the Northern Appalachians and their relationship to tectonic evolution in Wolf, K. H., editor, Handbook of strata-bound and stratiform ore deposits, V. 5, pp. 109-159.

10. Gates, O. and Moench, R. H. 1981. Bimodal Silurian and Lower Devonian volcanic rock assemblages in the Machias-Eastport area, Maine. U.S. Geol. Surv. Prof. Paper, 1184, 32 p.

11. Cathles, L. M., Dudas, F. O. and Lenagh, T., 1981. Exploration significance of a failed rift-hypothesis for the genesis of Kuroko-type massive sulphide deposits. Abstracts Geol. Assoc. Canada, Ann. Meeting.

12. Ruitenberg, A. A. 1972. Metallization episodes related to tectonic evolution, Rolling Dam and Mascarene-Nerepis Belts, New Brunswick. Econ. Geol. 67, pp. 434-444.

13. Ruitenberg, A. A. and Fyffe, L. R. 1982. Mineral deposits associated with granitoid intrusions and related subvolcanic stocks in New Brunswick and their relationship to Appalachian tectonic evolution. Can. Inst. of Mining and Metall. Bull. V. 75, pp. 83-97.

14. Ruitenberg, A. A. 1967. Stratigraphy, structure and metallization, Piskahegan-Rolling Dam area. Leidse Geologische Mededelingen 40, pp. 79-120.

15. Fyffe, L. R., Pajari, G. E. and Cherry, M. E. 1981. The Acadian plutonic rocks of New Brunswick. Maritime Sediments and Atlantic Geology, V. 17, pp. 23-26.

16. St. Peter, C. 1978. Geology of parts of Restigouche, Victoria and Madawaska counties, northwestern New Brunswick. Mineral Resources Branch, New Brunswick Department of Natural Resources, Report of Investigation 17, 69 p.

17. Ruitenberg, A. A. 1982. Gold potential evaluation of southern New Brunswick in Report of Activities, Mineral Resources Division, New Brunswick Department of Natural Resources, Information Circular 82-1, pp. 14-16.

18. Poole, W. H. 1973. Stratigraphic framework massive sulphide deposits, Northern Appalachian orogen, in Symposium on matavolcanic massive sulphides with reference to the Northern Appalachians, McGill University, Montreal, Quebec.

19. Carpenter, P. A. 1976. Metallic mineral deposits of the Carolina slate belt, Northern Carolina. Bull. 84, Geological Survey Section, North Carolina Department of Natural Resources and Community Development, 89 p.

20. Evans, A. M. 1976. Genesis of Irish base metal deposits, in Wolf, K. H., editor, Handbook of stratabound and strataform ore deposits. Elsevier Scientific Publishing Company, V. 5, pp. 231-256.

21. Brück, P. M., Gardiner, P. R. R., Max, M. D. and Stillman, C. J. 1978. Field Guide to the Caledonian and pre-Caledonian rocks of southeast Ireland. Geological Survey of Ireland, Guide Series No. 2, 86 p.

22. Strong, D. 1973. Plate tectonic setting of Appalachian-Caledonian mineral deposits as indicated by Newfoundland examples. Trans. A.I.M.E., 73-I (320):21.

23. Aye, Françoise. 1978. Les gisements a zinc-plomb-cuivre-argent de Bodennec et Porte-aux-Moines in Chronique de la recherche minière 445, Bureau de Recherches Géologiques et minières, pp. 47-69.

24. Young, 1962. Prospect evaluations Hancock County, Maine. Special Economic Studies Series No. 2. Maine Geological Survey, Augusta, Maine 113 p.

25. Young, 1963. Prospect evaluations, Washington County, Maine. Special Economic Studies Series No. 3, Maine Geological Survey, Augusta, Maine, 86 p.

26. Keppie, J. D., Gregory, D. J., Chatterjee, A. K. and Lyttle, N. A. 1979. Geological Map of Nova Scotia, Mineral Occurrence Edition. Department of Mines and Energy, Nova Scotia.

27. Tsu, D. N. 1976. Metallic mineral occurrences map of Newfoundland. Department of Mines and Energy, Newfoundland.

28. Dines, H. G. and Phemister, J. 1956. The metalliferous mining region of southwest England. Memoir of the Geological Survey of Great Britian, London, U.K.

EVIDENCE FOR THE ALLOCHTHONOUS NATURE OF THE DUNNAGE ZONE IN CENTRAL NEWFOUNDLAND

COLMAN-SAAD, S.P.

Department of Mines & Energy,
Government of Newfoundland & Labrador
St. John's, Newfoundland

Dismembered ophiolite complexes in central Newfoundland form part of an ophiolite-lined fault zone that has a roughly elliptical outcrop. The ellipse has a northeast trending long axis of 55 km and a short axis of 20 km. A complete ophiolite stratigraphy has been demonstrated in the Coy Pond and Pipestone Pond Complexes, which have steep dips and face east and west respectively, outwards from the centre of the ellipse.

The ophiolite belt is bounded on the outside by Lower to Middle Ordovician volcanogenic rocks (Davidsville and Baie d'Espoir Groups), and by Upper Ordovician to Silurian clastic sediments (Botwood Group). The volcanogenic rocks are part of the Central Newfoundland island arc and probably lay directly on the ophiolite. The whole sequence has been regionally deformed twice, forming folds with roughly horizontal axes, and is mainly metamorphosed in the greenschist facies.

The ophiolite belt surrounds an area underlain by metamorphosed shale, quartz-rich sandstone, and gneisses; a limestone occurrence contains shelly fossils of Lower to Middle Ordovician age. An early deformation formed folds with near-vertical axes, and subsequent metamorphism resulted in a progression from greenschist facies, inwards to sillimanite-cordierite gneisses and migmatites. All the rocks appear to belong to the same sedimentary sequence, except for amphibolite-bearing gneisses in the central part which may represent a conformably underlying unit or basement.

P. E. Schenk (ed.),
Regional Trends in the Geology of the Appalachian-Caledonian-Hercynian-Mauritanide Orogen, 375–376.

Sedimentary sequences, inside and outside the ophiolite belt, have contrasting compositions and facies but are of the same age, implying that one of them is allochthonous. Since the ophiolitic rocks face outwards, they and the overlying volcanogenic rocks form the upper sheet and the rocks inside the ophiolite belt represent a second layer of crust that is exposed by a window. The time of emplacement postdated the Silurian Botwood Group. Paleontologic and tectonic considerations suggest that the underlying crustal layer belonged to the eastern margin of Iapetus and is perhaps equivalent to the Gander Zone, and that the Dunnage Zone has been thrust eastwards over it.

THRUSTING IN THE NEW WORLD ISLAND - HAMILTON SOUND AREA OF NEWFOUNDLAND

WILLIAMS, P.F.*; KARLSTROM, K.E. and VAN DER PLUIJM, B.

Department of Geology, University of New Brunswick
Fredericton, New Brunswick, Canada E3B 5A3

Detailed structural mapping is in progress in the New World Island - Hamilton Sound area of the "North Shore" in Newfoundland. It is an area composed primarily of pillow lavas of island arc affinity (1), turbidites, conglomerates, mélange and thin Caradocian black shales and limestone. Deformation is intense and extremely complex. Throughout the area we recognize a minimum of four generations of folds and several generations of faults, both brittle and ductile. Parallelism of faults and axial surfaces and over- printing relationships show that there are faults broadly contemporaneous with each generation of folds. Consequently the faults are commonly folded by open to isoclinal folds.

Deformation is extremely heterogeneous and narrow zones characterized by intense and complex deformation alternating with broader zones of less deformed rocks in which sedimentary structures are well preserved. Some of these zones are first generation structures and are interpreted as thrust zones. They are characterized by variable dip and by reclined folds and parallel stretching lineations. Other similar zones are characterized by steep dips and a horizontal stretching lineation. They may be late, transcurrent, ductile faults but only preliminary work has been done on these zones to date and they are therefore not discussed further.

The first generation thrust zones contain as many as three generations of folds all of which are associated with development of the zone and which have no known representatives outside the zones. These folds appear to have developed as approximately

P. E. Schenk (ed.),
Regional Trends in the Geology of the Appalachian-Caledonian-Hercynian-Mauritanide Orogen, 377–378.

north east-south west trending horizontally plunging structures that were then rotated into a north west - south east reclined orientation. One sheath fold of the latter orientation has been recorded. Locally quartz veins are present in the zones and they also are involved in the folding indicating that at least some of the deformation associated with the zones is post lithification. Elsewhere however, the zones are, at least in part, olistostromal (2) and are associated with the emplacement by thrusting of lithified or partially lithified Ordovician rocks onto Middle Silurian sediments.

We interpret these observations in terms of an accretionary prism developing by thrusting of pillow lavas and overlying sediments while the sediments were in varying stages of lithification. Indications so far are that thrusting was towards the north west. Timing of the thrusting remains a problem. The olistostrome at Cobbs Arm (3) dates thrusting there as Middle Silurian but that does not limit the timing in any way since thrusts confined to older rocks could be older in age. Similar structures on Fogo Island are contemporary with the Fogo granodiorite which has been dated at 380 My so that there is evidence to suggest that the thrusting continued into the Devonian.

Thrusting was followed by the development of large, overturned tight folds which generally have a penetrative cleavage. It is these north east trending, north westly overturned folds that are responsible for the inverted sequence over much of New World Island. They have an approximately horizontal enveloping surface and therefore regionally give rise to repetition of the same stratigraphic units desite a preponderance of steep dips. They greatly complicate the F_1 thrusting picture by tightly folding the thrust.

F_3 folds are approximately upright and vary considerably in degree of development and in tightness. Their effect on the regional structure is quite minor in the area under consideration.

REFERENCES

[1] Arnott, R.J.: 1983, Sedimentology and stratigraphy of Upper Ordovician-Silurian sequences on New World Island, Newfoundland: Separate fault controlled basins? Can. J. Earth Sci., in press.

[2] McKerrow, W.S.: 1978, A lower Paleozoic trench-fill sequence, New World Island, Newfoundland. Geol. Soc. Am. Bull. 89, pp. 1121-1132.

[3] Strong, D.F., and Payne, J.G.: 1973, Early Paleozoic Volcanism and Metamorphism of the Moretons Harbour-Twillingate Area, Newfoundland. Can. J. Earth Sci. 10, pp. 1363-1379.

PLATE KINEMATIC ORIGINS OF CARBONIFEROUS AND OFFSHORE BASINS OF EASTERN CANADA, WITH IMPLICATIONS FOR TACONIAN AND LATER MOTIONS

OSMASTON, Miles F.

The White Cottage, Sendmarsh, Ripley
Woking, Surrey GU23 6JT U.K.

The configuration of basins developed by restricted plate separations in the course of orogen evolution offer a potentially powerful, and so far little used, source of evidence on the plate kinematic history of the orogen. Sea-floor spreading principles imply no obstacle to the separative detachment, and incorporation into the basin floor, of quite tiny fragments of pre-existing continental lithosphere. Vigorous contemporaneous sedimentation, however, will probably result in a predominantly gabbroic igneous construction of the basin's lower crust. Minor compression of such basins cannot erase their proneness to prolonged subsidence, and sedimentation (+ volcanism) will eventually result in a crustal thickness exceeding 26 km, depending on the incidence of burial metamorphism.

Much of the complex basin system extending from Chaleur Bay eastward to beyond the Miquelons and southward to the Orpheus Graben and the northern Bay of Fundy can be explained as follows. The 250-450 km westward offset of tectonic elements in the mainland Maritimes relative to those in Newfoundland is mainly due to subduction and compression along the Gaspé-Connecticut Valley Synclinorium in two stages; 195 km Taconian (O_2) and 130 km Acadian (D_2). Five well-constrained partly-separative along-the-belt plate motions were also involved, as was a late Carboniferous westward compression in the SE of the region. The Cape Ray and Fredericton Faults are accepted as lines of substantial closure.

P. E. Schenk (ed.),
Regional Trends in the Geology of the Appalachian-Caledonian-Hercynian-Mauritanide Orogen, 379–380.

At the start of these motions, the Maquereau Dome was in contact with the Port au Port Peninsula, while the Springdale-to-Cape Ray strip of crust had not yet been juxtaposed with the Long Range Peninsula. The Port aux Basques-to-Hermitage Bay rocks, on the other hand, lay immediately east of the Caledonia Highlands inlier, extending along the north of the Cobequid block to Cape North on a southwestward-restored (see below) Cape Breton Island (CBI). The Sydney area was occupied by Petit Miquelon (Langlade) and St. Pierre. During Llandovery time, northeastward separations by the eastern half of Newfoundland caused extensive partly-transcurrent movements and volcanism in coastal Maine and New Brunswick, extracted the now-buried Westmorland block from the Sussex Basin, complexly disrupted elements of CBI and extracted both it and the Antigonish block from the basins on the northwest side of the Meguma block (which at that time lay sufficiently further east than now).

Early Acadian (Ludlovian?) dextral drag, between the west-moving Meguma block and southern Newfoundland, moved the intervening CBI, Antigonish, and Cobequid blocks (<u>inter alia</u>) 250 km westward relative to the latter. Shortly prior to final (D_2) closure of the Cape Ray-Fredericton suture, E Newfoundland plus CBI and the Meguma moved NNE relative to Cobequid to form the basin W of the Hollow Fault.

The palinspastic position of CBI lies wholly SE of any possible extension of the Dover-Hermitage Fault, which is inferred to be a major suture of Wenlock age and probably joined the Fredericton closure zone via the Hermitage "Flexure" and Cape Ray. Perhaps that is where most of Iapetus disappeared.

The Cabot Fault "system", by contrast, appears to have disjoint origins.

EARLY PALEOZOIC TECTONIC DEVELOPMENT OF SOUTHWESTERN NEWFOUNDLAND

CHORLTON, Lesley

Memorial University of Newfoundland
St. John's, Newfoundland Canada

Southwestern Newfoundland is traversed by two major faults, the Long Range Fault and the Cape Ray Fault. Carboniferous sedimentary rocks are exposed northwest of the Long Range Fault, and an early Paleozoic terrane based on ophiolitic crust, imbricated in stages punctuated by widespread intrusion of tonalite and subsequently cut by dioritic to granitic plutons, underlies the southern Long Range Mountains southeast of the fault. The southwest end of the Cape Ray Fault separates the early Paleozoic ophiolite/granitoid terrane on its northwest side from a terrane characterized by polydeformed, Ordovician metasedimentary, metavolcanic, and amphibolitic rocks. Ophiolitic gabbro intruded by diorite occurs as remnants beneath metasedimentary and metavolcanic rocks southeast of the fault.

The terrane southeast of the Cape Ray Fault underwent four distinct phases of regional deformation:

(D_1) Fold nappes or thrusts affected the dated Middle Ordovician volcanosedimentary pile around the La Poile Bay area as well as the sedimentary and mafic igneous rocks to the north and west. Thrusting was followed by the attainment of a medium pressure amphibolite facies metamorphic peak in tectonically buried rocks, and by the generation of synmetamorphic granitoid rocks.

(D_2) Deformation under the ambient, amphibolite facies conditions is largely responsible for the prominent regional structural grain and mesoscopic structures. D_2 may have culminated in the initiation of sinistral wrench faults, the most

P. E. Schenk (ed.),
Regional Trends in the Geology of the Appalachian-Caledonian-Hercynian-Mauritanide Orogen, 381–382.

important being the Cape Ray Fault. Differential vertical movement accompanied strike-slip displacement on these faults.

The Emsian-Eifelian Windsor Point Group was deposited along the Cape Ray Fault, unconformably on its northwest side in the southwest and unconformably on both sides to the northeast. The group was possibly deformed during the last stages of D_2.

(D_3) The Windsor Point Group in the southwest was subsequently deformed by dextrally oblique, high angle reverse faulting which uplifted the buried and still hot amphibolite facies rocks.

(D_4) Late deformation was minor, and resulted only in small scale faulting and kink or chevron folding.

The geology southeast of the Cape Ray Fault has traditionally been incorporated into the Gander tectonostratigraphic zone, but differs from the northeastern end of the zone in two main features: (i) it includes arc volcanic rocks overlying fragments of ophiolitic crust, a characteristic of the Dunnage zone, and (ii) it displays a more extensively developed medium pressure, rather than low pressure, metamorphic sequence. In contrast to the Dunnage zone of northeastern Newfoundland, the geology exposed in the southwest represents deeper tectonic levels (up to approximately 25 km), and consequently the rocks are deformed in a more ductile manner and more highly metamorphosed and granitized.

LIST OF PARTICIPANTS

C A N A D A

BARR, SANDRA M., Acadia University, Wolfville, Nova Scotia BOP 1XO; tel.: 902-542-2201(340).

BLANCHARD, MARIE-CLAUDE, Dalhousie University, Department of Geology, Halifax, NS, B3H 3J5; tel.: 424-3438.

BLACK, WILFRIED, Urauelz Exploration and Mining Ltd., 3633 Sources Blvd., Suite 277, Dollardde, Ormeaux, Quebec, H9B 2K4; tel.: 683-9370 (574).

CHANDLER, F.W., Geology Survey of Canada, 223 Roger Road, Ottawa, K1H 5C5; tel.: 995-4003 (613).

DOUGLAS, MALCOLM, Soquip, P.O. Box 10650, 1175 Rue de Lavigerie, Sainte-Foy, Quebec, G1V 4P5; tel.: 651-9543 (418).

FRALICK, PHILIP W., University of Toronto, Department of Geology, Toronto, Canada, M5S 1A1; tel.: 978-8841 (416).

FYFFE, LESLIE, New Brunswick Geological Surveys Branch, P.O. Box 6000, Fredericton, NB, E3B 5H1; tel.: 453-2082.

GRANGER, BERNARD, Soquip, Place D'Iberville Deux, 1175 Rue de Lavigerie, C.P. 10650, Ste-Foy, Quebec, G1V 4P5; tel.: 651-9543 (418).

GRANTHAM, ROBERT G., Curator of Geology, Nova Scotia Museum, 1747 Summer Street, Halifax, NS, B3H 3A6; tel.: 429-4610 (902).

JAMIESON, R.A., Department of Geology, Dalhousie University, Halifax, NS; tel.: 424-3771 (902).

KEPPIE, J. DUNCAN, Nova Scotia Department of Mines and Energy, P.O. Box 1087, Halifax, NS, Canada, B3J 2X1; tel.: 424-4015 (5943).

KING, ART, Department of Geology, Memorial University, St. John's, Newfoundland; tel.: 737-7660 (709).

LAPOINTE, PIERRE, Earth Physics Branch, E.M.R., Geomagnetic Lab, Anderson Road, Ottawa, Ontario, K1A OY3; tel.: 824-1630 (613).

LEWRY, JOHN F., Department of Geology, University of Regina, Regina, Saskatchewan; tel.: 584-4270 (306).

MCGUGAN, ALAN, University of Calgary, Department of Geology and Geophysics, Calgary, Alberta, T2N 1N4; tel.: 284-5845.

NOBLE, J.P.A., University of New Brunswick, Fredericton, New Brunswick; tel.: 453-4804.

PARKER, STEPHEN, Project Geologist, Tantalum Mining Corp., Bernic Lake, Manitoba; tel.: 345-8658 (204).

POOLE, WILLIAM H., Geological Survey of Canada, 601 Booth Street, Ottawa, Ontario, K1A OE8; tel.: 995-4309 (613).

QUINLAN, GARRY, Atlantic Geoscience Centre, P.O. Box 1006, Dartmouth, NS, B2Y 4A2; tel.: 426-6048.

RAESIDE, ROBERT P., Department of Geology, Acadia University, Wolfville, NS, BOP 1XO,

REIMANN, CLEMENS, Project Geologist, Selco, 7071 Bayers Road, Suite 215; tel.: 455-8292.

REYNOLDS, P.H., Dalhousie University, Halifax, NS; tel.: 424-2325 (902).

ROY, JEAN, Earth Physics Branch, E.M.R., Geomagnetic Lab., Anderson Road, Ottawa, Ontario, K1A OY3; tel.: 824-1630 (613).

RUITENBERG, ARNE-A, Geological Surveys Branch, New Brunswick, P.O. Box 1519, Sussex, NB, Canada, EOE 1PO; tel.: 433-4317.

SCHENK, P.E., Department of Geology, Dalhousie University, Halifax, NS, B3H 3J5; tel.: 424-2365 (902).

SEAL, ROBERT, Queen's University, Department of Geological Sciences, Kingston, Ontario, K7L 3N6; tel.: 547-3103 (613).

SEQUIN, MAURICE, Department of Geology, Universite Laval; tel.: 656-2196 (418).

SMITH, PAUL K., Nova Scotia Department of Mines and Energy, P.O. Box 1087, 1690 Hollis Street, Halifax, NS; tel.: 424-5943 (902).

STRINGER, P., Department of Geology, University of New Brunswick, P.O. Box 4400, Fredericton, NS, Canada, E3B 5A6; tel.: 453-4804 (506).

STRONG, D.F., Department of Geology, Memorial University of Newfoundland, St.John's, Newfoundland; tel.: 737- 8384 (709).

TRZCIENSKI, WALTER E., Department of Geology, University of Montreal, Montreal, H3C 3I7; tel.: 344-4611 (504).

WILLIAMS, H., Department of Geology, Memorial University, St.John's, Newfoundland; tel.: 737-8396 (709).

WILLIAMS, PAUL F., University of New Brunswick, Fredericton, New Brunswick; tel.: 453-4803.

FRANCE

CABANIS, BRUNO, Universite Pierre et Marie Curie, Lab Geochimie Comparie et Systematique, 4 Plac Jussieu, 75230 Paris CDX 05; tel.: 542-35-62.

LECORCHE, JEAN-PAUL, Laboratoire Etudes Geologiques Quest Afircairres, Faculte des Sciences St.Jerome, 13397 Marseille cedex 13, France; tel.: (91) 989010 (p447).

LEFORT, JEAN-PIERRE, Institut de Geologie, Faculte des Sciences, Rennes, Campus de Beauliew, Avenue du general Leclere, 35042 Rennes-Cedex, France; tel.: (99) 36-48-15.

PIQUE, ALAIN, Universite de Stresboruz, 1'rue Blessig, 67084 Strasburg-Cedex, France; tel.: 35-66-03 (88).

ROLET, JOEL, Universite de Brest, Geologie Structurale-UBO-6 Ava Le Gorgeu, 29283 Brest, France; tel.: 03-16-94 porte 367.

SANTALLIER, DANIELLE S., Laboratoire de Geologie, 123 Avenue Albert Thomas, 87060 Limoges Cedex, France; tel.: 79-46-22 (55) (423).

THONON, PIERRE, Universite de Brest, 29283 Brest Cedex, France; tel.: 03-16-94 (98).

G E R M A N Y

VOGE, WOLFRAM, Utaueizbergbau-Gmbh, Koelnstrasse 367, 5300 Bonna, W-Germany; tel.: (0228) 5577.

G R E A T B R I T A I N

BROOK, MAUREEN, Isotope Geology Unit, Institute of Geological Sciences, 64-78 Grays Inn Road, London, WC1X 8NG, UK; tel.: 01-242-4531.

FETTES, DOUGLAS, Institute of Geological Sciences, Murchison House, West Mains Road, Edinburgh, EH9 3LA, UK; tel.: 031-667-1000.

FRANCIS, E.H., Department of Earth Sciences, University of Leeds, Leeds, LS2 9JT, UK; tel.: 431-751 (7259).

HARRIS, A.L., Department of Geology, University of Liverpool, UK; tel.: 051-709-6022 (2334/2343).

MCKERROW, W.S., University of Oxford, Department of Geology, Parks Road, Oxford, OX1 3PR, England; tel.: 54511.

OSMASTON, MILES F., The White Cottage, Sendmarsh, Ripley, Woking, Surrey, GU23 6JT, UK; tel.: 048-643-2138.

POWELL, DEREK, Department of Geology, Bedford College, University of London, Regents Park, London, NW1 4NS; tel.: 01-486-4400.

STONE, PHILIP, Institute of Geological Sciences, Murchison House, West Mains Road, Edinburgh, Scotland, EH9 3LA; tel.: 031-667-1000.

I R E L A N D

KENNAN, PADHRAIG S., University College, Dublin, Geology Department, Belfield, Dublin 4, Ireland; tel.: 693-244-326.

LONG, C. BARRY, Geological Survey of Ireland, 14 Hume Street, Dublin 2, Ireland; tel.: 01-760855.

STILLMANN, CHRIS J., Department of Geology, Trinity College, Dublin 2, Ireland; tel.: Dublin 772941.

M O R O C C O

BENSAID, MOHAMMED, Direction de La Geologie, Ministere de L'Energie et es Mines Rabat, Morocco; tel.: 728-24 74726.

DAHMANI ,MOHAMED , Chef de la Division de la Geologie Ministere de L'Energie et des Mines Rabat, Morocco; tel.: 889-13.

N O R W A Y

ANDERSEN, TORGEIR BJORGE, Department of Geology, AVD A, 5014 University of Beruen, Norway; tel.: 2100-40 (3519).

BRUTON, DAVID L., Paleontologisk Museum, University of Oslo, Norway, Sars Gate 1, Oslo 5, Norway; tel.: 68-69-60 (02).

FURNES, HARALD, Geologisk Institutt, AVD A, Allegt 41, 5014 Bergen, Norway; tel.: 212-040 (05) (3530).

ROBERTS, DAVID, Norges Geologiske Undersokelse, P.O. Box 3006, 7001 Trondheim, Norway; tel.: 075-75860.

STURT, BRIAN, University of Bergen, 5014 Bergen, Norway; tel.: 212-040 (05).

S W E D E N

ANDREASSON, PER-GUNNAR, Department of Mineralogy and Petrology, University of Lund, Sweden, Solvegatan 13, S-223 62 LUND, Sweden; tel.: Sweden 046-107872.

GEE, D.G., Geological Survey of Sweden, Box 670 S-75128 Uppsala, Sweden; tel.: 155-280 (018).

JOHANSSON, LEIF, Department of Geology, University of Lund, Solvegatan 13, 223 62 LUND, Sweden; tel.: 046-107892.

LINDQVIST, JAN-ERIK, Department of Geology, Lunds University, Solvegatan 13, S-223 62 Lund, Sweden; tel.: 046-107892.

STIGH, JIMMY, Department of Geology, Chalmers University of Technology, S-412 96 Goteborg, Sweden; tel.: 031-810-100-1637.

T U R K E Y

OGUZ, ARDA, Middle East Technical University, Ankara, Turkey, Geology Engineering Department; tel.: 237100/ 2679.

U . S . A .

BAGWELL, LAURA, University of South Carolina, Columbia, SC, 29208; tel.: 777-6823.

DRAKE, AVERY A., U.S. Geological Survey, 953 National Center, Reston, VA, 22092, USA; tel.: 860-6631 (703).

DREIER, RA NAYE, University of Texas-Austin, 2009 Arpdale, Austin, TX, 78704; tel.: 477-0341 (512).

ERIKSSON, KENNETH A., VPI & SU, Department of Geology, Blacksburg, VA, 724061; tel.: 961-4680 (703).

EVANS, NICK H., Orogenic Studies Laboratory, Department of Geology, VPI, Blackburg, Virginia, 24061, USA; tel.: 961-6700.

FISHER, GEORGE W., Department of Earth and Planetary Sciences, The Johns Hopkins University, Baltimore, Maryland, 21218, USA; tel.: 338-7038 (301).

FORBES, BILL, University of Maine, 181 Main Street, Presque Isle, Maine; tel.: 764-0311 (207).

GOLDSMITH, RICHARD, U.S. Geological Survey, National Center, STOP 925 Reston, VA, 22092, USA; tel.: 860-6406 (703).

GUIDOTTI, CHARLES V., Department of Geological Science, University of Maine at Orono, Orono, Maine, 04469; tel.: 581-2730 (207).

HANLEY, THOMAS B., Columbus College, Algonquin Drive; tel.: 568-2292 (404).

HATCH, NORMAN L., U.S. Geological Survey, MS-926, Reston, VA, 22092, USA; tel.: 860-6421 (703).

HERMES, O. DON, University of Rhode Island, Department of Geology, Kingston, RI, 02881, USA; tel.: 792-2265 (401).

HIBBARD, JIM, Newfoundland Department of Mines and Energy, P.O. Box 146, Sundown, New York, 12782; tel.: 985-7406 (914).

HOGAN, JOHN P., Vriginia Polytech, 615 Washington Street, Apt. 5, Blacksburg, NA, 24060; tel.: 552-0674 (703).

HORKOWITZ, JOHN, University of South Carolina, 2524 Kwah Avenue, Cola, SC, 29205; tel.: 777-6823.

HORTON, JAMES WRIGHT, U.S. Geological Survey, 928 National Center, Reston, Virginia, 22092, USA; tel.: 860-6595 (703).

HUSSEY II, A.M., Bowdoin College, Brunswick, Maine, 04011, USA; tel.: 725-8731 (207) (219).

JOHNSON, REX, University of Michigan, Department of Geological Science, 1006 CC Little Bld., Ann Arbor, MI, 48109; tel.: 763-2149 (313).

KANE, MARTIN F., U.S. Geological Survey, Box 25046 DFC, MS/964, Denver, CO, 80225; tel.: 234-4593 (303).

KIMBRELL, PHIL, University of South Carolina, Columbia, SC, 29208; tel.: 777-6823 (803).

NEUMAN, R.B., United States Geological Survey, E-501 U.S. National Museum, Washington, DC, 20244; tel.: 343-3319 (202).

OSBERG, PHILIP H., University of Maine at Orono, Department of Geological Science, Orono, Maine, USA; tel.: 581-2704 (207).

PAVILDES, LOUIS, U.S. Geological Survey, National Center, MS/928, Reston, VA, 22092; tel.: 860-6503 (703).

PENN, SHELDON H., Suffolk Community College, 16 Seward Drive, Dix Hills, NY, 11746, USA; tel.: 451-4302 (516).

RANKIN, DOUGLAS W., U.S. Geological Survey, Mail Stop 926-A, National Center, Reston, VA, 22092; tel.: 860-6404 (703).

ROBINSON, PETER, Department of Geology and Geography, University of Massachusetts, Amherst, Mass., 01003; tel.: 545-2593 or 2286 (413).

RODGERS, JOHN, Yale University, Department of Geology and Geophysics, Yale University, New Haven, Conn., 06511, USA; tel.: 436-0616 (203).

ROY, DAVID C., Boston College, Department of Geology and Geophysics, BC, Chestnut HILL, MA, 02167; tel.: 969-0100-3647 (617).

SAMSON, SAVA L., University of South Carolina, Columbia, SC; tel.: 777-0409 (803).

SCHOLTEN, ROBERT, The Pennsylvania State University, 334A Deike Building, University Park, Pennsylvania, 16802, USA; tel.: 865-6393 (814).

SCHUMACHER, JOHN C., Department of Geology and Geography, University of Massachusetts, Amherst, MA, USA, 01003; tel.: 545-2286 (413).

SECOR, DONALD T., Department of Geology, University of South Carolina, Columbia, SC, 29208; tel.: 777-4516 (803).

SHANKLE, LAURIE, Amselco Exploration Inc., 1112 Mill, P.O. Box 891, Camden, SC, 29020; tel.: 432-9829 (803).

SIZE, WILLIAM B., Emory University, Department of Geology, Atlanta, Georgia, 30322, USA; tel.: 329-6491 (404).

SKEHAN, JAMES W., Weston Observatory, 381 Concord Road, Weston, MA, 02193; tel.: 899-0950 (617).

SMITH, WILLIAM A., University of South Carolina, Columbia, SC; tel.: 777-6484.

TULL, JAMES F., Florida State University, Tallahassee, Fla., 32306, USA; tel.: 644-1448 (904).

WONES, DAVID R., Department of Geological Sciences, VPI & SV, Blacksburg, VA, 24061; tel.: 961-6521 (703).

SUBJECT INDEX